THE
DEEP
LEARNING
WORKSHOP

Learn the skills you need to develop your own next-generation deep learning models with TensorFlow and Keras

Mirza Rahim Baig, Thomas V. Joseph, Nipun Sadvilkar, Mohan Kumar Silaparasetty, and Anthony So

Packt>

THE DEEP LEARNING WORKSHOP

Authors: Mirza Rahim Baig, Thomas V. Joseph, Nipun Sadvilkar, Mohan Kumar Silaparasetty, and Anthony So

Reviewers: Ankit Bhatia, Akshay Chauhan, Francesco Mosconi, Nagendra Nagaraj, Bernard Ong, and Robert Ridley

Managing Editor: Abhishek Rane

Acquisitions Editors: Royluis Rodrigues, Sneha Shinde, and Archie Vankar

Production Editor: Roshan Kawale

Editorial Board: Megan Carlisle, Samuel Christa, Mahesh Dhyani, Heather Gopsill, Manasa Kumar, Alex Mazonowicz, Monesh Mirpuri, Bridget Neale, Dominic Pereira, Shiny Poojary, Abhishek Rane, Brendan Rodrigues, Erol Staveley, Ankita Thakur, Nitesh Thakur, and Jonathan Wray

First Published: July 2020

Production Reference: 2240221

ISBN: 978-1-83921-985-6

Published by Packt Publishing Ltd.

Livery Place, 35 Livery Street

Birmingham B3 2PB, UK

WHY LEARN WITH A PACKT WORKSHOP?

LEARN BY DOING

Packt Workshops are built around the idea that the best way to learn something new is by getting hands-on experience. We know that learning a language or technology isn't just an academic pursuit. It's a journey towards the effective use of a new tool—whether that's to kickstart your career, automate repetitive tasks, or just build some cool stuff.

That's why Workshops are designed to get you writing code from the very beginning. You'll start fairly small—learning how to implement some basic functionality—but once you've completed that, you'll have the confidence and understanding to move onto something slightly more advanced.

As you work through each chapter, you'll build your understanding in a coherent, logical way, adding new skills to your toolkit and working on increasingly complex and challenging problems.

CONTEXT IS KEY

All new concepts are introduced in the context of realistic use-cases, and then demonstrated practically with guided exercises. At the end of each chapter, you'll find an activity that challenges you to draw together what you've learned and apply your new skills to solve a problem or build something new.

We believe this is the most effective way of building your understanding and confidence. Experiencing real applications of the code will help you get used to the syntax and see how the tools and techniques are applied in real projects.

BUILD REAL-WORLD UNDERSTANDING

Of course, you do need some theory. But unlike many tutorials, which force you to wade through pages and pages of dry technical explanations and assume too much prior knowledge, Workshops only tell you what you actually need to know to be able to get started making things. Explanations are clear, simple, and to-the-point. So you don't need to worry about how everything works under the hood; you can just get on and use it.

Written by industry professionals, you'll see how concepts are relevant to real-world work, helping to get you beyond "Hello, world!" and build relevant, productive skills. Whether you're studying web development, data science, or a core programming language, you'll start to think like a problem solver and build your understanding and confidence through contextual, targeted practice.

ENJOY THE JOURNEY

Learning something new is a journey from where you are now to where you want to be, and this Workshop is just a vehicle to get you there. We hope that you find it to be a productive and enjoyable learning experience.

Packt has a wide range of different Workshops available, covering the following topic areas:

- Programming languages

- Web development

- Data science, machine learning, and artificial intelligence

- Containers

Once you've worked your way through this Workshop, why not continue your journey with another? You can find the full range online at http://packt.live/2MNkuyl.

If you could leave us a review while you're there, that would be great. We value all feedback. It helps us to continually improve and make better books for our readers, and also helps prospective customers make an informed decision about their purchase.

Thank you,
The Packt Workshop Team

Table of Contents

Chapter 2: Neural Networks 45

Chapter 4: Deep Learning for Text – Embeddings

Chapter 6: LSTMs, GRUs, and Advanced RNNs   265

PREFACE

ABOUT THE BOOK

Are you fascinated by how deep learning powers intelligent applications such as self-driving cars, virtual assistants, facial recognition devices, and chatbots to process data and solve complex problems? Whether you are familiar with machine learning or are new to this domain, *The Deep Learning Workshop* will make it easy for you to understand deep learning with the help of interesting examples and exercises throughout.

The book starts by highlighting the relationship between deep learning, machine learning, and artificial intelligence and helps you get comfortable with the TensorFlow 2.0 programming structure using hands-on exercises. You'll understand neural networks, the structure of a perceptron, and how to use TensorFlow to create and train models. The book will then let you explore the fundamentals of computer vision by performing image recognition exercises with **Convolutional Neural Networks (CNNs)** using Keras. As you advance, you'll be able to make your model more powerful by implementing text embedding and sequencing the data using popular deep learning solutions. Finally, you'll get to grips with bidirectional **Recurrent Neural Networks (RNNs)** and build **Generative Adversarial Networks (GANs)** for image synthesis.

By the end of this deep learning book, you'll have learned the skills essential for building deep learning models with TensorFlow and Keras.

AUDIENCE

If you are interested in machine learning and want to create and train deep learning models using TensorFlow and Keras, this workshop is for you. A solid understanding of Python and its packages, along with basic machine learning concepts, will help you to learn the topics quickly.

ABOUT THE CHAPTERS

Chapter 1, Building Blocks of Deep Learning, discusses the practical applications of deep learning. One such application includes a hands-on code demo you can run right away to recognize an image from the internet. Through practical exercises, you'll also learn the key code implementations of TensorFlow 2.0 that will help you build exciting neural network models in the coming chapters.

Chapter 2, Neural Networks, teaches you the structure of artificial neural networks. Using TensorFlow 2.0, you will not only implement a neural network, but also train it. You will later build multiple deep neural networks with different configurations, thereby experiencing the neural network training process first-hand.

Chapter 3, Image Classification with Convolutional Neural Networks (CNNs), covers image processing, how it works, and how that knowledge can be applied to Convolutional Neural Networks (CNNs). Through practical exercises, you will create and train CNN models that can be used to recognize images of handwritten digits and even fruits. You'll also learn some key concepts such as pooling layers, data augmentation, and transfer learning.

Chapter 4, Deep Learning for Text – Embeddings, introduces you to the world of Natural Language Processing. You will first perform text preprocessing, an important skill when dealing with raw text data. You will implement classical approaches to text representation, such as one-hot encoding and the TF-IDF approach. Later in the chapter, you will learn about embeddings, and use the Skip-gram and Continuous Bag of Words algorithms to generate your own word embeddings.

Chapter 5, Deep Learning for Sequences, shows you how to work on a classic sequence processing task—stock price prediction. You will first create a model based on Recurrent Neural Networks (RNNs), then implement a 1D convolutions-based model and compare its performance with that RNN model. You will combine both approaches by combining RNNs with 1D convolutions in a hybrid model.

Chapter 6, LSTMs, GRUs, and Advanced RNNs, reviews RNNs' practical drawbacks and how Long Short Term Memory (LSTM) models help overcome them. You will build a model that analyzes sentiments in movie reviews and study the inner workings of Gated Recurring Units (GRUs). Throughout the chapter, you will create models based on plain RNNs, LSTMs, and GRUs and, at the end of the chapter, compare their performance.

Chapter 7, Generative Adversarial Networks, introduces you to generative adversarial networks (GANs) and their basic components. Through practical exercises, you will use GANs to generate a distribution that mimics a data distribution produced by a sine function. You will also learn about deep convolutional GANs and implement them in an exercise. Toward the end of the chapter, you will create GANs that are able to replicate images with convincing accuracy.

CONVENTIONS

Code words in text, database table names, folder names, filenames, file extensions, pathnames, dummy URLs, user input, and Twitter handles are shown as follows:

"Load the MNIST dataset using **`mnist.load_data()`**"

Words that you see on the screen (for example, in menus or dialog boxes) appear in the same format.

A block of code is set as follows:

```
from sklearn.preprocessing import MinMaxScaler
scaler = MinMaxScaler()
train_scaled = scaler.fit_transform(train_data)
test_scaled = scaler.transform(test_data)
```

New terms and important words are shown like this:

The first step in preprocessing is inevitably **tokenization**—splitting the raw input text sequence into **tokens**. Long code snippets are truncated and the corresponding names of the code files on GitHub are placed at the top of the truncated code. The permalinks to the entire code are placed below the code snippet. It should look as follows:

Exercise7.04.ipynb

```
# Function to generate real samples
def realData(loc,batch):
    """
    loc is the random location or mean
    around which samples are centered
    """
    # Generate numbers to right of the random point
    xr = np.arange(loc,loc+(0.1*batch/2),0.1)
    xr = xr[0:int(batch/2)]
    # Generate numbers to left of the random point
    xl = np.arange(loc-(0.1*batch/2),loc,0.1)
```

The complete code for this step can be found at https://packt.live/3iIJHVS.

CODE PRESENTATION

Lines of code that span multiple lines are split using a backslash (\ ). When the code is executed, Python will ignore the backslash, and treat the code on the next line as a direct continuation of the current line.

For example:

```
history = model.fit(X, y, epochs=100, batch_size=5, verbose=1, \
                    validation_split=0.2, shuffle=False)
```

Comments are added into code to help explain specific bits of logic. Single-line comments are denoted using the **#** symbol, as follows:

```
# Print the sizes of the dataset
print("Number of Examples in the Dataset = ", X.shape[0])
print("Number of Features for each example = ", X.shape[1])
```

Multi-line comments are enclosed by triple quotes, as shown below:

```
"""
Define a seed for the random number generator to ensure the
result will be reproducible
"""
seed = 1
np.random.seed(seed)
random.set_seed(seed)
```

SETTING UP YOUR ENVIRONMENT

Before we explore the book in detail, we need to set up specific software and tools. In the following section, we shall see how to do that.

HARDWARE REQUIREMENTS

For the optimal user experience, we recommend 8 GB RAM.

INSTALLING ANACONDA ON YOUR SYSTEM

All the exercises and activities in this book will be executed in Jupyter Notebooks. To install Jupyter on Windows, macOS, or Linux we first need to install Anaconda. Installing Anaconda will also install Python.

1. Head to https://www.anaconda.com/distribution/ to install the Anaconda Navigator, which is an interface through which you can access your local Jupyter Notebook.

2. Now, based on your operating system (Windows, macOS, or Linux) you need to download the Anaconda Installer. Select your operating system first and then choose the Python version. For this book, it is recommended that you use the latest version of Python.

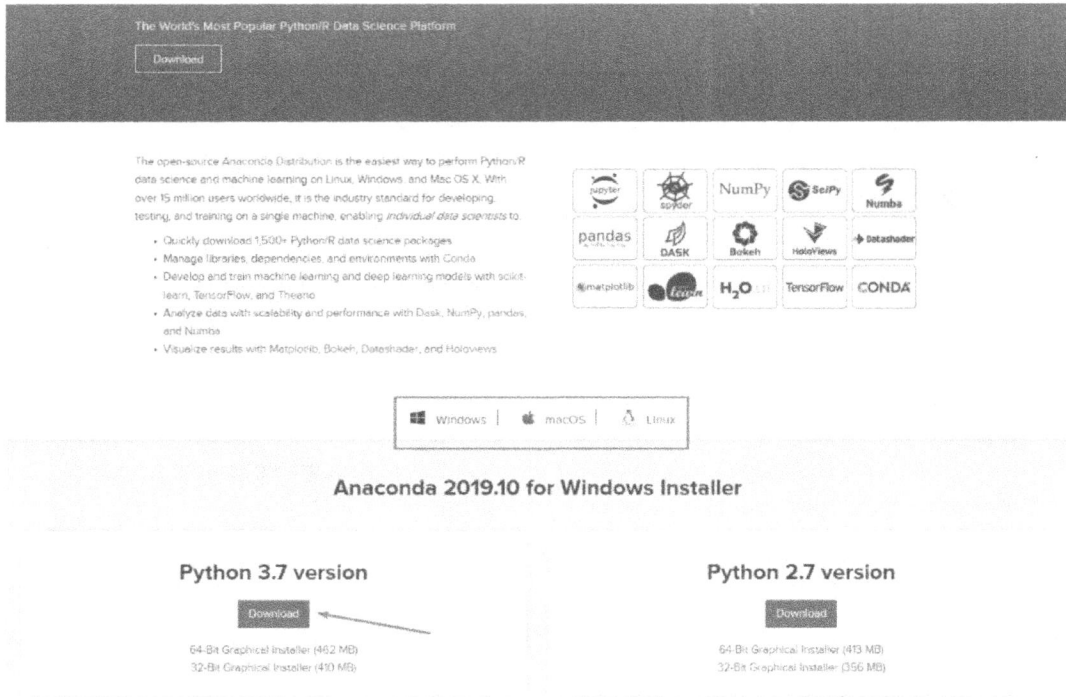

The open-source Anaconda Distribution is the easiest way to perform Python/R data science and machine learning on Linux, Windows, and Mac OS X. With over 15 million users worldwide, it is the industry standard for developing, testing, and training on a single machine, enabling *individual data scientists* to:

- Quickly download 1,500+ Python/R data science packages
- Manage libraries, dependencies, and environments with Conda
- Develop and train machine learning and deep learning models with scikit-learn, TensorFlow, and Theano
- Analyze data with scalability and performance with Dask, NumPy, pandas, and Numba
- Visualize results with Matplotlib, Bokeh, Datashader, and Holoviews

Windows | macOS | Linux

Anaconda 2019.10 for Windows Installer

Python 3.7 version

Download

64-Bit Graphical Installer (462 MB)
32-Bit Graphical Installer (410 MB)

Python 2.7 version

Download

64-Bit Graphical Installer (413 MB)
32-Bit Graphical Installer (356 MB)

Figure 0.1: The Anaconda home screen

3. To check if Anaconda Navigator is correctly installed, look for **Anaconda Navigator** in your applications. Look for the icon shown below. However, note that the icon's aesthetics may vary slightly depending on your operating system.

Anaconda Navigator (Anaconda3)

Figure 0.2 Anaconda Navigator icon

4. Click the icon to open Anaconda Navigator. It may take a while to load for the first time, but upon successful installation, you should see a similar screen:

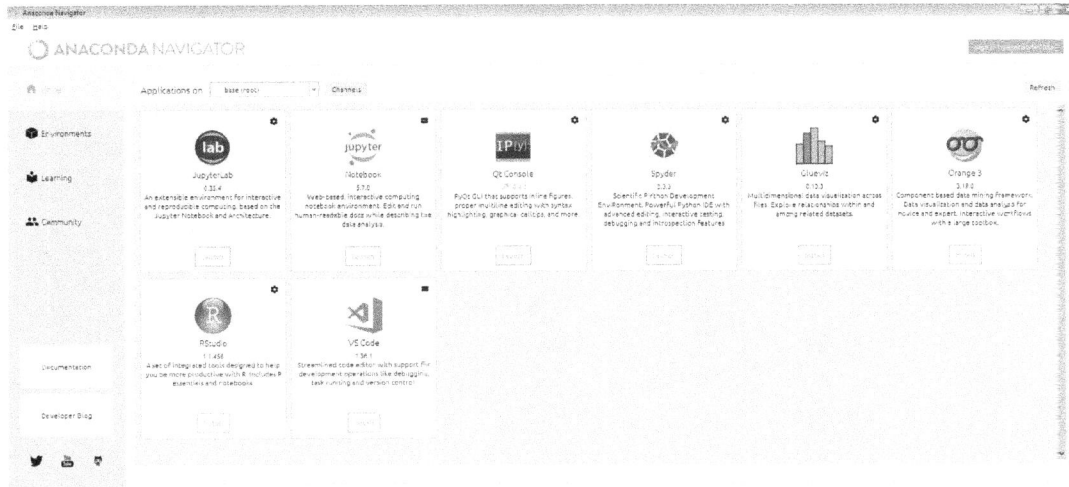

Figure C.3 Anaconda Navigator icon

LAUNCHING JUPYTER NOTEBOOK

To launch Jupyter Notebook from the Anaconda Navigator, follow these steps:

1. Open Anaconda Navigator. You should see the following screen:

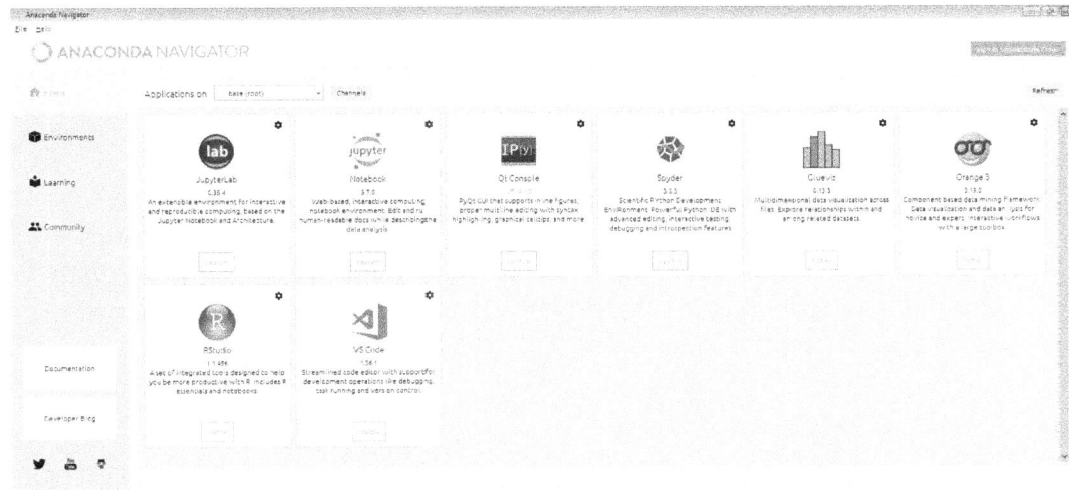

Figure 0.4: Anaconda installation screen

2. Now, click **Launch** under the `Jupyter Notebook` panel to start the notebook on your local system:

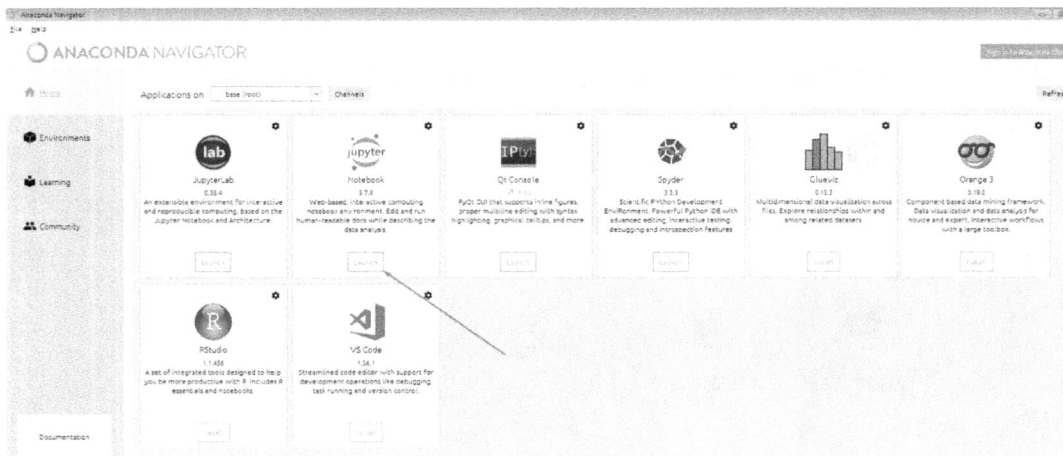

Figure 0.5: Jupyter Notebook launch option

You have successfully installed Jupyter Notebook onto your system. You can also open a Jupyter Notebook by simply running the command `jupyter notebook` in the Terminal or Anaconda Prompt.

INSTALLING LIBRARIES

`pip` comes pre-installed with Anaconda. Once Anaconda is installed on your machine, all the required libraries can be installed using `pip`, for example, `pip install numpy`. Alternatively, you can install all the required libraries using `pip install -r requirements.txt`. You can find the `requirements.txt` file at https://packt.live/303E4dD.

The exercises and activities will be executed in Jupyter Notebooks. Jupyter is a Python library and can be installed in the same way as the other Python libraries – that is, with `pip install jupyter`, but fortunately, it comes pre-installed with Anaconda. To open a notebook, simply run the command `jupyter notebook` in the Terminal or Command Prompt.

INSTALLING TENSORFLOW 2.0

Before installing TensorFlow 2.0, ensure that you have the latest version of **pip** installed on your system. You can check that by using the following command:

```
pip --version
```

To install TensorFlow 2.0, the version of **pip** on your system must be greater than **19.0**. You can upgrade your version of **pip** using the following command on Windows, Linux, or macOS:

```
pip install --upgrade pip
```

Once upgraded, use the following command to install TensorFlow on Windows, Linux, or macOS:

```
pip install --upgrade tensorflow
```

On Linux and macOS, if elevated rights are required, use the following command:

```
sudo pip install --upgrade tensorflow
```

> **NOTE**
>
> TensorFlow is not supported on Windows with Python 2.7.

INSTALLING KERAS

To install Keras on Windows, macOS, or Linux, use the following command:

```
pip install keras
```

On Linux and macOS, if elevated rights are required, use the following command:

```
sudo pip install keras
```

ACCESSING THE CODE FILES

You can find the complete code files of this book at https://packt.live/3edmwj4. You can also run many activities and exercises directly in your web browser by using the interactive lab environment at https://packt.live/2CGCWUz.

We've tried to support interactive versions of all activities and exercises, but we recommend a local installation as well for instances where this support isn't available.

NOTE

This book contains certain code snippets that read data from CSV files. It is assumed that the CSV files are stored in the same folder as the Jupyter Notebook. In case you have stored them elsewhere, you'll have to modify the path.

If you have any issues or questions about installation, please email us at **workshops@packt.com**.

1

BUILDING BLOCKS OF DEEP LEARNING

INTRODUCTION

In this chapter, you will be introduced to deep learning and its relationship with artificial intelligence and machine learning. We will also learn about some of the important deep learning architectures, such as the multi-layer perceptron, convolutional neural networks, recurrent neural networks, and generative adversarial networks. As we progress, we will get hands-on experience with the TensorFlow framework and use it to implement a few linear algebra operations. Finally, we will be introduced to the concept of optimizers. We will understand their role in deep learning by utilizing them to solve a quadratic equation. By the end of this chapter, you will have a good understanding of what deep learning is and how programming with TensorFlow works.

INTRODUCTION

You have just come back from your yearly vacation. Being an avid social media user, you are busy uploading your photographs to your favorite social media app. When the photos get uploaded, you notice that the app automatically identifies your face and tags you in them almost instantly. In fact, it does that even in group photos. Even in some poorly lit photos, you notice that the app has, most of the time, tagged you correctly. How does the app learn how to do that?

To identify a person in a picture, the app requires accurate information on the person's facial structure, bone structure, eye color, and many other details. But when you used that photo app, you didn't have to feed all these details explicitly to the app. All you did was upload your photos, and the app automatically began identifying you in them. How did the app know all these details?

When you uploaded your first photo to the app, the app would have asked you to tag yourself. When you manually tagged yourself, the app automatically "learned" all the information it needed to know about your face. Then, every time you upload a photo, the app uses the information it learned to identify you. It improves when you manually tag yourself in photos in which the app incorrectly tagged you.

This ability of the app to learn new details and improve itself with minimal human intervention is possible due to the power of **deep learning** (**DL**). Deep learning is a part of **artificial intelligence** (**AI**) that helps a machine learn by recognizing patterns from labeled data. But wait a minute, isn't that what **machine learning** (**ML**) does? Then what is the difference between deep learning and machine learning? What is the point of confluence among domains such as AI, machine learning, and deep learning? Let's take a quick look.

AI, MACHINE LEARNING, AND DEEP LEARNING

Artificial intelligence is the branch of computer science aimed at developing machines that can simulate human intelligence. Human intelligence, in a simplified manner, can be explained as decisions that are taken based on the inputs received from our five senses – sight, hearing, touch, smell, and taste. AI is not a new field and has been in vogue since the 1950s. Since then, there have been multiple waves of ecstasy and agony within this domain. The 21st century has seen a resurgence in AI following the big strides made in computing, the availability of data, and a better understanding of theoretical underpinnings. Machine learning and deep learning are subfields of AI and are increasingly used interchangeably.

The following figure depicts the relationship between AI, ML, and DL:

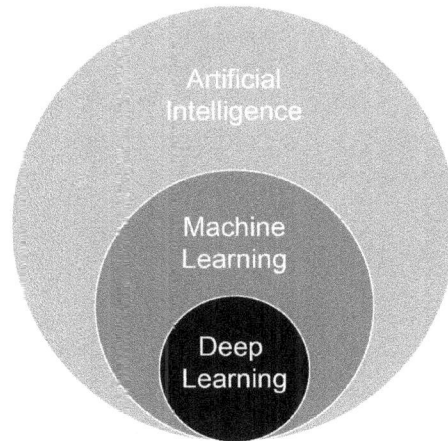

Figure 1.1: Relationship between AI, ML, and DL

MACHINE LEARNING

Machine learning is the subset of AI that performs specific tasks by identifying patterns within data and extracting inferences. The inferences that are derived from data are then used to predict outcomes on unseen data. Machine learning differs from traditional computer programming in its approach to solving specific tasks. In traditional computer programming, we write and execute specific business rules and heuristics to get the desired outcomes. However, in machine learning, the rules and heuristics are not explicitly written. These rules and heuristics are learned by providing a dataset. The dataset provided for learning the rules and heuristics is called a **training dataset**. The whole process of learning and inferring is called **training**.

Learning rules and heuristics is done using different algorithms that use statistical models for that purpose. These algorithms make use of many representations of data for learning. Each such representation of data is called an **example**. Each element within an example is called a **feature**. The following is an example of the famous IRIS dataset (https://archive.ics.uci.edu/ml/datasets/Iris). This dataset is a representation of different species of iris flowers based on different characteristics, such as the length and width of their sepals and petals:

sepal_length	sepal_width	petal_length	petal_width	species
5.1	3.5	1.4	0.2	setosa
4.9	3.0	1.4	0.2	setosa
4.7	3.2	1.3	0.2	setosa
4.6	3.1	1.5	0.2	setosa
5.0	3.6	1.4	0.2	setosa
5.4	3.9	1.7	0.4	setosa

Figure 1.2: Sample data from the IRIS dataset

In the preceding dataset, each row of data represents an example, and each column is a feature. Machine learning algorithms make use of these features to draw inferences from the data. The veracity of the models, and thereby the outcomes that are predicted, depend a lot on the features of the data. If the features provided to the machine learning algorithm are a good representation of the problem statement, the chances of getting a good result are high. Some examples of machine learning algorithms are *linear regression*, *logistic regression*, *support vector machines*, *random forest*, and *XGBoost*.

Even though traditional machine learning algorithms are useful for a lot of use cases, they have a lot of dependence on the quality of the features to get superior outcomes. The creation of features is a time-consuming art and requires a lot of domain knowledge. However, even with comprehensive domain knowledge, there are still limitations on transferring that knowledge to derive features, thereby encapsulating the nuances of the data generating processes. Also, with the increasing complexity of the problems that are tackled with machine learning, particularly with the advent of unstructured data (images, voice, text, and so on), it can be almost impossible to create features that represent the complex functions, which, in turn, generate data. As a result, there is often a need to find a different approach to solving complex problems; that is where deep learning comes into play.

DEEP LEARNING

Deep learning is a subset of machine learning and an extension of a certain kind of algorithm called Artificial Neural Networks (ANNs). Neural networks are not a new phenomenon. Neural networks were created in the first half of the 1940s. The development of neural networks was inspired by the knowledge of how the human brain works. Since then, there have been several ups and downs in this field. One defining moment that renewed enthusiasm around neural networks was the introduction of an algorithm called backpropagation by stalwarts in the field such as Geoffrey Hinton. For this reason, Hinton is widely regarded as the 'Godfather of Deep Learning'. We will be discussing neural networks in depth in *Chapter 2, Neural Networks*.

ANNs with multiple (deep) layers lie at the heart of deep learning. One defining characteristic of deep learning models is their ability to learn features from the input data. Unlike traditional machine learning, where there is the need to create features, deep learning excels in learning different hierarchies of features across multiple layers. Say, for example, we are using a deep learning model to detect faces. The initial layers of the model will learn low-level approximations of a face, such as the edges of the face, as shown in *Figure 1.3*. Each succeeding layer takes the lower layers' features and puts them together to form more complex features. In the case of face detection, if the initial layer has learned to detect edges, the subsequent layers will put these edges together to form parts of a face such as the nose or eyes. This process continues with each successive layer, with the final layer generating an image of a human face:

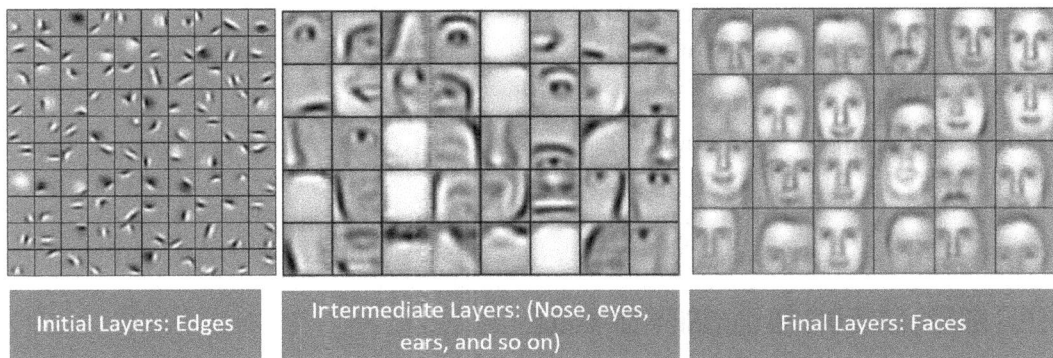

| Initial Layers: Edges | Intermediate Layers: (Nose, eyes, ears, and so on) | Final Layers: Faces |

Figure 1.3: Deep learning model for detecting faces

> **NOTE**
>
> The preceding image is sourced from the popular research paper:
> Lee, Honglak & Grosse, Roger & Ranganath, Rajesh & Ng, Andrew.
> (2011). *Unsupervised Learning of Hierarchical Representations with
> Convolutional Deep Belief Networks.* Commun. ACM. 54. 95-103.
> 10.1145/2001269.2001295.

Deep learning techniques have made great strides over the past decade. There are different factors that have led to the exponential rise of deep learning techniques. At the top of the list is the availability of large quantities of data. The digital age, with its increasing web of connected devices, has generated lots of data, especially unstructured data. This, in turn, has fueled the large-scale adoption of deep learning techniques as they are well-suited to handle large unstructured data.

Another major factor that has led to the rise in deep learning is the strides that have been made in computing infrastructure. Deep learning models that have large numbers of layers and millions of parameters necessitate great computing power. The advances in computing layers such as **Graphical Processing Units** (**GPUs**) and **Tensor Processing Units** (**TPUs**) at an affordable cost has led to the large-scale adoption of deep learning.

The pervasiveness of deep learning was also accelerated by open sourcing different frameworks in order to build and implement deep learning models. In 2015, the Google Brain team open sourced the TensorFlow framework and since then TensorFlow has grown to be one of the most popular frameworks for deep learning. The other major frameworks available are PyTorch, MXNet, and Caffe. We will be using the TensorFlow framework in this book.

Before we dive deep into the building blocks of deep learning, let's get our hands dirty with a quick demo that illustrates the power of deep learning models. You don't need to know any of the code that is presented in this demo. Simply follow the instructions and you'll be able to get a quick glimpse of the basic capabilities of deep learning.

USING DEEP LEARNING TO CLASSIFY AN IMAGE

In the exercise that follows, we will classify an image of a pizza and convert the resulting class text into speech. To classify the image, we will be using a pre-trained model. The conversion of text into speech will be done using a freely available API called **Google Text-to-Speech** (**gTTS**). Before we get into it, let's understand some of the key building blocks of this demo.

PRE-TRAINED MODELS

Training a deep learning model requires a lot of computing infrastructure and time, with big datasets. However, to aid with research and learning, the deep learning community has also made models that have been trained on large datasets available. These pre-trained models can be downloaded and used for predictions or can be used for further training. In this demo, we will be using a pre-trained model called **ResNet50**. This model is available along with the Keras package. This pre-trained model can predict 1,000 different classes of objects that we encounter in our daily lives, such as birds, animals, automobiles, and more.

THE GOOGLE TEXT-TO-SPEECH API

Google has made its Text-to-Speech algorithm available for limited use. We will be using this algorithm to convert the predicted text into speech.

PREREQUISITE PACKAGES FOR THE DEMO

For this demo to work, you will need the following packages installed on your machine:

- TensorFlow 2.0

- Keras

- gTTS

Please refer to the *Preface* to understand the process of installing the first two packages. Installing gTTS will be shown in the exercise. Let's dig into the demo.

EXERCISE 1.01: IMAGE AND SPEECH RECOGNITION DEMO

In this exercise, we will demonstrate image recognition and speech-to-text conversion using deep learning models. At this point, you will not be able to understand each and every line of the code. This will be explained later. For now, just execute the code and find out how easy it is to build deep learning and AI applications with TensorFlow. Follow these steps to complete this exercise:

1. Open a Jupyter Notebook and name it *Exercise 1.01*. For details on how to start a Jupyter Notebook, please refer to the preface.

2. Import all the required libraries:

```
from tensorflow.keras.preprocessing.image import load_img
from tensorflow.keras.preprocessing.image import img_to_array
```

```
from tensorflow.keras.applications.resnet50 import ResNet50
from tensorflow.keras.preprocessing import image
from tensorflow.keras.applications.resnet50 \
import preprocess_input
from tensorflow.keras.applications.resnet50 \
import decode_predictions
```

> **NOTE**
>
> The code snippet shown here uses a backslash (\ ) to split the logic across multiple lines. When the code is executed, Python will ignore the backslash, and treat the code on the next line as a direct continuation of the current line.

Here is a brief description of the packages we'll be importing:

load_img: Loads the image into the Jupyter Notebook

img_to_array: Converts the image into a NumPy array, which is the desired format for Keras

preprocess_input: Converts the input into a format that's acceptable for the model

decode_predictions: Converts the numeric output of the model prediction into text labels

Resnet50: This is the pre-trained image classification model

3. Create an instance of the pre-trained **Resnet** model:

```
mymodel = ResNet50()
```

You should get a message similar to the following as it downloads:

```
Downloading data from https://github.com/keras-team/keras-applications/release
s/download/resnet/resnet50_weights_tf_dim_ordering_tf_kernels.h5
 92823552/102967424 [============================>...] - ETA: 4s
```

Figure 1.4: Loading Resnet50

Resnet50 is a pre-trained image classification model. For first-time users, it will take some time to download the model into your environment.

4. Download an image of a pizza from the internet and store it in the same folder that you are running the Jupyter Notebook in. Name the image **im1.jpg**.

> **NOTE**
>
> You can also use the image we are using by downloading it from this link:
> https://packt.live/2AHTAC9

5. Load the image to be classified using the following command:

```
myimage = load_img 'im1.jpg', target_size=(224, 224))
```

If you are storing the image in another folder, the complete path of the location where the image is located has to be given in place of the **im1.jpg** command. For example, if the image is stored in **D:/projects/demo**, the code should be as follows:

```
myimage = load_img 'D:/projects/demo/im1.jpg', \
                    target_size=(224, 224))
```

6. Let's display the image using the following command:

```
myimage
```

The output of the preceding command will be as follows:

Figure 1.5: Output displayed after loading the image

7. Convert the image into a **numpy** array as the model expects it in this format:

```
myimage = img_to_array(myimage)
```

8. Reshape the image into a four-dimensional format since that's what is expected by the model:

```
myimage = myimage.reshape((1, 224, 224, 3))
```

9. Prepare the image for submission by running the **preprocess_input()** function:

```
myimage = preprocess_input(myimage)
```

10. Run the prediction:

```
myresult = mymodel.predict(myimage)
```

11. The prediction results in a number that needs to be converted into the corresponding label in text format:

```
mylabel = decode_predictions(myresult)
```

12. Next, type in the following code to display the label:

```
mylabel = mylabel[0][0]
```

13. Print the label using the following code:

```
print("This is a : " + mylabel[1])
```

If you have followed the steps correctly so far, the output will be as follows:

```
This is a : pizza
```

The model has successfully identified our image. Interesting, isn't it? In the next few steps, we'll take this a step further and convert this result into speech.

> **TIP**
>
> While we have used an image of a pizza here, you can use just about any image with this model. We urge you to try out this exercise multiple times with different images.

14. Prepare the text to be converted into speech:

```
sayit="This is a "+mylabel[1]
```

15. Install the **gtts** package, which is required for converting text into speech. This can be implemented in the Jupyter Notebook, as follows:

```
!pip install gtts
```

16. Import the required libraries:

```
from gtts import gTTS
import os
```

The preceding code will import two libraries. One is **gTTS**, that is, Google Text-to-Speech, which is a cloud-based open source API for converting text into speech. Another is the **os** library that is used to play the resulting audio file.

17. Call the **gTTS** API and pass the text as a parameter:

```
myobj = gTTS(text=sayit)
```

> **NOTE**
>
> You need to be online while running the preceding step.

18. Save the resulting audio file. This file will be saved in the home directory where the Jupyter Notebook is being run.

```
myobj.save("prediction.mp3")
```

> **NOTE**
>
> You can also specify the path where you want it to be saved by including the absolute path in front of the name; for example, **(myobj.save('D:/projects/prediction.mp3')**.

19. Play the audio file:

```
os.system("prediction.mp3")
```

If you have correctly followed the preceding steps, you will hear the words **This is a pizza** being spoken.

> **NOTE**
>
> To access the source code for this specific section, please refer to https://packt.live/2ZPZx8B.
>
> You can also run this example online at https://packt.live/326cRIu. You must execute the entire Notebook in order to get the desired result.

In this exercise, we learned how to build a deep learning model by making use of publicly available models using a few lines of code in TensorFlow. Now that you have got a taste of deep learning, let's move forward and learn about the different building blocks of deep learning.

DEEP LEARNING MODELS

At the heart of most of the popular deep learning models are ANNs, which are inspired by our knowledge of how the brain works. Even though no single model can be called perfect, different models perform better in different scenarios. In the sections that follow, we will learn about some of the most prominent models.

THE MULTI-LAYER PERCEPTRON

The **multi-layer perceptron** (**MLP**) is a basic type of neural network. An MLP is also known as a feed-forward network. A representation of an MLP can be seen in the following figure:

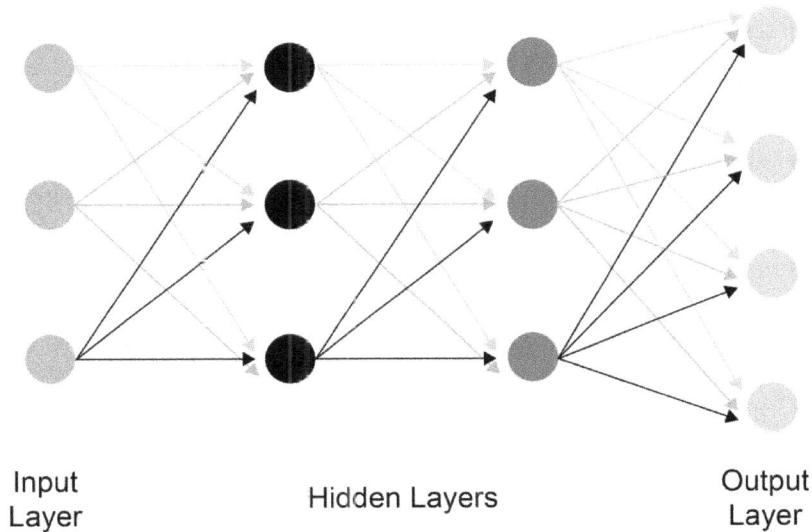

Input
Layer   Hidden Layers   Output
Layer

Figure 1.6: MLP representation

One of the basic building blocks of an MLP (or any neural network) is a neuron. A network consists of multiple neurons connected to successive layers. At a very basic level, an MLP will consist of an input layer, a hidden layer, and an output layer. The input layer will have neurons equal to the input data. Each input neuron will have a connection to all the neurons of the hidden layer. The final hidden layer will be connected to the output layer. The MLP is a very useful model and can be tried out on various classification and regression problems. The concept of an MLP will be covered in detail in *Chapter 2, Neural Networks*.

CONVOLUTIONAL NEURAL NETWORKS

A convolutional neural network (CNN) is a class of deep learning model that is predominantly used for image recognition. When we discussed the MLP, we saw that each neuron in a layer is connected to every other neuron in the subsequent layer. However, CNNs adopt a different approach and do not resort to such a fully connected architecture. Instead, CNNs extract local features from images, which are then fed to the subsequent layers.

CNNs rose to prominence in 2012 when an architecture called AlexNet won a premier competition called the **ImageNet Large-Scale Visual Recognition Challenge (ILSVRC)**. ILSVRC is a large-scale computer vision competition where teams from around the globe compete for the prize of the best computer vision model. Through the 2012 research paper titled *ImageNet Classification with Deep Convolutional Neural Networks* (https://papers.nips.cc/paper/4824-imagenet-classification-with-deep-convolutional-neural-networks), Alex Krizhevsky, et al. (University of Toronto) showcased the true power of CNN architectures, which eventually won them the 2012 ILSVRC challenge. The following figure depicts the structure of the *AlexNet* model, a CNN model whose high performance catapulted CNNs to prominence in the deep learning domain. While the structure of this model may look complicated to you, in *Chapter 3, Image Classification with Convolutional Neural Networks*, the working of such CNN networks will be explained to you in detail:

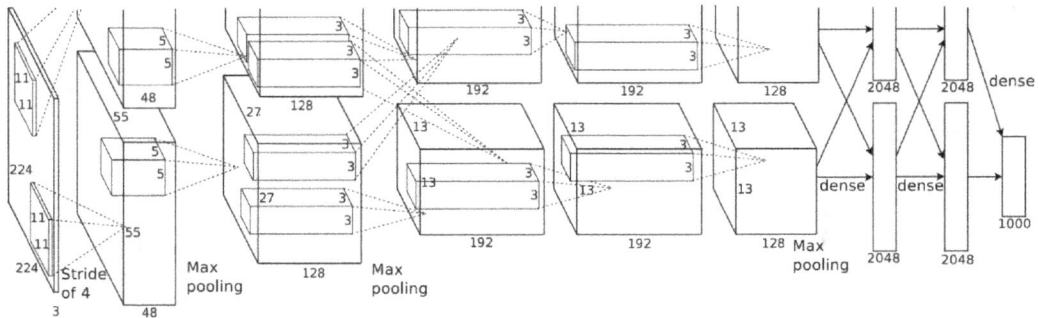

Figure 1.7: CNN architecture of the AlexNet model

> **NOTE**
>
> The aforementioned diagram is sourced from the popular research paper: Krizhevsky, Alex & Sutskever, Ilya & Hinton, Geoffrey. (2012). *ImageNet Classification with Deep Convolutional Neural Networks.* Neural Information Processing Systems. 25. 10.1145/3065386.

Since 2012, there have been many breakthrough CNN architectures expanding the possibilities for computer vision. Some of the prominent architectures are ZFNet, Inception (GoogLeNet), VGG, and ResNet.

Some of the most prominent use cases where CNNs are put to use are as follows:

- Image recognition and **optical character recognition (OCR)**
- Face recognition on social media
- Text classification
- Object detection for self-driving cars
- Image analysis for health care

Another great benefit of working with deep learning is that you needn't always build your models from scratch – you could use models built by others and use them for your own applications. This is known as "transfer learning", and it allows you to benefit from the active deep learning community.

We will apply transfer learning to image processing and learn about CNNs and their dynamics in detail in *Chapter 3, Image Classification with Convolutional Neural Networks*.

RECURRENT NEURAL NETWORKS

In traditional neural networks, the inputs are independent of the outputs. However, in cases such as language translation, where there is dependence on the words preceding and succeeding a word, there is a need to understand the dynamics of the sequences in which words appear. This problem was solved by a class of networks called **recurrent neural networks** (**RNNs**). RNNs are a class of deep learning networks where the output from the previous step is sent as input to the current step. A distinct characteristic of an RNN is a hidden layer, which remembers the information of other inputs in a sequence. A high-level representation of an RNN can be seen in the following figure. You'll learn more about the inner workings of these networks in *Chapter 5, Deep Learning for Sequences*:

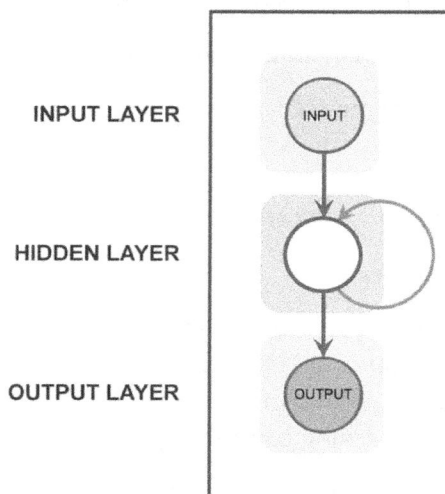

Figure 1.8: Structure of RNNs

There are different types of RNN architecture. Some of the most prominent ones are **long short-term memory** (**LSTM**) and **gated recurrent units** (**GRU**).

Some of the important use cases for RNNs are as follows:

- Language modeling and text generation
- Machine translation
- Speech recognition
- Generating image descriptions

RNNs will be covered in detail in *Chapter 5, Deep Learning for Sequences*, and *Chapter 6, LSTMs, GRUs, and Advanced RNNs*.

GENERATIVE ADVERSARIAL NETWORKS

Generative adversarial networks (GANs) are networks that are capable of generating data distributions similar to any real data distributions. One of the pioneers of deep learning, Yann LeCun, described GANs as one of the most promising ideas in deep learning in the last decade.

To give you an example, suppose we want to generate images of dogs from random noise data. For this, we train a GAN network with real images of dogs and the noise data until we generate data that looks like the real images of dogs. The following diagram explains the concept behind GANs. At this stage, you might not fully understand this concept. It will be explained in detail in *Chapter 7, Generative Adversarial Networks*.

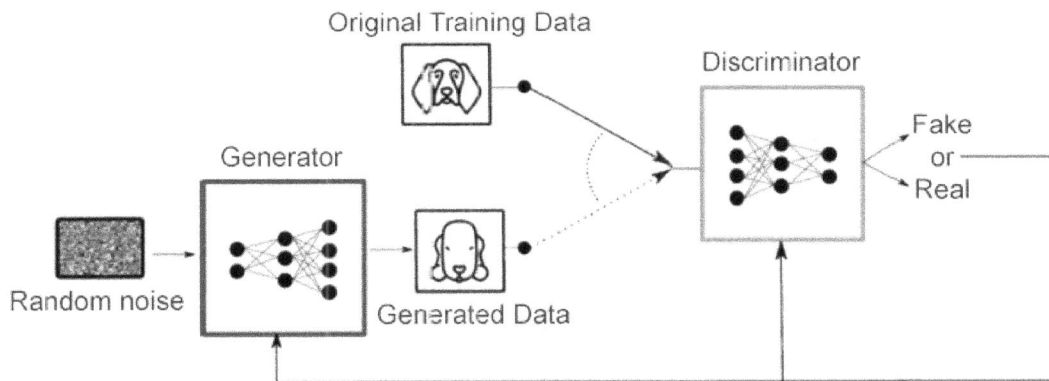

Figure 1.9: Structure of GANs

NOTE

The aforementioned diagram is sourced from the popular research paper: Barrios, Buldain, Comech, Gilbert & Orue (2019). *Partial Discharge Classification Using Deep Learning Methods—Survey of Recent Progress* (https://doi.org/10.3390/en12132485).

GANs are a big area of research, and there are many use cases for them. Some of the useful applications of GANs are as follows:

- Image translation
- Text to image synthesis
- Generating videos
- The restoration of art

GANs will be covered in detail in *Chapter 7, Generative Adversarial Networks*.

The possibilities and promises of deep learning are huge. Deep learning applications have become ubiquitous in our daily lives. Some notable examples are as follows:

- Chatbots
- Robots
- Smart speakers (such as Alexa)
- Virtual assistants
- Recommendation engines
- Drones
- Self-driving cars or autonomous vehicles

This ever-expanding canvas of possibilities makes it a great toolset in the arsenal of a data scientist. This book will progressively introduce you to the amazing world of deep learning and make you adept at applying it to real-world scenarios.

INTRODUCTION TO TENSORFLOW

TensorFlow is a deep learning library developed by Google. At the time of writing this book, TensorFlow is by far the most popular deep learning library. It was originally developed by a team within Google called the Google Brain team for their internal use and was subsequently open sourced in 2015. The Google Brain team has developed popular applications such as Google Photos and Google Cloud Speech-to-Text, which are deep learning applications based on TensorFlow. TensorFlow 1.0 was released in 2017, and within a short period of time, it became the most popular deep learning library ahead of other existing libraries, such as Caffe, Theano, and PyTorch. It is considered the industry standard, and almost every organization that is doing something in the deep learning space has adopted it. Some of the key features of TensorFlow are as follows:

- It can be used with all common programming languages, such as Python, Java, and R

- It can be deployed on multiple platforms, including Android and Raspberry Pi

- It can run in a highly distributed mode and hence is highly scalable

After being in Alpha/Beta release for a long time, the final version of TensorFlow 2.0 was released on September 30, 2019. The focus of TF2.0 was to make the development of deep learning applications easier. Let's go ahead and understand the basics of the TensorFlow 2.0 framework.

Tensors

Inside the TensorFlow program, every data element is called a **tensor**. A tensor is a representation of vectors and matrices in higher dimensions. The rank of a tensor denotes its dimensions. Some of the common data forms represented as tensors are as follows.

Scalar

A scalar is a tensor of rank 0, which only has magnitude.

For example, [**12**] is a scalar of magnitude 12.

Vector

A vector is a tensor of rank 1.

For example, [**10** , **11**, **12**, **13**].

Matrix

A matrix is a tensor of rank 2.

For example, [**[10,11]** , **[12,13]**]. This tensor has two rows and two columns.

Tensor of rank 3

This is a tensor in three dimensions. For example, image data is predominantly a three-dimensional tensor with width, height, and the number of channels as its three dimensions. The following is an example of a tensor with three dimensions, that is, it has two rows, three columns, and three channels:

```
array([[[1, 1, 1],
        [1, 1, 1]],

       [[1, 1, 1],
        [1, 1, 1]],

       [[1, 1, 1],
        [1, 1, 1]]])>
```

Figure 1.10: Tensor with three dimensions

The shape of a tensor is represented by an array and indicates the number of elements in each dimension. For example, if the shape of a tensor is [2,3,5], it means the tensor has three dimensions. If this were to be image data, this shape would mean that this tensor has two rows, three columns, and five channels. We can also get the rank from the shape. In this example, the rank of the tensor is three, since there are three dimensions. This is further illustrated in the following diagram:

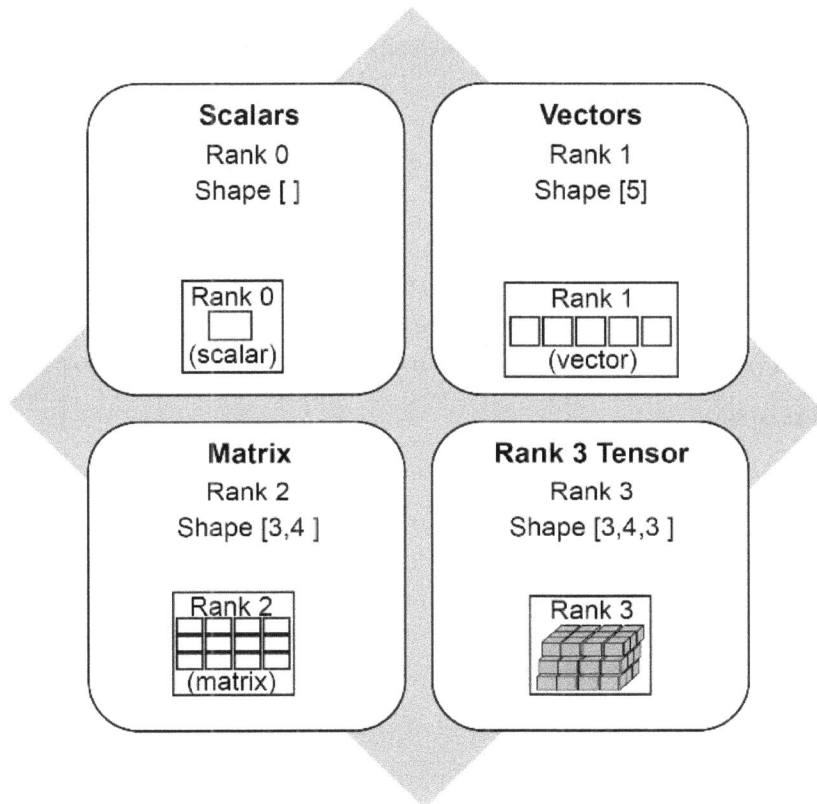

Figure 1.11: Examples of Tensor rank and shape

CONSTANTS

Constants are used to store values that are not changed or modified during the course of the program. There are multiple ways in which a constant can be created, but the simplest way is as follows:

```
a = tf.constant (10)
```

This creates a tensor initialized to 10. Keep in mind that a constant's value cannot be updated or modified by reassigning a new value to it. Another example is as follows:

```
s = tf.constant("Hello")
```

In this line, we are instantiating a string as a constant.

VARIABLES

A variable is used to store data that can be updated and modified during the course of the program. We will look at this in more detail in *Chapter 2, Neural Networks*. There are multiple ways of creating a variable, but the simplest way is as follows:

```
b=tf.Variable(20)
```

In the preceding code, the variable **b** is initialized to **20**. Note that in TensorFlow, unlike constants, the term **Variable** is written with an uppercase **V**.

A variable can be reassigned a different value during the course of the program. Variables can be used to assign any type of object, including scalars, vectors, and multi-dimensional arrays. The following is an example of how an array whose dimensions are 3 x 3 can be created in TensorFlow:

```
C = tf.Variable([[1,2,3],[4,5,6],[7,8,9]])
```

This variable can be initialized to a 3 x 3 matrix, as follows:

1	2	3
4	5	6
7	8	9

Figure 1.12: 3 x 3 matrix

Now that we know some of the basic concepts of TensorFlow, let's learn how to put them into practice.

DEFINING FUNCTIONS IN TENSORFLOW

A function can be created in Python using the following syntax:

```
def myfunc(x,y,c):
    Z=x*x*y+y+c
    return Z
```

A function is initiated using the special operator **def**, followed by the name of the function, **myfunc**, and the arguments for the function. In the preceding example, the body of the function is in the second line, and the last line returns the output.

In the following exercise, we will learn how to implement a small function using the variables and constants we defined earlier.

EXERCISE 1.02: IMPLEMENTING A MATHEMATICAL EQUATION

In this exercise, we will solve the following mathematical equation using TensorFlow:

$$Z = X^2 Y + Y + 2$$

Figure 1.13: Mathematical equation to be solved using TensorFlow

We will use TensorFlow to solve it, as follows:

```
X=3
Y=4
```

While there are multiple ways of doing this, we will only explore one of the ways in this exercise. Follow these steps to complete this exercise:

1. Open a new Jupyter Notebook and rename it *Exercise 1.02*.

2. Import the TensorFlow library using the following command:

    ```
    import tensorflow as tf
    ```

3. Now, let's solve the equation. For that, you will need to create two variables, **X** and **Y**, and initialize them to the given values of **3** and **4**, respectively:

    ```
    X=tf.Variable(3)
    Y=tf.Variable(4)
    ```

4. In our equation, the value of **2** isn't changing, so we'll store it as a constant by typing the following code:

```
C=tf.constant(2)
```

5. Define the function that will solve our equation:

```
def myfunc(x,y,c):
    Z=x*x*y+y+c
    return Z
```

6. Call the function by passing **X**, **Y**, and **C** as parameters. We'll be storing the output of this function in a variable called **result**:

```
result=myfunc(X,Y,C)
```

7. Print the result using the **tf.print()** function:

```
tf.print(result)
```

The output will be as follows:

```
42
```

> **NOTE**
>
> To access the source code for this specific section, please refer to https://packt.live/2CIXKjj.
>
> You can also run this example online at https://packt.live/2ZOIN1C. You must execute the entire Notebook in order to get the desired result.

In this exercise, we learned how to define and use a function. Those familiar with Python programming will notice that it is not a lot different from normal Python code.

In the rest of this chapter, we will prepare ourselves by looking at some basic linear algebra and familiarize ourselves with some of the common vector operations, so that understanding neural networks in the next chapter will be much easier.

LINEAR ALGEBRA WITH TENSORFLOW

The most important linear algebra topic that will be used in neural networks is matrix multiplication. In this section, we will explain how matrix multiplication works and then use TensorFlow's built-in functions to solve some matrix multiplication examples. This is essential in preparation for neural networks in the next chapter.

How does matrix multiplication work? You might have studied this as part of high school, but let's do a quick recap.

Let's say we have to perform a matrix multiplication between two matrices, A and B, where we have the following:

$$A = \begin{bmatrix} 1 & 2 & 3 \\ 4 & 5 & 6 \end{bmatrix}$$

Figure 1.14: Matrix A

$$B = \begin{bmatrix} 7 & 8 \\ 9 & 10 \\ 11 & 12 \end{bmatrix}$$

Figure 1.15: Matrix B

The first step would be to check whether multiplying a 2 x 3 matrix by a 3 x 2 matrix is possible. There is a prerequisite for matrix multiplication. Remember that C=R, that is, the number of columns (C) in the first matrix should be equal to the number of rows (R) in the second matrix. And remember the sequence matters here, and that's why, A x B is not equal to B x A. In this example, C=3 and R=3. So, multiplication is possible.

The resultant matrix would have the number of rows equal to that in A and the number of columns equal to that in B. So, in this case, the result would be a 2 x 2 matrix.

To begin multiplying the two matrices, take the elements of the first row of A (R_1) and the elements of the first column of B (C_1):

$$A(R_1) = \begin{vmatrix} 1 & 2 & 3 \end{vmatrix}$$

Figure 1.16: Matrix A(R_1)

$$B(C_1) = \begin{vmatrix} 7 \\ 9 \\ 11 \end{vmatrix}$$

Figure 1.17: Matrix B(C_1)

Get the sum of the element-wise products, that is, (1 x 7) + (2 x 9) + (3 x 11) = 58. This will be the first element in the resultant 2 x 2 matrix. We'll call this incomplete matrix D(i) for now:

$$D(i) = \begin{vmatrix} 58 \end{vmatrix}$$

Figure 1.18: Incomplete matrix D(i)

Repeat this with the first row of A(R_1) and the second column of B (C_2):

$$A(R_1) = \begin{vmatrix} 1 & 2 & 3 \end{vmatrix}$$

Figure 1.19: First row of matrix A

$$B(C_2) = \begin{vmatrix} 8 \\ 10 \\ 12 \end{vmatrix}$$

Figure 1.20: Second column of matrix B

Get the sum of the products of the corresponding elements, that is, (1 x 8) + (2 x 10) + (3 x 12) = 64. This will be the second element in the resultant matrix:

$$D(i) = \begin{vmatrix} 58 & 64 \end{vmatrix}$$

Figure 1.21: Second element of matrix D(i)

Repeat the same with the second row to get the final result:

$$D = \begin{vmatrix} 58 & 64 \\ 139 & 154 \end{vmatrix}$$

Figure 1.22: Matrix D

The same matrix multiplication can be performed in TensorFlow using a built-in method called **tf.matmul()**. The matrices that need to be multiplied must be supplied to the model as variables, as shown in the following example:

```
C = tf.matmul(A,B)
```

In the preceding case, A and B are the matrices that we want to multiply. Let's practice this method by using TensorFlow to multiply the two matrices we multiplied manually.

EXERCISE 1.03: MATRIX MULTIPLICATION USING TENSORFLOW

In this exercise, we will use the **tf.matmul()** method to multiply two matrices using **tensorflow**. Follow these steps to complete this exercise:

1. Open a new Jupyter Notebook and rename it *Exercise 1.03*.

2. Import the **tensorflow** library and create two variables, **X** and **Y**, as matrices. **X** is a 2 x 3 matrix and **Y** is a 3 x 2 matrix:

```
import tensorflow as tf
X=tf.Variable([[1,2,3],[4,5,6]])
Y=tf.Variable([[7,8],[9,10],[11,12]])
```

3. Print and display the values of **X** and **Y** to make sure the matrices are created correctly. We'll start by printing the value of **X**:

```
tf.print(X)
```

The output will be as follows:

```
[[1 2 3]
 [4 5 6]]
```

Now, let's print the value of **Y**:

```
tf.print(Y)
```

The output will be as follows:

```
[[7 8]
 [9 10]
 [11 12]]
```

4. Perform matrix multiplication by calling the TensorFlow **tf.matmul()** function:

```
c1=tf.matmul(X,Y)
```

To display the result, print the value of **c1**:

```
tf.print(c1)
```

The output will be as follows:

```
[[58 64]
 [139 154]]
```

5. Let's perform matrix multiplication by changing the order of the matrices:

```
c2=tf.matmul(Y,X)
```

To display the result, let's print the value of **c2**:

```
tf.print(c2)
```

The resulting output will be as follows.

```
[[39 54 69]
 [49 68 87]
 [59 82 105]]
```

Note that the results are different since we changed the order.

> **NOTE**
>
> To access the source code for this specific section, please refer to https://packt.live/3eevyw4.
>
> You can also run this example online at https://packt.live/2CfGGvE. You must execute the entire Notebook in order to get the desired result.

In this exercise, we learned how to create matrices in TensorFlow and how to perform matrix multiplication. This will come in handy when we create our own neural networks.

THE RESHAPE FUNCTION

Reshape, as the name suggests, changes the shape of a tensor from its current shape to a new shape. For example, you can reshape a 2 × 3 matrix to a 3 × 2 matrix, as shown here:

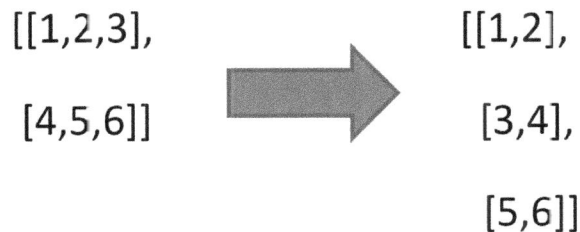

$$[[1,2,3],$$
$$[4,5,6]] \Rightarrow \begin{array}{l} [[1,2], \\ [3,4], \\ [5,6]] \end{array}$$

Figure 1.23: Reshaped matrix

Let's consider the following 2 × 3 matrix, which we defined as follows in the previous exercise:

```
X=tf.Variable([[1,2,3],[4,5,6]])
```

We can print the shape of the matrix using the following code:

```
X.shape
```

From the following output, we can see the shape, which we already know:

```
TensorShape([2, 3])
```

Now, to reshape **X** into a 3 × 2 matrix, TensorFlow provides a handy function called **tf.reshape()**. The function is implemented with the following arguments:

```
tf.reshape(X, [3,2])
```

In the preceding code, **X** is the matrix that needs to be reshaped, and **[3,2]** is the new shape that the **X** matrix has to be reshaped to.

Reshaping matrices is a handy operation when implementing neural networks. For example, a prerequisite when working with images using CNNs is that the image has to be of rank 3, that is, it has to have three dimensions: width, height, and depth. If our image is a grayscale image that has only two dimensions, the **reshape** operation will come in handy to add a third dimension. In this case, the third dimension will be 1:

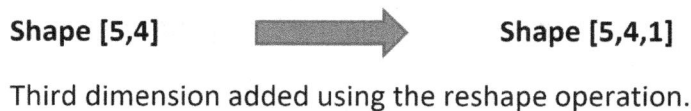

Shape [5,4] ⟹ **Shape [5,4,1]**

Third dimension added using the reshape operation.

Figure 1.24: Changing the dimension using reshape()

In the preceding figure, we are reshaping a matrix of shape **[5,4]** to a matrix of shape **[5,4,1]**. In the exercise that follows, we will be using the **reshape()** function to reshape a **[5,4]** matrix.

There are some important considerations when implementing the **reshape()** function:

- The total number of elements in the new shape should be equal to the total number of elements in the original shape. For example, you can reshape a 2 × 3 matrix (a total of 6 elements) to a 3 × 2 matrix since the new shape also has 6 elements. However, you cannot reshape it to 3 × 3 or 3 × 4.

- The **reshape()** function should not be confused with **transpose()**. In **reshape()**, the sequence of the elements of the matrix is retained and the elements are rearranged in the new shape in the same sequence. However, in the case of **transpose()**, the rows become columns and the columns become rows. Hence the sequence of the elements will change.

- The **reshape()** function will not change the original matrix unless you assign the new shape to it. Otherwise, it simply displays the new shape without actually changing the original variable. For example, let's say **x** has shape [2,3] and you simply run **tf.reshape(x, [3,2])**. When you check the shape of **x** again, it will remain as [2,3]. In order to actually change the shape, you need to assign the new shape to it, like this:

```
x=tf.reshape(x,[3,2])
```

Let's try implementing **reshape()** in TensorFlow in the exercise that follows.

EXERCISE 1.04: RESHAPING MATRICES USING THE RESHAPE() FUNCTION IN TENSORFLOW

In this exercise, we will reshape a **[5,4]** matrix into the shape of **[5,4,1]** using the **reshape()** function. This exercise will help us understand how **reshape()** can be used to change the rank of a tensor. Follow these steps to complete this exercise:

1. Open a Jupyter Notebook and rename it *Exercise 1.04*. Then, import **tensorflow** and create the matrix we want to reshape:

```
import tensorflow as tf
A=tf.Variable([[1,2,3,4], \
               [5,6,7,8], \
               [9,10,11,12], \
               [13,14,15,16], \
               [17,18,19,20]])
```

2. First, we'll print the variable **A** to check whether it is created correctly, using the following command:

```
tf.print(A)
```

The output will be as follows:

```
[[1 2 3 4]
 [5 6 7 8]
 [9 10 11 12]
 [13 14 15 16]
 [17 18 19 20]]
```

3. Let's print the shape of **A**, just to be sure:

```
A.shape
```

The output will be as follows:

```
TensorShape([5, 4])
```

Currently, it has a rank of 2. We'll be using the **reshape()** function to change its rank to 3.

4. Now, we will reshape **A** to the shape [5,4,1] using the following command. We've thrown in the **print** command just to see what the output looks like:

```
tf.print(tf.reshape(A, [5,4,1]))
```

We'll get the following output:

```
[[[1]
  [2]
  [3]
  [4]]

 [[5]
  [6]
  [7]
  [8]]

 [[9]
  [10]
  [11]
  [12]]

 [[13]
  [14]
  [15]
  [16]]

 [[17]
  [18]
  [19]
  [20]]]]
```

That worked as expected.

5. Let's see the new shape of **A**:

```
A.shape
```

The output will be as follows:

```
TensorShape([5, 4])
```

We can see that **A** still has the same shape. Remember that we discussed that in order to save the new shape, we need to assign it to itself. Let's do that in the next step.

6. Here, we'll assign the new shape to **A**:

```
A = tf.reshape(A, [5,4,1])
```

7. Let's check the new shape of **A** once again:

```
A.shape
```

We will see the following output:

```
TensorShape([5, 4, 1])
```

With that, we have not just reshaped the matrix but also changed its rank from 2 to 3. In the next step, let's print out the contents of **A** just to be sure.

8. Let's see what **A** contains now:

```
tf.print(A)
```

The output, as expected, will be as follows:

```
[[[1]
  [2]
  [3]
  [4]]

 [[5]
  [6]
  [7]
  [8]]

 [[9]
  [10]
  [11]
  [12]]
```

```
[[13]
 [14]
 [15]
 [16]]


[[17]
 [18]
 [19]
 [20]]]
```

> **NOTE**
>
> To access the source code for this specific section, please refer
> to https://packt.live/3gHvyGQ.
>
> You can also run this example online at https://packt.live/2ZdjdUY.
> You must execute the entire Notebook in order to get the desired result.

In this exercise, we saw how to use the **reshape()** function. Using **reshape()**, we can change the rank and shape of tensors. We also learned that reshaping a matrix changes the shape of the matrix without changing the order of the elements within the matrix. Another important thing that we learned was that the reshape dimension has to align with the number of elements in the matrix. Having learned about the **reshape** function, we will go ahead and learn about the next function, which is Argmax.

THE ARGMAX FUNCTION

Now, let's understand the **argmax** function, which is frequently used in neural networks. **Argmax** returns the position of the maximum value along a particular axis in a matrix or tensor. It must be noted that it does not return the maximum value, but rather the index position of the maximum value.

For example, if **x = [1,10,3,5]**, then **tf.argmax(x)** will return 1 since the maximum value (which in this case is 10) is in the index position 1.

> **NOTE**
>
> In Python, the index starts with 0. So, considering the preceding example of **x**, the element 1 will have an index of 0, 10 will have an index of 1, and so on.

Now, let's say we have the following:

$$x = \begin{vmatrix} 1 & 5 & 3 \\ 2 & 4 & 8 \\ 9 & 2 & 3 \end{vmatrix}$$

Figure 1.25: An example matrix

In this case, **argmax** has to be used with the **axis** parameter. When **axis** equals 0, it returns the position of the maximum value in each column, as shown in the following figure:

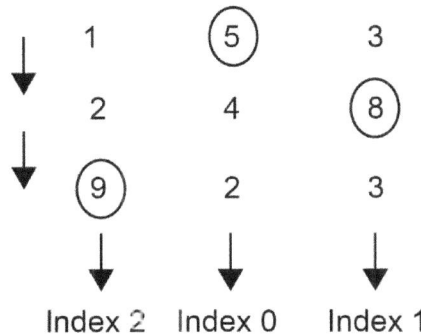

Axis 0, downward along the columns

Index 2   Index 0   Index 1

Figure 1.26: The argmax operation along axis 0

As you can see, the maximum value in the first column is 9, so the index, in this case, will be 2. Similarly, if we move along to the second column, the maximum value is 5, which has an index of 0. In the third column, the maximum value is 8, and hence the index is 1. If we were to run the **argmax** function on the preceding matrix with the **axis** as 0, we would get the following output:

```
[2,0,1]
```

When **axis** = 1, **argmax** returns the position of the maximum value across each row, like this:

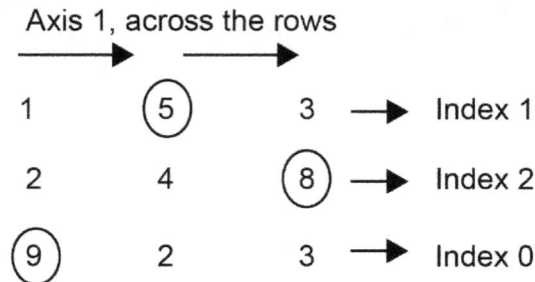

Axis 1, across the rows

1   (5)   3 ⟶ Index 1

2   4   (8) ⟶ Index 2

(9)   2   3 ⟶ Index 0

Figure 1.27: The argmax operation along axis 1

Moving along the rows, we have 5 at index 1, 8 at index 2, and 9 at index 0. If we were to run the **argmax** function on the preceding matrix with the axis as 1, we would get the following output:

```
[1,2,0]
```

With that, let's try and implement **argmax** on a matrix.

EXERCISE 1.05: IMPLEMENTING THE ARGMAX() FUNCTION

In this exercise, we are going to use the **argmax** function to find the position of the maximum value in a given matrix along axes 0 and 1. Follow these steps to complete this exercise:

1. Import **tensorflow** and create the following matrix:

```
import tensorflow as tf
X=tf.Variable([[91,12,15], [11,88,21],[90, 87,75]])
```

2. Let's print **X** and see what the matrix looks like:

```
tf.print(X)
```

The output will be as follows:

```
[[91 12 15]
 [11 88 21]
 [90 87 75]]
```

3. Print the shape of **X**:

```
X.shape
```

The output will be as follows:

```
TensorShape([3, 3])
```

4. Now, let's use **argmax** to find the positions of the maximum values while keeping **axis** as **0**:

```
tf.print(tf.argmax(X,axis=0))
```

The output will be as follows:

```
[0 1 2]
```

Referring to the matrix in *Step 2*, we can see that, moving across the columns, the index of the maximum value (91) in the first column is 0. Similarly, the index of the maximum value along the second column (88) is 1. And finally, the maximum value across the third column (75) has index 2. Hence, we have the aforementioned output.

5. Now, let's change the **axis** to **1**:

```
tf.print(tf.argmax(X,axis=1))
```

The output will be as follows:

```
[0 1 0]
```

Again, referring to the matrix in *Step 2*, if we move along the rows, the maximum value along the first row is 91, which is at index 0. Similarly, the maximum value along the second row is 88, which is at index 1. Finally, the third row is at index 0 again, with a maximum value of 75.

> **NOTE**
>
> To access the source code for this specific section, please refer to https://packt.live/2ZR5q5p.
>
> You can also run this example online at https://packt.live/3eewhNO. You must execute the entire Notebook in order to get the desired result.

In this exercise, we learned how to use the **argmax** function to find the position of the maximum value along a given axis of a tensor. This will be used in the subsequent chapters when we perform classification using neural networks.

OPTIMIZERS

Before we look at neural networks, let's learn about one more important concept, and that is optimizers. Optimizers are extensively used for training neural networks, so it is important to understand their application. In this chapter, let's get a basic introduction to the concept of an optimizer. As you might already be aware, the purpose of machine learning is to find a function (along with its parameters) that maps inputs to outputs.

For example, let's say the original function of a data distribution is a linear function (linear regression) of the following form:

```
Y = mX + b
```

Here, **Y** is the dependent variable (label), **X** the independent variable (features), and **m** and **b** are the parameters of the model. Solving this problem with machine learning would entail learning the parameters **m** and **b** and thereby the form of the function that connects **X** to **Y**. Once the parameters have been learned, if we are given a new value for **X**, we can calculate or predict the value of **Y**. It is in learning these parameters that optimizers come into play. The learning process entails the following steps:

1. Assume some arbitrary random values for the parameters **m** and **b**.

2. With these assumed parameters, for a given dataset, estimate the values of **Y** for each **X** variable.

3. Find the difference between the predicted value of **Y** and the actual value of **Y** associated with the **X** variable. This difference is called the **loss function** or **cost function**. The magnitude of loss will depend on the parameter values we initially assumed. If the assumptions were way off the actual values, then the loss will be high. The way to get toward the right parameter is by changing or altering the initial assumed values of the parameters in such a way that the loss function is minimized. This task of changing the values of the parameters to reduce the loss function is called optimization.

There are different types of optimizers that are used in deep learning. Some of the most popular ones are stochastic gradient descent, Adam, and RMSprop. The detailed functionality and the internal workings of optimizers will be described in *Chapter 2, Neural Networks*, but here, we will see how they are applied in solving certain common problems, such as simple linear regression. In this chapter, we will be using an optimizer called Adam, which is a very popular optimizer. We can define the Adam optimizer in TensorFlow using the following code:

```
tf.optimizers.Adam()
```

Once an optimizer has been defined, we can use it to minimize the loss using the following code:

```
optimizer.minimize(loss,[m,b])
```

The terms **[m,b]** are the parameters that will be changed during the optimization process. Now, let's use an optimizer to train a simple linear regression model using TensorFlow.

EXERCISE 1.06: USING AN OPTIMIZER FOR A SIMPLE LINEAR REGRESSION

In this exercise, we are going to see how to use an optimizer to train a simple linear regression model. We will start off by assuming arbitrary values for the parameters (**w** and **b**) in a linear equation **w*x + b**. Using the optimizer, we will observe how the values of the parameters change to get to the right parameter values, thus mapping the relationship between the input values (**x**) and output (**y**). Using the optimized parameter values, we will predict the output (**y**) for some given input values (**x**). After completing this exercise, we will see that the linear output, which is predicted by the optimized parameters, is very close to the real values of the output values. Follow these steps to complete this exercise:

1. Open a Jupyter Notebook and rename it *Exercise 1.06*.

2. Import **tensorflow**, create the variables, and initialize them to 0. Here, our assumed values are zero for both these parameters:

```
import tensorflow as tf
w=tf.Variable(0.0)
b=tf.Variable(0.0)
```

3. Define a function for the linear regression model. We learned how to create functions in TensorFlow earlier:

```
def regression(x):
    model=w*x+b
    return model
```

4. Prepare the data in the form of features (**x**) and labels (**y**):

```
x=[1,2,3,4]
y=[0,-1,-2,-3]
```

5. Define the **loss** function. In this case, this is the absolute value of the difference between the predicted value and the label:

```
loss=lambda:abs(regression(x)-y)
```

6. Create an **Adam** optimizer instance with a learning rate of .**01**. The learning rate defines at what rate the optimizer should change the assumed parameters. We will discuss the learning rate in subsequent chapters:

```
optimizer=tf.optimizers.Adam(.01)
```

7. Train the model by running the optimizer for 1,000 iterations to minimize the loss:

```
for i in range(1000):
    optimizer.minimize(loss, [w,b])
```

8. Print the trained values of the **w** and **b** parameters:

```
tf.print(w,b)
```

The output will be as follows:

```
-1.00371706 0.999803364
```

We can see that the values of the **w** and **b** parameters have been changed from their original values of 0, which were assumed. This is what is done during the optimizing process. These updated parameter values will be used for predicting the values of **Y**.

> **NOTE**
>
> The optimization process is stochastic in nature (having a random probability distribution), and you might get values for **w** and **b** that are different to the value that was printed here.

9. Use the trained model to predict the output by passing in the **x** values. The model predicts the values, which are very close to the label values (**y**), which means the model was trained to a high level of accuracy:

```
tf.print(regression([1,2,3,4]))
```

The output of the preceding command will be as follows:

```
[-0.00391370058 -1.00763083 -2.01134801 -3.01506495]
```

> **NOTE**
>
> To access the source code for this specific section, please refer to https://packt.live/3gSBs8b.
>
> You can also run this example online at https://packt.live/2OaFs7C. You must execute the entire Notebook in order to get the desired result.

In this exercise, we saw how to use an optimizer to train a simple linear regression model. During this exercise, we saw how the initially assumed values of the parameters were updated to get the true values. Using the true values of the parameters, we were able to get the predictions close to the actual values. Understanding how to apply the optimizer will help you later with training neural network models.

Now that we have seen the use of an optimizer, let's take what we've learned and apply the optimization function to solve a quadratic equation in the next activity.

ACTIVITY 1.01: SOLVING A QUADRATIC EQUATION USING AN OPTIMIZER

In this activity, you will use an optimizer to solve the following quadratic equation:

$$x^2 - 10x + 25 = 0$$

Figure 1.28: A quadratic equation

Here are the high-level steps you need to follow to complete this activity:

1. Open a new Jupyter Notebook and import the necessary packages, just as we did in the previous exercises.

2. Initialize the variable. Please note that, in this example, **x** is the variable that you will need to initialize. You can initialize it to a value of 0.

3. Construct the **loss** function using the **lambda** function. The **loss** function will be the quadratic equation that you are trying to solve.

4. Use the **Adam** optimizer with a learning rate of **.01**.

5. Run the optimizer for different iterations and minimize the loss. You can start the number of iterations at 1,000 and then increase it in subsequent trials until you get the result you desire.

6. Print the optimized value of **x**.

The expected output is as follows:

```
4.99919891
```

Please note that while your actual output might be a little different, it should be a value close to 5.

> **NOTE**
>
> The detailed steps for this activity, along with the solutions and additional commentary, are presented on page 388.

SUMMARY

That brings us to the end of this chapter. Let's revisit what we have learned so far. We started off by looking at the relationship between AI, machine learning, and deep learning. Then, we implemented a demo of deep learning by classifying an image and then implementing a text to speech conversion using a Google API. This was followed by a brief description of different use cases and types of deep learning, such as MLP, CNN, RNN, and GANs.

In the next section, we were introduced to the TensorFlow framework and understood some of the basic building blocks, such as tensors and their rank and shape. We also implemented different linear algebra operations using TensorFlow, such as matrix multiplication. Later in the chapter, we performed some useful operations such as **reshape** and **argmax**. Finally, we were introduced to the concept of optimizers and implemented solutions for mathematical expressions using optimizers.

Now that we have laid the foundations for deep learning and introduced you to the TensorFlow framework, the stage has been set for you to take a deep dive into the fascinating world of neural networks. In the next chapter, you will be introduced to neural networks, and in the successive chapters, we will take a look at more in-depth deep learning concepts. We hope you enjoy this fascinating journey.

2

NEURAL NETWORKS

OVERVIEW

This chapter starts with an introduction to biological neurons; we see how an artificial neural network is inspired by biological neural networks. We will examine the structure and inner workings of a simple single-layer neuron called a perceptron and learn how to implement it in TensorFlow. We will move on to building multilayer neural networks to solve more complex multiclass classification tasks and discuss the practical considerations of designing a neural network. As we build deep neural networks, we will move on to Keras to build modular and easy-to-customize neural network models in Python. By the end of this chapter, you'll be adept at building neural networks to solve complex problems.

INTRODUCTION

In the previous chapter, we learned how to implement basic mathematical concepts such as quadratic equations, linear algebra, and matrix multiplication in TensorFlow. Now that we have learned the basics, let's dive into **Artificial Neural Networks (ANNs)**, which are central to artificial intelligence and deep learning.

Deep learning is a subset of machine learning. In supervised learning, we often use traditional machine learning techniques, such as support vector machines or tree-based models, where features are explicitly engineered by humans. However, in deep learning, the model explores and identifies the important features of a labeled dataset without human intervention. ANNs, inspired by biological neurons, have a layered representation, which helps them learn labels incrementally—from the minute details to the complex ones. Consider the example of image recognition: in a given image, an ANN would just as easily identify basic details such as light and dark areas as it would identify more complex structures such as shapes. Though neural network techniques are tremendously successful at tasks such as identifying objects in images, how they do so is a black box, as the features are learned implicitly. Deep learning techniques have turned out to be powerful at tackling very complex problems, such as speech/image recognition, and hence are used across industry in building self-driving cars, Google Now, and many more applications.

Now that we know the importance of deep learning techniques, we will take a pragmatic step-by-step approach to understanding a mix of theory and practical considerations in building deep-learning-based solutions. We will start with the smallest component of a neural network, which is an artificial neuron, also referred to as a perceptron, and incrementally increase the complexity to explore **Multi-Layer Perceptrons (MLPs)** and advanced models such as **Recurrent Neural Networks (RNNs)** and **Convolutional Neural Networks (CNNs)**.

NEURAL NETWORKS AND THE STRUCTURE OF PERCEPTRONS

A neuron is a basic building block of the human nervous system, which relays electric signals across the body. The human brain consists of billions of interconnected biological neurons, and they are constantly communicating with each other by sending minute electrical binary signals by turning themselves on or off. The general meaning of a neural network is a network of interconnected neurons. In the current context, we are referring to ANNs, which are actually modeled on a biological neural network. The term artificial intelligence is derived from the fact that natural intelligence exists in the human brain (or any brain for that matter), and we humans are trying to simulate this natural intelligence artificially. Though ANNs are inspired by biological neurons, some of the advanced neural network architectures, such as CNNs and RNNs, do not actually mimic the behavior of a biological neuron. However, for ease of understanding, we will begin by drawing an analogy between the biological neuron and an artificial neuron (perceptron).

A simplified version of a biological neuron is represented in *Figure 2.1*:

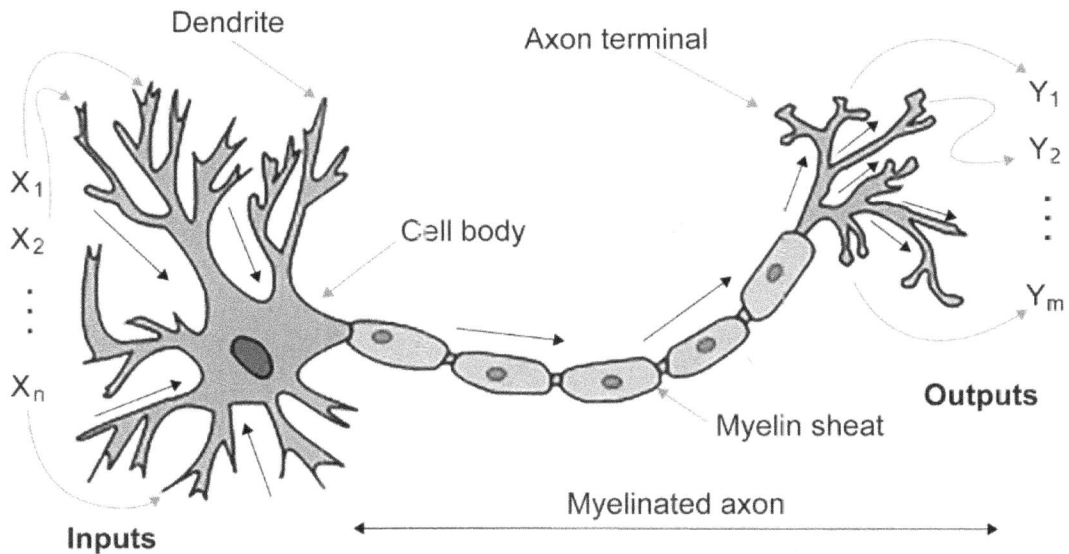

Figure 2.1: Biological neuron

This is a highly simplified representation. There are three main components:

- The dendrites, which receive the input signals

- The cell body, where the signal is processed in some form

- The tail-like axon, through which the neuron transfers the signal out to the next neuron

A perceptron can also be represented in a similar way, although it is not a physical entity but a mathematical model. *Figure 2.2* shows a high-level representation of an artificial neuron:

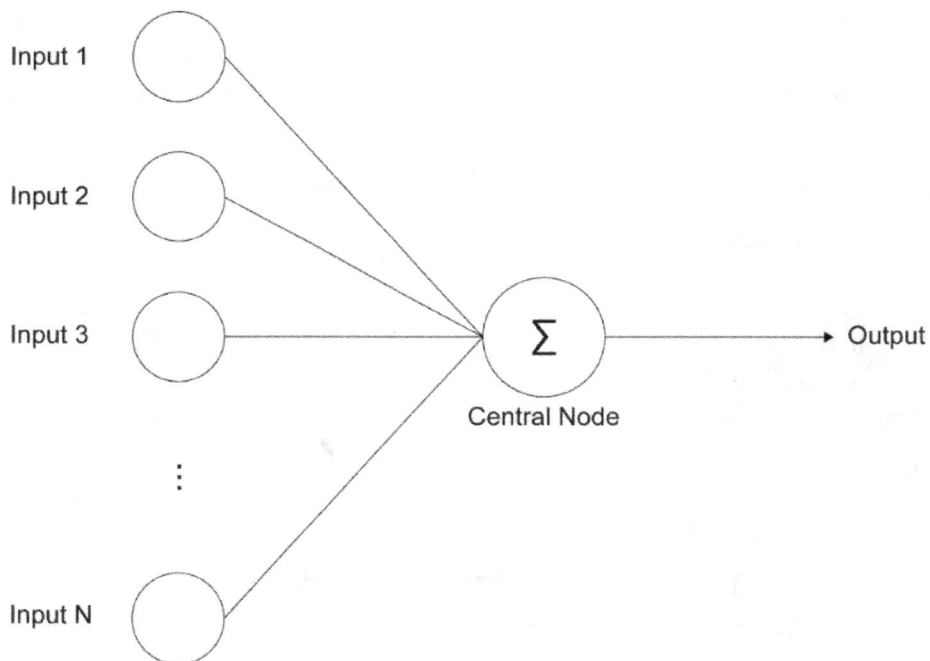

Figure 2.2: Representation of an artificial neuron

In an artificial neuron, as in a biological one, there is an input signal. The central node conflates all the signals and fires the output signal if it is above a certain threshold. A more detailed representation of a perceptron is shown in *Figure 2.3*. Each component of this perceptron is explained in the sections that follow:

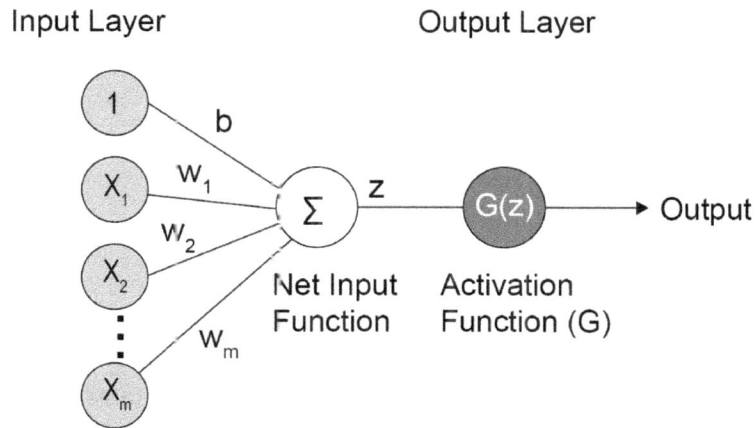

Figure 2.3: Representation of a perceptron

A perceptron has the following components:

- Input layer
- Weights
- Bias
- Net input function
- Activation function

Let's look at these components and their TensorFlow implementations in detail by considering an **OR** table dataset.

INPUT LAYER

Each example of input data is fed through the input layer. Referring to the representation shown in *Figure 2.3* depending on the size of the input example, the number of nodes will vary from x_1 to x_m. The input data can be structured data (such as a CSV file) or unstructured data, such as an image. These inputs, x_1 to x_m, are called features (**m** refers to the number of features). Let's illustrate this with an example.

Let's say the data is in the form of a table as follows:

x1	x2	y
0	0	0
0	1	1
1	0	1
1	1	1

Figure 2.4: Sample input and output data – OR table

Here, the inputs to the neuron are the columns x_1 and x_2, which correspond to one row. At this point, it may be difficult to comprehend, but for now, accept it that the data is fed one row at a time in an iterative manner during training. We will represent the input data and the true labels (output **y**) with the TensorFlow **Variable** class as follows:

```
X = tf.Variable([[0.,0.],[0.,1.],\
                [1.,0.],[1.,1.]], \
             tf.float32)
y = tf.Variable([0, 1, 1, 1], tf.float32)
```

WEIGHTS

Weights are associated with each neuron, and the input features dictate how much influence each of the input features should have in computing the next node. Each neuron will be connected to all the input features. In the example, since there were two inputs (x_1 and x_2) and the input layer is connected to one neuron, there will be two weights associated with it: w_1 and w_2. A weight is a real number; it can be positive or negative and is mathematically represented as **R**. When we say that a neural network is learning, what is happening is that the network is adjusting its weights and biases to get the correct predictions by adjusting to the error feedback. We will see this in more detail in the sections that follow. For now, we will initialize the weights as zeros and use the same TensorFlow **Variable** class as follows:

```
number_of_features = x.shape[1]
number_of_units = 1
Weight = tf.Variable(tf.zeros([number_of_features, \
                            number_of_units]), \
                        tf.float32)
```

Weights would be of the following dimension: *number of input features × output size*.

BIAS

In *Figure 2.3*, bias is represented by *b*, which is called additive bias. Every neuron has one bias. When *x* is zero, that is, no information is coming from the independent variables, then the output should be biased to just *b*. Like the weights, the bias also a real number, and the network has to learn the bias value to get the correct predictions.

In TensorFlow, bias is the same size as the output size and can be represented as follows:

```
B = tf.Variable(tf.zeros([1, 1]), tf.float32)
```

NET INPUT FUNCTION

The net input function, also commonly referred to as the input function, can be described as the sum of the products of the inputs and their corresponding weights plus the bias. Mathematically, it is represented as follows:

$$z = \sum_{i=1}^{m} w_i x_i + b$$

Figure 2.5: Net input function in mathematical form

Here:

- x_i: input data—x_1 to x_m
- w_i: weights—w_1 to w_m
- *b*: additive bias

As you can see, this formula involves inputs and their associated weights and biases. This can be written in vectorized form, and we can use matrix multiplication, which we learned about in *Chapter 1, Building Blocks of Deep Learning*. We will see this when we start the code demo. Since all the variables are numbers, the result of the net input function is just a number, a real number. The net input function can be easily implemented using the TensorFlow **matmul** functionality as follows:

```
z = tf.add(tf.matmul(X, W), B)
```

w stands for weight, **X** stands for input, and **B** stands for bias.

ACTIVATION FUNCTION (G)

The output of the net input function (**z**) is fed as input to the activation function. The activation function squashes the output of the net input function (**z**) into a new output range depending on the choice of activation function. There are a variety of activation functions, such as sigmoid (logistic), ReLU, and tanh. Each activation function has its own pros and cons. We will take a deep dive into activation functions later in the chapter. For now, we will start with a sigmoid activation function, also known as a logistic function. With the sigmoid activation function, the linear output **z** is squashed into a new output range of (0,1). The activation function provides non-linearity between layers, which gives neural networks the ability to approximate any continuous function.

The mathematical equation of the sigmoid function is as follows, where *G(z)* is the sigmoid function and the right-hand equation details the derivative with respect to *z*:

$$G(z) = \frac{1}{1 + e^{-z}}$$

Figure 2.6: Mathematical form of the sigmoid function

As you can see in *Figure 2.7*, the sigmoid function is a more or less S-shaped curve with values between 0 and 1, no matter what the input is:

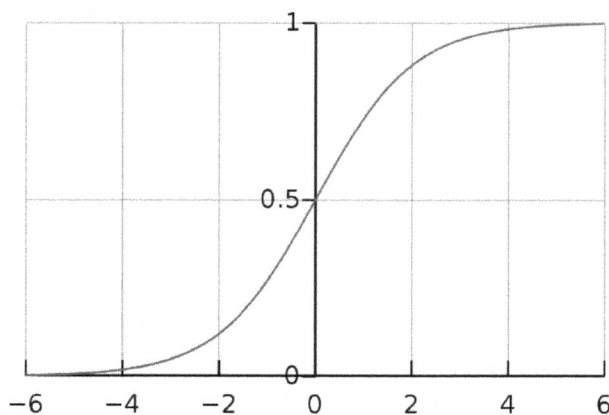

Figure 2.7: Sigmoid curve

And if we set a threshold (say **0.5**), we can convert this into a binary output. Any output greater than or equal to **.5** is considered **1**, and any value less than **.5** is considered **0**.

Activation functions such as sigmoid are provided out of the box in TensorFlow. A sigmoid function can be implemented in TensorFlow as follows:

```
output = tf.sigmoid(z)
```

Now that we have seen the structure of a perceptron and its code representation in TensorFlow, let's put all the components together to make a perceptron.

PERCEPTRONS IN TENSORFLOW

In TensorFlow, a perceptron can be implemented just by defining a simple function, as follows:

```
def perceptron(X):
    z = tf.add(tf.matmul(X, W), B)
    output = tf.sigmoid(z)
    return output
```

At a very high level, we can see that the input data passes through the net input function. The output of the net input function is passed to the activation function, which, in turn, gives us the predicted output. Now, let's look at each line of the code:

```
z = tf.add(tf.matmul(X, W), B)
```

The output of the net input function is stored in **z**. Let's see how we got that result by breaking it down further into two parts, that is, the matrix multiplication part contained in **tf.matmul** and the addition contained in **tf.add**.

Let's say we're storing the result of the matrix multiplication of **X** and **W** in a variable called **m**:

```
m = tf.matmul(X, W)
```

Now, let's consider how we got that result. For example, let's say **X** is a row matrix, like [X_1 X_2], and **W** is a column matrix, as follows:

$$\begin{bmatrix} w_1 \\ w_2 \end{bmatrix}$$

Figure 2.8: Column matrix

Recall from the previous chapter that **tf.matmul** will perform matrix multiplication. So, the result is this:

```
m = x1*w1 + x2*w2
```

And then, we add the output, **m**, to the bias, **B**, as follows:

```
z = tf.add(m, B)
```

Note that what we do in the preceding step is the same as the mere addition of the two variables **m** and **b**:

```
m + b
```

Hence, the final output is:

```
z = x1*w1 + x2*w2 + b
```

z would be the output of the net input function.

Now, let's consider the next line:

```
output= tf.sigmoid(z)
```

As we learned earlier, **tf.sigmoid** is a readily available implementation of the sigmoid function. The net input function's output (**z**) computed in the previous line is fed as input to the sigmoid function. The result of the sigmoid function is the output of the perceptron, which is in the range of 0 to 1. During training, which will be explained later in the chapter, we will feed the data in batches to this function, which will calculate the predicted values.

EXERCISE 2.01: PERCEPTRON IMPLEMENTATION

In this exercise, we will implement the perceptron in TensorFlow for an **OR** table. Let's set the input data in TensorFlow and freeze the design parameters of perceptron:

1. Let's import the necessary package, which, in our case, is **tensorflow**:

```
import tensorflow as tf
```

2. Set the input data and labels of the **OR** table data in TensorFlow:

```
X = tf.Variable([[0.,0.],[0.,1.],\
                [1.,0.],[1.,1.]], \
                dtype=tf.float32)
print(X)
```

As you can see in the output, we will have a 4 × 2 matrix of input data:

```
<tf.Variable 'Variable:0' shape=(4, 2) dtype=float32,
numpy=array([[0., 0.],
            [0., 1.],
            [1., 0.],
            [1., 1.]], dtype=float32)>
```

3. We will set the actual labels in TensorFlow and use the **reshape()** function to reshape the **y** vector into a 4 × 1 matrix:

```
y = tf.Variable([0, 1, 1, 1], dtype=tf.float32)
y = tf.reshape(y, [4,1])
print(y)
```

The output is a 4 × 1 matrix, as follows:

```
tf.Tensor(
[[0.]
 [1.]
 [1.]
 [1.]], shape=(4, 1), dtype=float32)
```

4. Now let's design parameters of a perceptron.

   Number of neurons (units) = 1

   Number of features (inputs) = 2 (number of examples × number of features)

   The activation function will be the sigmoid function, since we are doing binary classification:

   ```
   NUM_FEATURES = X.shape[1]
   OUTPUT_SIZE = 1
   ```

   In the preceding code, **X.shape[1]** will equal **2** (since the indices start with zero, **1** refers to the second index, which is **2**).

5. Define the connections weight matrix in TensorFlow:

   ```
   W = tf.Variable(tf.zeros([NUM_FEATURES, \
                             OUTPUT_SIZE]), \
                             dtype=tf.float32)
   print(W)
   ```

   The weight matrix would essentially be a columnar matrix as shown in the following figure. It will have the following dimension: *number of features (columns) × output size*:

   $$\begin{bmatrix} w_1 \\ w_2 \end{bmatrix}$$

   Figure 2.9: A columnar matrix

   The output size will be dependent on the number of neurons—in this case, it is **1**. So, if you are developing a layer of 10 neurons with two features, the shape of this matrix will be [2,10]. The **tf.zeros** function creates a tensor with the given shape and initializes all the elements to zeros.

   So, this will result in a zero columnar matrix like this:

   ```
   <tf.Variable 'Variable:0' shape=(2, 1) dtype=float32, \
   numpy=array([[0.], [0.]], dtype=float32)>
   ```

6. Now create the variable for the bias:

   ```
   B = tf.Variable(tf.zeros([OUTPUT_SIZE, 1]), dtype=tf.float32)
   print(B)
   ```

There is only one bias per neuron, so in this case, the bias is just one number in the form of a single-element array. However, if we had a layer of 10 neurons, then it would be an array of 10 numbers—1 for each neuron.

This will result in a 0-row matrix with a single element like this:

```
<tf.Variable 'Variable:0' shape=(1, 1) dtype=float32,
numpy=array([[0.]], dtype=float32)>
```

7. Now that we have the weights and bias, the next step is to perform the computation to get the net input function, feed it to the activation function, and then get the final output. Let's define a function called **perceptron** to get the output:

```
def perceptron(X):
    z = tf.add(tf.matmul(X, W), B)
    output = tf.sigmoid(z)
    return output


print(perceptron(X))
```

The output will be a 4 × 1 array that contains the predictions by our perceptron:

```
tf.Tensor(
[[0.5]
 [0.5]
 [0.5]
 [0.5]], shape=(4, 1), dtype=float32)
```

As we can see, the predictions are not quite accurate. We will learn how to improve the results in the sections that follow.

> **NOTE**
>
> To access the source code for this specific section, please refer to https://packt.live/3feF7MC.
>
> You can also run this example online at https://packt.live/2CkMiEE. You must execute the entire Notebook in order to get the desired result.

In this exercise, we implemented a perceptron, which is a mathematical implementation of a single artificial neuron. Keep in mind that it is just the implementation of the model; we have not done any training. In the next section, we will see how to train the perceptron.

TRAINING A PERCEPTRON

To train a perceptron, we need the following components:

- Data representation
- Layers
- Neural network representation
- Loss function
- Optimizer
- Training loop

In the previous section, we covered most of the preceding components: the **data representation** of the input data and the true labels in TensorFlow. For **layers**, we have the linear layer and the activation functions, which we saw in the form of the net input function and the sigmoid function respectively. For the **neural network representation**, we made a function called `perceptron()`, which uses a linear layer and a sigmoid layer to perform predictions. What we did in the previous section using input data and initial weights and biases is called **forward propagation**. The actual neural network training involves two stages: forward propagation and backward propagation. We will explore them in detail in the next few steps. Let's look at the training process at a higher level:

- A training iteration where the neural network goes through all the training examples is called an Epoch. This is one of the hyperparameters to be tweaked in order to train a neural network.

- In each pass, a neural network does forward propagation, where data travels from the input to the output. As seen in *Exercise 2.01, Perceptron Implementation*, inputs are fed to the perceptron. Input data passes through the net input function and the activation function to produce the predicted output. The predicted output is compared with the labels or the ground truth, and the error or loss is calculated.

- In order to make a neural network learn, learning being the adjustment of weights and biases in order to make correct predictions, there needs to be a **loss function**, which will calculate the error between an actual label and the predicted label.

- To minimize the error in the neural network, the training loop needs an **optimizer**, which will minimize the loss on the basis of a loss function.

- Once the error is calculated, the neural network then sees which nodes of the network contributed to the error and by how much. This is essential in order to make the predictions better in the next epoch. This way of propagating the error backward is called **backward propagation** (backpropagation). Backpropagation uses the chain rule from calculus to propagate the error (the error gradient) in reverse order until it reaches the input layer. As it propagates the error back through the network, it uses gradient descent to make fine adjustments to the weights and biases in the network by utilizing the error gradient calculated before.

This cycle continues until the loss is minimized.

Let's implement the theory we have discussed in TensorFlow. Revisit the code in *Exercise 2.01, Perceptron Implementation,* where the perceptron we created just did one forward pass. We got the following predictions, and we saw that our perceptron had not learned anything:

```
tf.Tensor(
[[0.5]
 [0.5]
 [0.5]
 [0.5]], shape=(4, 1), dtype=float32)
```

In order to make our perceptron learn, we need additional components, such as a training loop, a loss function, and an optimizer. Let's see how to implement these components in TensorFlow.

PERCEPTRON TRAINING PROCESS IN TENSORFLOW

In the next exercise, when we train our model, we will use a **Stochastic Gradient Descent (SGD)** optimizer to minimize the loss. There are a few more advanced optimizers available and provided by TensorFlow out of the box. We will look at the pros and cons of each of them in later sections. The following code will instantiate a stochastic gradient descent optimizer using TensorFlow:

```
learning_rate = 0.01
optimizer = tf.optimizers.SGD(learning_rate)
```

The **perceptron** function takes care of the forward propagation. For the backpropagation of the error, we have used an optimizer. **Tf.optimizers. SGD** creates an instance of an optimizer. SGD will update the parameters of the networks—weights and biases—on each example from the input data. We will discuss the functioning of the gradient descent optimizer in greater detail later in this chapter. We will also discuss the significance of the **0.01** parameter, which is known as the learning rate. The learning rate is the magnitude by which SGD takes a step in order to reach the global optimum of the loss function. The learning rate is another hyperparameter that needs to be tweaked in order to train a neural network.

The following code can be used to define the epochs, training loop, and loss function:

```
no_of_epochs = 1000

for n in range(no_of_epochs):
    loss = lambda:abs(tf.reduce_mean(tf.nn.\
            sigmoid_cross_entropy_with_logits\
            (labels=y,logits=perceptron(X))))
    optimizer.minimize(loss, [W, B])
```

Inside the training loop, the loss is calculated using the loss function, which is defined as a lambda function.

The **tf.nn.sigmoid_cross_entropy_with_logits** function calculates the loss value of each observation. It takes two parameters: **Labels = y** and **logit = perceptron(x)**.

perceptron(X) returns the predicted value, which is the result of the forward propagation of the input, **x**. This is compared with the corresponding label value stored in **y**. The mean value is calculated using **Tf.reduce_mean**, and the magnitude is taken. The sign is ignored using the **abs** function. **Optimizer. minimize** takes the loss value and adjusts the weights and bias as a part of the backward propagation of the error.

The forward propagation is executed again with the new values of weights and bias. And this forward and backward process continues for the number of iterations we define.

During the backpropagation, the weights and biases are updated only if the loss is less than the previous cycle. Otherwise, the weights and biases remain unchanged. In this way, the optimizer ensures that even though it loops through the required number of iterations, it only stores the values of **w** and **b** for which the loss is minimal.

We have set the number of epochs for the training to 1,000 iterations. There is no rule of thumb for setting the number of epochs since the number of epochs is a hyperparameter. But how do we know when training has taken place successfully?

When we can see that the values of weights and biases have changed, we can conclude the training has taken place. Let's say we used a training loop for the **OR** data we saw in *Exercise 2.01, Perceptron Implementation*, we would see weights somewhat equal to the following:

```
[[0.412449151]
 [0.412449151]]
```

And the bias would be something like this:

```
0.236065879
```

When the network has learned, that is, the weights and biases have been updated, we can see whether it is making accurate predictions using **accuracy_score** from the **scikit-learn** package. We can use it to measure the accuracy of the predictions as follows:

```
from sklearn.metrics import accuracy_score
print(accuracy_score(y, ypred))
```

Here, **accuracy_score** takes two parameters—the label values (**y**) and the predicted values (**ypred**)—and measures the accuracy. Let's say the result is **1.0**. This means the perceptron is 100% accurate.

In the next exercise, we will train our perceptron to perform a binary classification.

EXERCISE 2.02: PERCEPTRON AS A BINARY CLASSIFIER

In the previous section, we learned how to train a perceptron. In this exercise, we will train our perceptron to approximate a slightly more complicated function. We will be using randomly generated external data with two classes: class **0** and class **1**. Our trained perceptron should be able to classify the random numbers based on their class:

> **NOTE**
>
> The data is in a CSV file called **data.csv**. You can download the file from GitHub by visiting https://packt.live/2BVtxlf.

1. Import the required libraries:

```
import tensorflow as tf
import pandas as pd
from sklearn.metrics import confusion_matrix
from sklearn.metrics import accuracy_score
import matplotlib.pyplot as plt
%matplotlib inline
```

Apart from **tensorflow**, we will need **pandas** to read the data from the CSV file, **confusion_matrix** and **accuracy_score** to measure the accuracy of our perceptron after the training, and **matplotlib** to visualize the data.

2. Read the data from the **data.csv** file. It should be in the same path as the Jupyter Notebook file in which you are running this exercise's code. Otherwise, you will have to change the path in the code before executing it:

```
df = pd.read_csv('data.csv')
```

3. Examine the data:

```
df.head()
```

The output will be as follows

	label	x1	x2
0	1	2.6487	4.5192
1	1	1.5438	2.4443
2	1	1.8990	4.2409
3	1	2.4711	5.8097
4	1	3.3590	6.4423

Figure 2.10: Contents of the DataFrame

As you can see, the data has three columns. **x1** and **x2** are the features, and the **label** column contains the labels **0** or **1** for each observation. The best way to see this kind of data s through a scatter plot.

4. Visualize the data by plotting it using **matplotlib**:

```
plt.scatter(df[df['label'] == 0]['x1'], \
            df[df['label'] == 0]['x2'], \
            marker='*')
plt.scatter(df[df['label'] == 1]['x1'], \
            df[df['label'] == 1]['x2'], marker='<')
```

The output will be as follows:

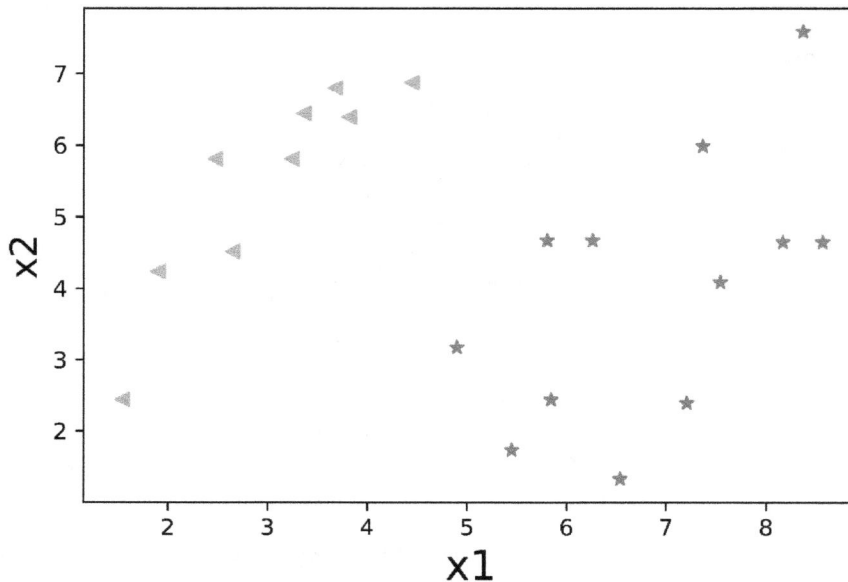

Figure 2.11: Scatter plot of external data

This shows the two distinct classes of the data shown by the two different shapes. Data with the label **0** is represented by a star, while data with the label **1** is represented by a triangle.

5. Prepare the data. This step is not unique to neural networks; you must have seen it in regular machine learning as well. Before submitting the data to a model for training, you split it into features and labels:

```
X_input = df[['x1','x2']].values
y_label = df[['label']].values
```

x_input contains the features, **x1** and **x2**. The values at the end convert it into matrix format, which is what is expected as input when the tensors are created. **y_label** contains the labels in matrix format.

6. Create TensorFlow variables for features and labels and typecast them to **float**:

```
x = tf.Variable(X_input, dtype=tf.float32)
y = tf.Variable(y_label, dtype=tf.float32)
```

7. The rest of the code is for the training of the perceptron, which we saw in *Exercise 2.01, Perceptron Implementation*:

Exercise2.02.ipynb

```
Number_of_features = 2
Number_of_units = 1
learning_rate = 0.01

# weights and bias
weight = tf.Variable(tf.zeros([Number_of_features, \
                              Number_of_units]))
bias = tf.Variable(tf.zeros([Number_of_units]))

#optimizer
optimizer = tf.optimizers.SGD(learning_rate)

def perceptron(x):
    z = tf.add(tf.matmul(x,weight),bias)
    output = tf.sigmoid(z)
    return output
```

The complete code for this step can be found at https://packt.live/3gJ73bY.

> **NOTE**
>
> The **#** symbol in the code snippet above denotes a code comment.
> Comments are added into code to help explain specific bits of logic.

8. Display the values of **weight** and **bias** to show that the perceptron has been trained:

```
tf.print(weight, bias)
```

The output is as follows:

```
[[-0.844034135]
 [0.673354745]] [0.0593947917]
```

9. Pass the input data to check whether the perceptron classifies it correctly:

```
ypred = perceptron(x)
```

10. Round off the output to convert it into binary format:

```
ypred = tf.round(ypred)
```

11. Measure the accuracy using the **accuracy_score** method, as we did in the previous exercise:

```
acc = accuracy_score(y.numpy(), ypred.numpy())
print(acc)
```

The output is as follows:

```
1.0
```

The perceptron gives 100% accuracy.

12. The confusion matrix helps to get the performance measurement of a model. We will plot the confusion matrix using the **scikit-learn** package.

```
cnf_matrix = confusion_matrix(y.numpy(), \
                              ypred.numpy())
print(cnf_matrix)
```

The output will be as follows:

```
[[12  0]
 [ 0  9]]
```

All the numbers are along the diagonal, that is, 12 values corresponding to class 0 and 9 values corresponding to class 1 are properly classified by our trained perceptron (which has achieved 100% accuracy).

> **NOTE**
>
> To access the source code for this specific section, please refer to https://packt.live/3gJ73bY.
>
> You can also run this example online at https://packt.live/2DhelFw.
> You must execute the entire Notebook in order to get the desired result.

In this exercise, we trained our perceptron into a binary classifier, and it has done pretty well. In the next exercise, we will see how to create a multiclass classifier.

MULTICLASS CLASSIFIER

A classifier that can handle two classes is known as a **binary classifier**, like the one we saw in the preceding exercise. A classifier that can handle more than two classes is known as a **multiclass classifier**. We cannot build a multiclass classifier with a single neuron. Now we move from one neuron to one layer of multiple neurons, which is required for multiclass classifiers.

A single layer of multiple neurons can be trained to be a multiclass classifier. Some of the key points are detailed here. You need as many neurons as the number of classes; that is, for a 3-class classifier, you need 3 neurons; for a 10-class classifier you need 10 neurons, and so on.

As we saw in binary classification, we used sigmoid (logistic layer) to get predictions in the range of 0 to 1. In multiclass classification, we use a special type of activation function called the **Softmax** activation function to get probabilities across each class that sums to 1. With the sigmoid function in a multiclass setting, the probabilities do not necessarily add up to 1, so Softmax is preferred.

Before we implement the multiclass classifier, let's explore the Softmax activation function.

THE SOFTMAX ACTIVATION FUNCTION

The Softmax function is also known as the **normalized exponential function**. As the word **normalized** suggests, the Softmax function normalizes the input into a probability distribution that sums to 1. Mathematically, it is represented as follows:

$$\mathrm{Softmax}(x_i) = \frac{\exp(x_i)}{\sum_j \exp(x_j)}$$

Figure 2.12: Mathematical form of the Softmax function

To understand what Softmax does, let's use TensorFlow's built-in **softmax** function and see the output.

So, for the following code:

```
values = tf.Variable([3,1,7,2,4,5], dtype=tf.float32)
output = tf.nn.softmax(values)
tf.print(output)
```

The output will be:

```
[0.0151037546 0.00204407098 0.824637055
 0.00555636082 0.0410562605 0.111602485]
```

As you can see in the output, the **values** input is mapped to a probability distribution that sums to 1. Note that **7** (the highest value in the original input values) received the highest weight, **0.824637055**. This is what the Softmax function is mainly used for: to focus on the largest values and suppress values that are below the maximum value. Also, if we sum the output, it adds up to ~ 1.

Illustrating the example in more detail, let's say we want to build a multiclass classifier with 3 classes. We will need 3 neurons connected to a Softmax activation function:

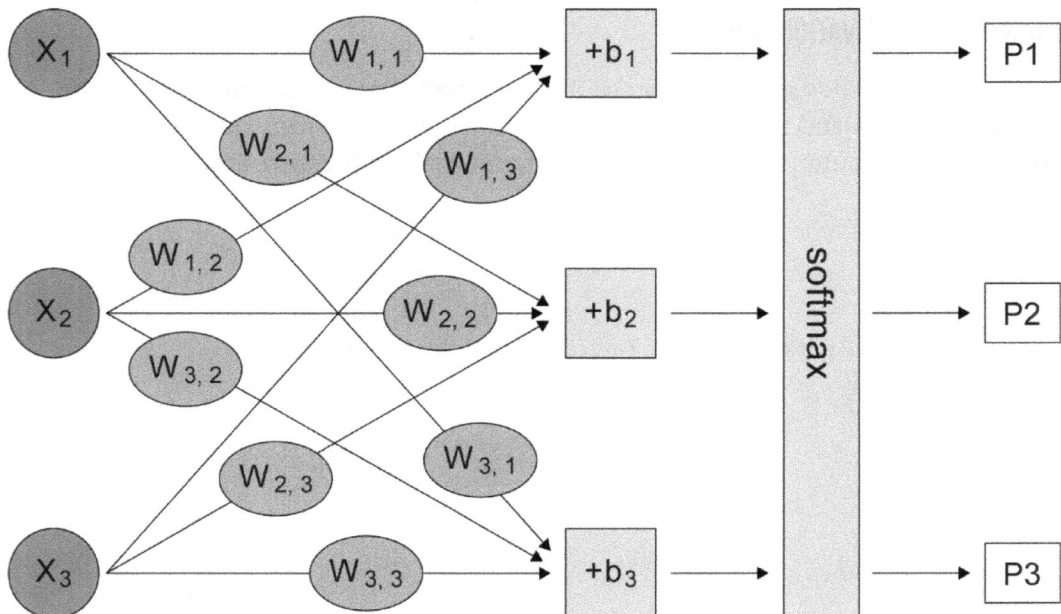

Figure 2.13: Softmax activation function used in a multiclass classification setting

As seen in *Figure 2.13*, $\mathbf{x}_1$, $\mathbf{x}_2$, and $\mathbf{x}_3$ are the input features, which go through the net input function of each of the three neurons, which have the weights and biases ($\mathbf{W}_{i,j}$ and $\mathbf{b}_i$) associated with it. Lastly, the output of the neuron is fed to the common Softmax activation function instead of the individual sigmoid functions. The Softmax activation function spits out the probabilities of the 3 classes: **P1**, **P2**, and **P3**. The sum of these three probabilities will add to 1 because of the Softmax layer.

As we saw in the previous section, Softmax highlights the maximum value and suppresses the rest of the values. Suppose a neural network is trained to classify the input into three classes, and for a given set of inputs, the output is class 2; then it would say that **P2** has the highest value since it is passed through a Softmax layer. As you can see in the following figure, **P2** has the highest value, which means the prediction is correct:

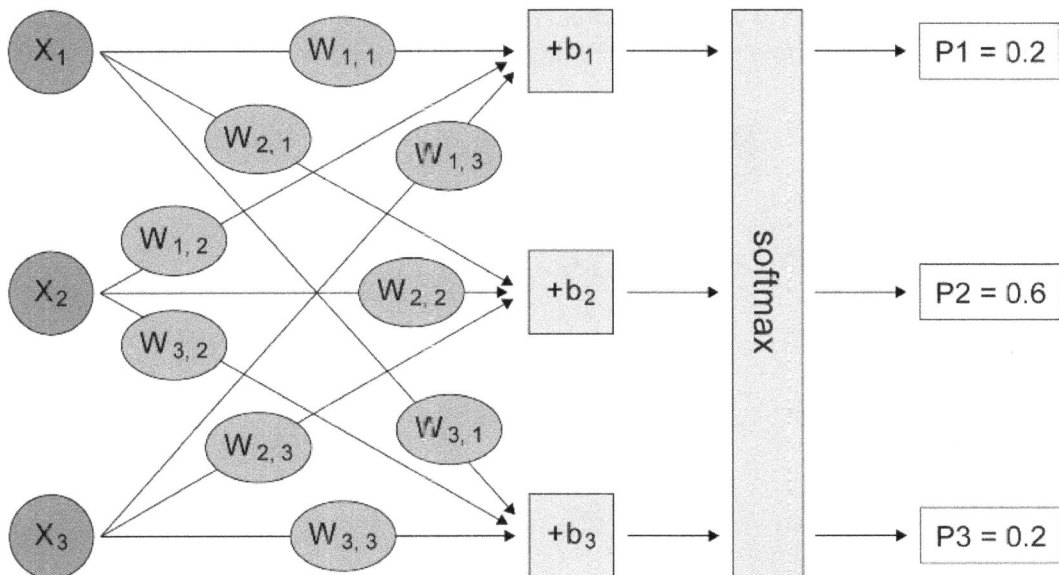

Figure 2.14: Probability P2 is the highest

An associated concept is one-hot encoding. As we have three different classes, **class1**, **class2**, and **class3**, we need to encode the class labels into a format that we can work with more easily; so, after applying one-hot encoding, we would see the following output:

	class 1	class 2	class 3
0	1	0	0
1	0	1	0
2	0	0	1

Figure 2.15: One-hot encoded data for three classes

This makes the results quick and easy to interpret. In this case, the output that has the highest value is set to 1, and all others are set to 0. The one-hot encoded output of the preceding example would be like this:

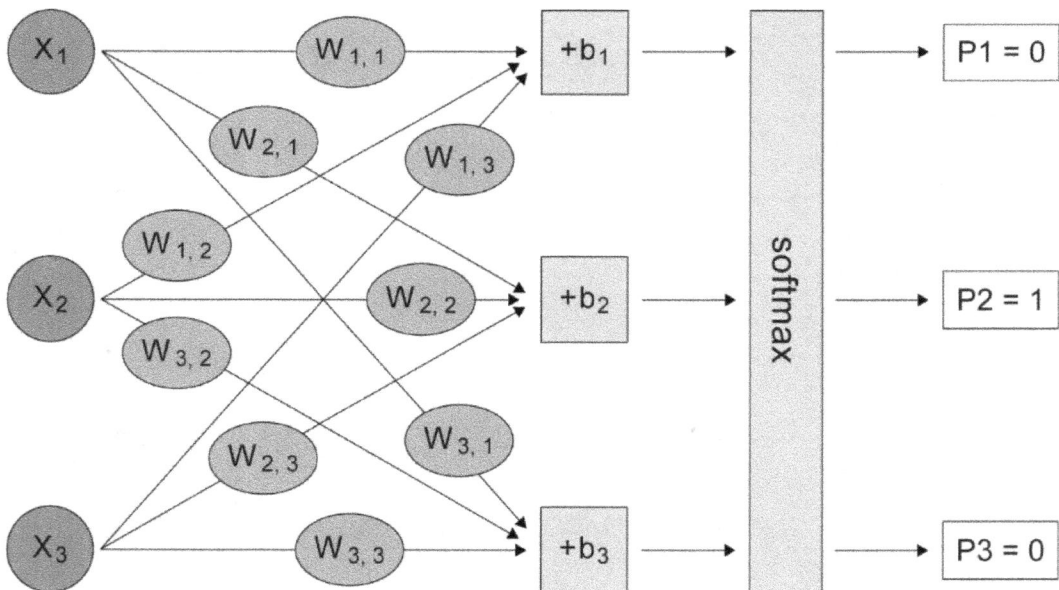

Figure 2.16: One-hot encoded output probabilities

The labels of the training data also need to be one-hot encoded. And if they have a different format, they need to be converted into one-hot-encoded format before training the model. Let's do an exercise on multiclass classification with one-hot encoding.

EXERCISE 2.03: MULTICLASS CLASSIFICATION USING A PERCEPTRON

To perform multiclass classification, we will be using the Iris dataset (https://archive.ics.uci.edu/ml/datasets/Iris), which has 3 classes of 50 instances each, where each class refers to a type of Iris. We will have a single layer of three neurons using the Softmax activation function:

> **NOTE**
>
> You can download the dataset from GitHub using this link:
> https://packt.live/3ekiBBf.

1. Import the required libraries:

```
import tensorflow as tf
import pandas as pd

from sklearn.metrics import confusion_matrix
from sklearn.metrics import accuracy_score

import matplotlib.pyplot as plt
%matplotlib inline

from pandas import get_dummies
```

You must be familiar with all of these imports as they were used in the previous exercise, except for **get_dummies**. This function converts a given label data into the corresponding one-hot-encoded format.

2. Load the **iris.csv** data:

```
df = pd.read_csv('iris.csv')
```

3. Let's examine the first five rows of the data:

```
df.head()
```

The output will be as follows:

	petallength	petalwidth	sepallength	sepalwidth	species
0	5.1	3.5	1.4	0.2	0
1	4.9	3.0	1.4	0.2	0
2	4.7	3.2	1.3	0.2	0
3	4.6	3.1	1.5	0.2	0
4	5.0	3.6	1.4	0.2	0

Figure 2.17: Contents of the DataFrame

4. Visualize the data by using a scatter plot:

```
plt.scatter(df[df['species'] == 0]['sepallength'],\
            df[df['species'] == 0]['sepalwidth'], marker='*')
plt.scatter(df[df['species'] == 1]['sepallength'],\
            df[df['species'] == 1]['sepalwidth'], marker='<')
plt.scatter(df[df['species'] == 2]['sepallength'], \
            df[df['species'] == 2]['sepalwidth'], marker='o')
```

The resulting plot will be as follows. The x axis denotes the sepal length and the y axis denotes the sepal width. The shapes in the plot represent the three species of Iris, setosa (star), versicolor (triangle), and virginica (circle):

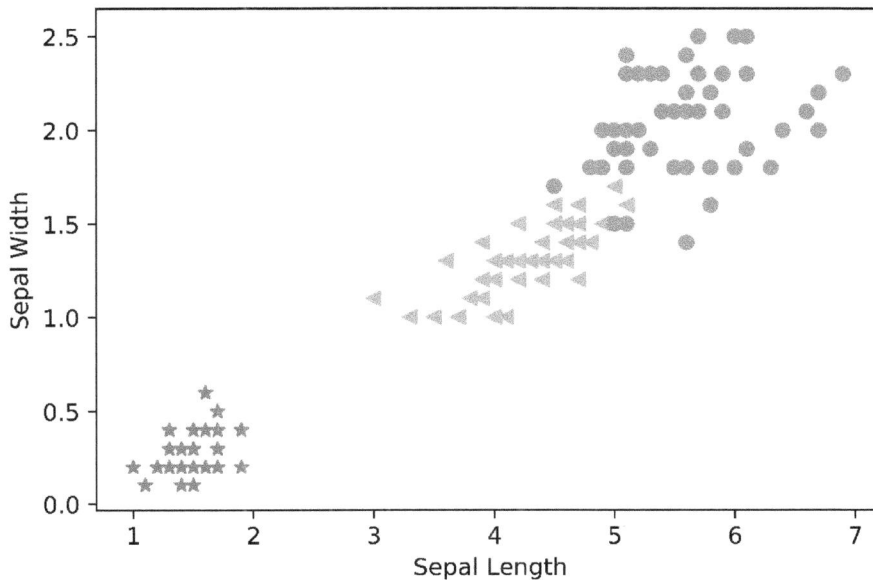

Figure 2.18: Iris data scatter plot

There are three classes, as can be seen in the visualization, denoted by different shapes.

5. Separate the features and the labels:

```
x = df[['petallength', 'petalwidth', \
        'sepallength', 'sepalwidth']].values
y = df['species'].values
```

values will transform the features into matrix format.

6. Prepare the data by doing one-hot encoding on the classes:

```
y = get_dummies(y)
y = y.values
```

get_dummies(y) will convert the labels into one-hot-encoded format.

7. Create a variable to load the features and typecast it to **float32**:

```
x = tf.Variable(x, dtype=tf.float32)
```

8. Implement the **perceptron** layer with three neurons:

```
Number_of_features = 4
Number_of_units = 3


# weights and bias
weight = tf.Variable(tf.zeros([Number_of_features, \
                                Number_of_units]))
bias = tf.Variable(tf.zeros([Number_of_units]))
def perceptron(x):
    z = tf.add(tf.matmul(x, weight), bias)
    output = tf.nn.softmax(z)
    return output
```

The code looks very similar to the single perceptron implementation. Only the **Number_of_units** parameter is set to **3**. Therefore, the weight matrix will be 4 x 3 and the bias matrix will be 1 x 3.

The other change is in the activation function:

Output=tf.nn.softmax(x)

We are using **softmax** instead of **sigmoid**.

9. Create an instance of the **optimizer**. We will be using the **Adam** optimizer. At this point, you can think of **Adam** as an improved version of gradient descent that converges faster. We will cover it in detail later in the chapter:

```
optimizer = tf.optimizers.Adam(.01)
```

10. Define the training function:

```
def train(i):
    for n in range(i):
        loss=lambda: abs(tf.reduce_mean\
                    (tf.nn.softmax_cross_entropy_with_logits(\
                    labels=y, logits=perceptron(x))))
        optimizer.minimize(loss, [weight, bias])
```

Again, the code looks very similar to the single-neuron implementation except for the loss function. Instead of **sigmoid_cross_entropy_with_logits**, we use **softmax_cross_entropy_with_logits**.

11. Run the training for **1000** iterations:

```
train(1000)
```

12. Print the values of the weights to see if they have changed. This is also an indication that our perceptron is learning:

```
tf.print(weight)
```

The output shows the learned weights of our perceptron:

```
[[0.684310317 0.895633 -1.0132345]
 [2.6424644 -1.13437736 -3.20665336]
 [-2.96634197 -0.129377216 3.2572844]
 [-2.97383809 -3.13501668 3.2313652]]
```

13. To test the accuracy, we feed the features to predict the output and then calculate the accuracy using **accuracy_score**, like in the previous exercise:

```
ypred=perceptron(x)
ypred=tf.round(ypred)
accuracy_score(y, ypred)
```

The output is:

```
0.98
```

It has given 98% accuracy, which is pretty good.

> **NOTE**
>
> To access the source code for this specific section, please refer to https://packt.live/2Dhes3U.
>
> You can also run this example online at https://packt.live/3iIJKkm. You must execute the entire Notebook in order to get the desired result.

In this exercise, we performed multiclass classification using our perceptron. Let's do a more complex and interesting case study of the handwritten digit recognition dataset in the next section.

MNIST CASE STUDY

Now that we have seen how to train a single neuron and a single layer of neurons, let's take a look at more realistic data. MNIST is a famous case study. In the next exercise, we will create a 10-class classifier to classify the MNIST dataset. However, before that, you should get a good understanding of the MNIST dataset.

Modified National Institute of Standards and Technology (**MNIST**) refers to the modified dataset that the team led by Yann LeCun worked with at NIST. This project was aimed at handwritten digit recognition using neural networks.

We need to understand the dataset before we get into writing the code. The MNIST dataset is integrated into the TensorFlow library. It consists of 70,000 handwritten images of the digits 0 to 9:

Figure 2.19: Handwritten digits

When we say images, you might think these are JPEG files, but they are not. They are actually stored in the form of pixel values. As far as the computer is concerned, an image is a bunch of numbers. These numbers are pixel values ranging from 0 to 255. The dimension of each of these images is 28 x 28. The images are stored in the form of a 28 x 28 matrix, each cell containing real numbers ranging from 0 to 255. These are grayscale images (commonly known as black and white). 0 indicates white and 1 indicates complete black, and values in between indicate a certain shade of gray. The MNIST dataset is split into 60,000 training images and 10,000 test images.

Each image has a label associated with it ranging from 0 to 9. In the next exercise, let's build a 10-class classifier to classify the handwritten MNIST images.

EXERCISE 2.04: CLASSIFYING HANDWRITTEN DIGITS

In this exercise, we will build a single-layer 10-class classifier consisting of 10 neurons with the Softmax activation function. It will have an input layer of 784 pixels:

1. Import the required libraries and packages just like we did in the earlier exercise:

```
import tensorflow as tf
import pandas as pd
from sklearn.metrics import accuracy_score

import matplotlib.pyplot as plt
%matplotlib inline

from pandas import get_dummies
```

2. Create an instance of the MNIST dataset:

```
mnist = tf.keras.datasets.mnist
```

3. Load the MNIST dataset's **train** and **test** data:

```
(train_features, train_labels), (test_features, test_labels) = \
mnist.load_data()
```

4. Normalize the data:

```
train_features, test_features = train_features / 255.0, \
                                test_features / 255.0
```

5. Flatten the 2-dimensional images into row matrices. So, a 28 × 28 pixel gets flattened to **784** using the **reshape** function:

```
x = tf.reshape(train_features, [60000, 784])
```

6. Create a **Variable** with the features and typecast it to **float32**:

```
x = tf.Variable(x)
x = tf.cast(x, tf.float32)
```

7. Create a one-hot encoding of the labels and transform it into a matrix:

```
y_hot = get_dummies(train_labels)
y = y_hot.values
```

8. Create the single-layer neural network with **10** neurons and train it for **1000** iterations:

Exercise2.04.ipynb

```
#defining the parameters
Number_of_features = 784
Number_of_units = 10

# weights and bias
weight = tf.Variable(tf.zeros([Number_of_features, \
                              Number_of_units]))
bias = tf.Variable(tf.zeros([Number_of_units]))
```

The complete code for this step can be accessed from https://packt.live/3efd7Yh.

9. Prepare the test data to measure the accuracy:

```
# Prepare the test data to measure the accuracy.
test = tf.reshape(test_features, [10000, 784])
test = tf.Variable(test)
test = tf.cast(test, tf.float32)

test_hot = get_dummies(test_labels)
test_matrix = test_hot.values
```

10. Run the predictions by passing the test data through the network:

```
ypred = perceptron(test)
ypred = tf.round(ypred)
```

11. Calculate the accuracy:

```
accuracy_score(test_hot, ypred)
```

The predicted accuracy is:

```
0.9304
```

> **NOTE**
>
> To access the source code for this specific section, please refer to https://packt.live/3efd7Yh.
>
> You can also run this example online at https://packt.live/2Oc83ZW.
> You must execute the entire Notebook in order to get the desired result.

In this exercise, we saw how to create a single-layer multi-neuron neural network and train it as a multiclass classifier.

The next step is to build a multilayer neural network. However, before we do that, we must learn about the Keras API, since we use Keras to build dense neural networks.

KERAS AS A HIGH-LEVEL API

In TensorFlow 1.0, there were several APIs, such as Estimator, Contrib, and layers. In TensorFlow 2.0, Keras is very tightly integrated with TensorFlow, and it provides a high-level API that is user-friendly, modular, composable, and easy to extend in order to build and train deep learning models. This also makes developing code for neural networks much easier. Let's see how it works.

EXERCISE 2.05: BINARY CLASSIFICATION USING KERAS

In this exercise, we will implement a very simple binary classifier with a single neuron using the Keras API. We will use the same **data.csv** file that we used in *Exercise 2.02, Perceptron as a Binary Classifier*:

> **NOTE**
>
> The dataset can be downloaded from GitHub by accessing the following GitHub link: https://packt.live/2BVtxIf.

1. Import the required libraries:

```
import tensorflow as tf

import pandas as pd

import matplotlib.pyplot as plt
%matplotlib inline

# Import Keras libraries
from tensorflow.keras.models import Sequential
from tensorflow.keras.layers import Dense
```

In the code, **Sequential** is the type of Keras model that we will be using because it is very easy to add layers to it. **Dense** is the type of layer that will be added. These are the regular neural network layers as opposed to the convolutional layers or pooling layers that will be used later on.

2. Import the data:

```
df = pd.read_csv('data.csv')
```

3. Inspect the data:

```
df.head()
```

The following will be the output:

	label	x1	x2
0	1	2.6487	4.5192
1	1	1.5438	2.4443
2	1	1.8990	4.2409
3	1	2.4711	5.8097
4	1	3.3590	6.4423

Figure 2.20: Contents of the DataFrame

4. Visualize the data using a scatter plot:

```
plt.scatter(df[df['label'] == 0]['x1'], \
            df[df['label'] == 0]['x2'], marker='*')
plt.scatter(df[df['label'] == 1]['x1'], \
            df[df['label'] == 1]['x2'], marker='<')
```

The resulting plot is as follows, with the *x* axis denoting **x1** values and the y-axis denoting **x2** values:

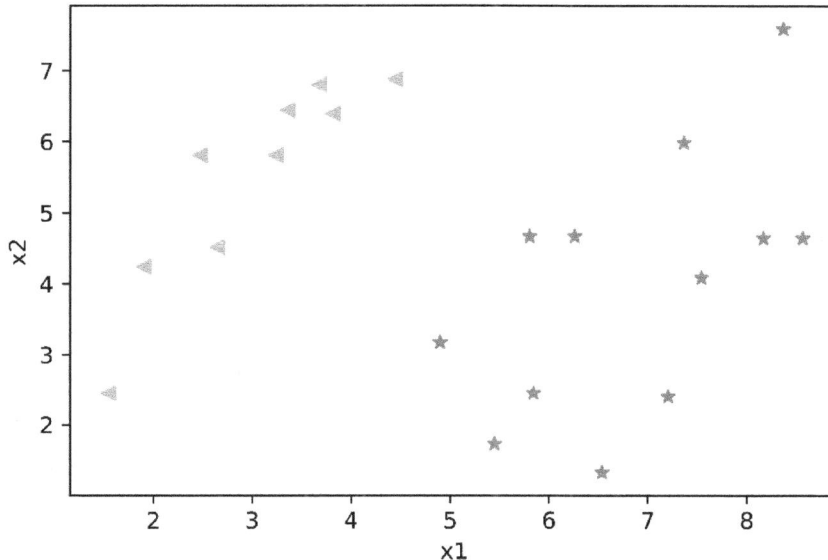

Figure 2.21: Scatter plot of the data

5. Prepare the data by separating the features and labels and setting the **tf** variables:

```
x_input = df[['x1','x2']].values
y_label = df[['label']].values
```

6. Create a neural network model consisting of a single layer with a neuron and a sigmoid activation function:

```
model = Sequential()
model.add(Dense(units=1, input_dim=2, activation='sigmoid'))
```

The parameters in **mymodel.add(Dense())** are as follows: **units** is the number of neurons in the layer; **input_dim** is the number of features, which in this case is **2**; and **activation** is **sigmoid**.

7. Once the model is created, we use the **compile** method to pass the additional parameters that are needed for training, such as the type of the optimizer, the loss function, and so on:

```
model.compile(optimizer='adam', \
              loss='binary_crossentropy',\
              metrics=['accuracy'])
```

In this case, we are using the **adam** optimizer, which is an enhanced version of the gradient descent optimizer, and the loss function is **binary_ crossentropy**, since this is a binary classifier.

The **metrics** parameter is almost always set to **['accuracy']**, which is used to display information such as the number of epochs, the training loss, the training accuracy, the test loss, and the test accuracy during the training process.

8. The model is now ready to be trained. However, it is a good idea to check the configuration of the model by using the **summary** function:

```
model.summary()
```

The output will be as follows:

```
Model: "sequential"
_____
Layer (type)                 Output Shape              Param #
=================================================================
dense (Dense)                (None, 1)                 3
=================================================================
Total params: 3
Trainable params: 3
Non-trainable params: 0
_____
```

Figure 2.22: Summary of the sequential model

9. Train the model by calling the **fit()** method:

```
model.fit(x_input, y_label, epochs=1000)
```

It takes the features and labels as the data parameters along with the number of epochs, which in this case is **1000**. The model will start training and will continuously provide the status as shown here:

```
Epoch 994/1000
21/21 [==============================] - 0s 144us/sample - loss: 0.2458 - accuracy: 1.0000
Epoch 995/1000
21/21 [==============================] - 0s 144us/sample - loss: 0.2456 - accuracy: 1.0000
Epoch 996/1000
21/21 [==============================] - 0s 123us/sample - loss: 0.2454 - accuracy: 1.0000
Epoch 997/1000
21/21 [==============================] - 0s 146us/sample - loss: 0.2451 - accuracy: 1.0000
Epoch 998/1000
21/21 [==============================] - 0s 143us/sample - loss: 0.2449 - accuracy: 1.0000
Epoch 999/1000
21/21 [==============================] - 0s 134us/sample - loss: 0.2447 - accuracy: 1.0000
Epoch 1000/1000
21/21 [==============================] - 0s 174us/sample - loss: 0.2444 - accuracy: 1.0000
```

Figure 2.23: Model training logs using Keras

10. We will evaluate our model using Keras's **evaluate** functionality:

```
model.evaluate(x_input, y_label)
```

The output is as follows:

```
21/21 [==============================] - 0s 611us/sample - loss:
   0.2442 - accuracy: 1.0000
[0.24421504139900208, 1.0]
```

As you can see, our Keras model is able to train well, as our accuracy is 100%.

> **NOTE**
>
> To access the source code for this specific section, please refer to https://packt.live/2ZVV1VY
>
> You can also run this example online at https://packt.live/38CzhTc.
> You must execute the entire Notebook in order to get the desired result.

In this exercise, we have learned how to build a perceptron using Keras. As you have seen, Keras makes the code more modular and more readable, and the parameters easier to tweak. In the next section, we will see how to build a multilayer or deep neural network using Keras.

MULTILAYER NEURAL NETWORK OR DEEP NEURAL NETWORK

In the previous example, we developed a single-layer neural network, often referred to as a shallow neural network. A diagram of this follows:

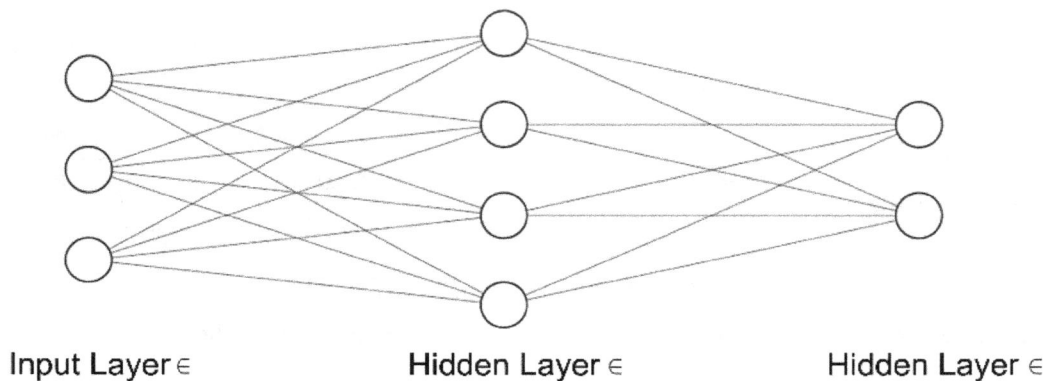

Input Layer ∈   **Hidden Layer** ∈   **Hidden Layer** ∈

Figure 2.24: Shallow neural network

One layer of neurons is not sufficient to solve more complex problems, such as face recognition or object detection. You need to stack up multiple layers. This is often referred to as creating a deep neural network. A diagram of this follows:

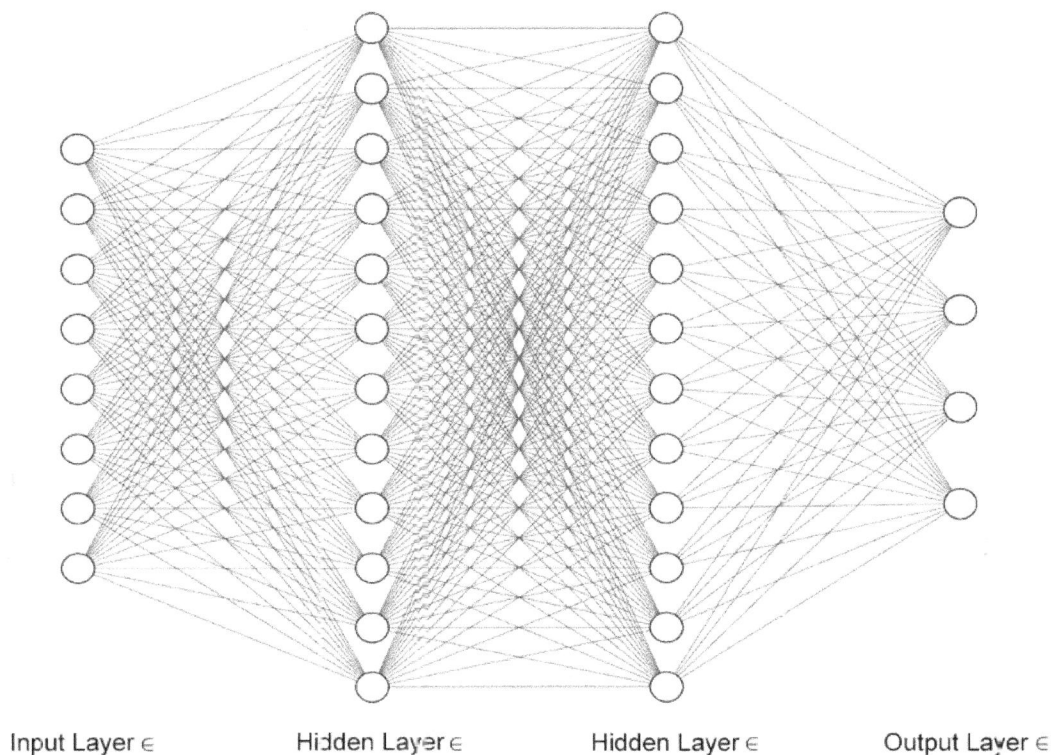

Input Layer ∈   Hidden Layer ∈   Hidden Layer ∈   Output Layer ∈

Figure 2.25: Deep neural network

Before we jump into the code, let's try to understand how this works. Input data is fed to the neurons in the first layer. It must be noted that every input is fed to every neuron in the first layer, and every neuron has one output. The output from each neuron in the first layer is fed to every neuron in the second layer. The output of each neuron in the second layer is fed to every neuron in the third layer, and so on.

That is why this kind of network is also referred to as a dense neural network or a fully connected neural network. There are other types of neural networks with different workings, such as CNNs, but that is something we will discuss in the next chapter. There is no set rule about the number of neurons in each layer. This is usually determined by trial and error in a process known as hyperparameter tuning (which we'll learn about later in the chapter). However, when it comes to the number of neurons in the last layers, there are some restrictions. The configuration of the last layer is determined as follows:

Binary classification	Multiclass classification
No. of Neurons = 1	No. of Neurons = number of classes
Activation function – Sigmoid	Activation function - Softmax

Figure 2.26: Last layer configuration

RELU ACTIVATION FUNCTION

One last thing to do before we implement the code for deep neural networks is learn about the ReLU activation function. This is one of the most popular activation functions used in multilayer neural networks.

ReLU is a shortened form of **Rectified Linear Unit**. The output of the ReLU function is always a non-negative value that is greater than or equal to 0:

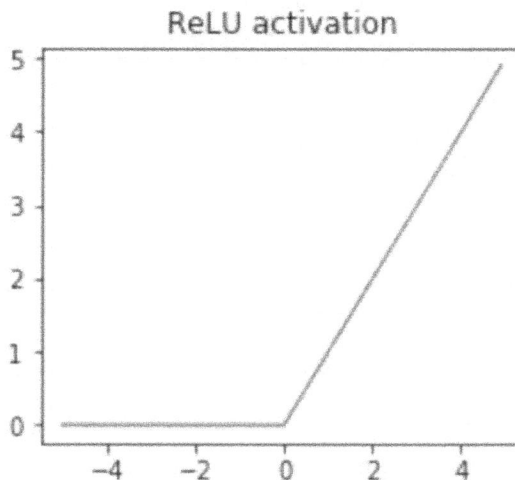

Figure 2.27: ReLU activation function

The mathematical expression for ReLU is:

$$f(x) = max(0, x)$$

Figure 2.28: ReLU activation function

ReLU converges much more quickly than the sigmoid activation function, and therefore it is by far the most widely used activation function. ReLU is used in almost every deep neural network. It is used in all the layers except the last layer, where either sigmoid or Softmax is used.

The ReLU activation function is provided by TensorFlow out of the box. To see how it is implemented, let's give some sample input values to a ReLU function and see the output:

```
values = tf.Variable([1.0, -2., 0., 0.3, -1.5], dtype=tf.float32)
output = tf.nn.relu(values)
tf.print(output)
```

The output is as follows:

```
[1 0 0 0.3 0]
```

As you can see, all the positive values are retained, and the negative values are suppressed to zero. Let's use this ReLU activation function in the next exercise to do a multilayer binary classification task.

EXERCISE 2.06: MULTILAYER BINARY CLASSIFIER

In this exercise, we will implement a multilayer binary classifier using the **data.csv** file that we used in *Exercise 2.02, Perceptron as a Binary Classifier*.

We will build a binary classifier with a deep neural network of the following configuration. There will be an input layer with 2 nodes and 2 hidden layers, the first with 50 neurons and the second with 20 neurons, and lastly a single neuron to do the final prediction belonging to any binary class:

> **NOTE**
>
> The dataset can be downloaded from GitHub using the following link:
> https://packt.live/2BVtxIf .

1. Import the required libraries and packages:

```
import tensorflow as tf
import pandas as pd

import matplotlib.pyplot as plt
%matplotlib inline

##Import Keras libraries
from tensorflow.keras.models import Sequential
from tensorflow.keras.layers import Dense
```

2. Import and inspect the data:

```
df = pd.read_csv('data.csv')
df.head()
```

The output is as follows:

	label	x1	x2
0	1	2.6487	4.5192
1	1	1.5438	2.4443
2	1	1.8990	4.2409
3	1	2.4711	5.8097
4	1	3.3590	6.4423

Figure 2.29: The first five rows of the data

3. Visualize the data using a scatter plot:

```
plt.scatter(df[df['label'] == 0]['x1'], \
            df[df['label'] == 0]['x2'], marker='*')
plt.scatter(df[df['label'] == 1]['x1'], \
            df[df['label'] == 1]['x2'], marker='<')
```

The resulting output is as follows, with the *x* axis showing **x1** values and the *y* axis showing **x2** values:

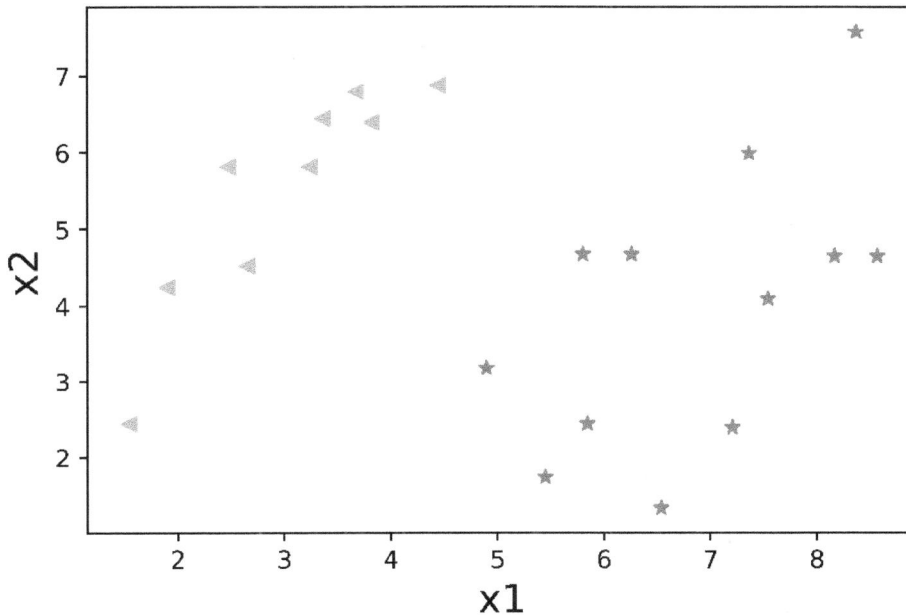

Figure 2.30: Scatter plot for given data

4. Prepare the data by separating the features and labels and setting the **tf** variables:

```
x_input = df[['x1','x2']].values
y_label = df[['label']].values
```

5. Build the **Sequential** model:

```
model = Sequential()
model.add(Dense(units = 50,input_dim=2, activation = 'relu'))
model.add(Dense(units = 20 , activation = 'relu'))
model.add(Dense(units = 1,input_dim=2, activation = 'sigmoid'))
```

Here are a couple of points to consider. We provide the input details for the first layer, then use the ReLU activation function for all the intermediate layers, as discussed earlier. Furthermore, the last layer has only one neuron with a sigmoid activation function for binary classifiers.

6. Provide the training parameters using the **compile** method:

```
model.compile(optimizer='adam', \
              loss='binary_crossentropy', metrics=['accuracy'])
```

7. Inspect the **model** configuration using the **summary** function:

```
model.summary()
```

The output will be as follows:

```
Model: "sequential"
_____
Layer (type)                 Output Shape              Param #
=================================================================
dense (Dense)                (None, 50)                150
_____
dense_1 (Dense)              (None, 20)                1020
_____
dense_2 (Dense)              (None, 1)                 21
=================================================================
Total params: 1,191
Trainable params: 1,191
Non-trainable params: 0
_____
```

Figure 2.31: Deep neural network model summary using Keras

In the model summary, we can see that there are a total of **1191** parameters—weights and biases—to learn across the hidden layers to the output layer.

8. Train the model by calling the **fit()** method:

```
model.fit(x_input, y_label, epochs=50)
```

Notice that, in this case, the model reaches 100% accuracy within **50** epochs, unlike the single-layer model, which needed about 1,000 epochs:

```
Epoch 8/50
21/21 [==============================] - 0s 285us/sample - loss: 0.4644 - accuracy: 0.5714
Epoch 9/50
21/21 [==============================] - 0s 285us/sample - loss: 0.4502 - accuracy: 0.5714
Epoch 10/50
21/21 [==============================] - 0s 380us/sample - loss: 0.4378 - accuracy: 0.5714
Epoch 11/50
21/21 [==============================] - 0s 380us/sample - loss: 0.4267 - accuracy: 0.7143
Epoch 12/50
21/21 [==============================] - 0s 380us/sample - loss: 0.4163 - accuracy: 0.8571
Epoch 13/50
21/21 [==============================] - 0s 285us/sample - loss: 0.4059 - accuracy: 0.9524
Epoch 14/50
21/21 [==============================] - 0s 190us/sample - loss: 0.3960 - accuracy: 0.9524
Epoch 15/50
21/21 [==============================] - 0s 238us/sample - loss: 0.3863 - accuracy: 1.0000
Epoch 16/50
21/21 [==============================] - 0s 332us/sample - loss: 0.3769 - accuracy: 1.0000
Epoch 17/50
21/21 [==============================] - 0s 380us/sample - loss: 0.3677 - accuracy: 1.0000
Epoch 18/50
21/21 [==============================] - 0s 237us/sample - loss: 0.3585 - accuracy: 1.0000
Epoch 19/50
21/21 [==============================] - 0s 236us/sample - loss: 0.3495 - accuracy: 1.0000
```

Figure 2.32: Multilayer model train logs

9. Let's evaluate the model's performance:

```
model.evaluate(x_input, y_label)
```

The output is as follows:

```
21/21 [==============================] - 0s 6ms/sample - loss:
  0.1038 - accuracy: 1.0000
[0.1037961095571518, 1.0]
```

Our model has now been trained and demonstrates 100% accuracy.

> **NOTE**
>
> To access the source code for this specific section, please refer
> to https://packt.live/2ZUkM94.
>
> You can also run this example online at https://packt.live/3iKsD1W.
> You must execute the entire Notebook in order to get the desired result.

In this exercise, we learned how to build a multilayer neural network using Keras. This is a binary classifier. In the next exercise, we will build a deep neural network for a multiclass classifier with the MNIST dataset.

EXERCISE 2.07: DEEP NEURAL NETWORK ON MNIST USING KERAS

In this exercise, we will perform a multiclass classification by implementing a deep neural network (multi-layer) for the MNIST dataset where our input layer comprises 28 × 28 pixel images flattened to 784 input nodes followed by 2 hidden layers, the first with 50 neurons and the second with 20 neurons. Lastly, there will be a Softmax layer consisting of 10 neurons since we are classifying the handwritten digits into 10 classes:

1. Import the required libraries and packages:

```
import tensorflow as tf
import pandas as pd

import matplotlib.pyplot as plt
%matplotlib inline

# Import Keras libraries
from tensorflow.keras.models import Sequential
from tensorflow.keras.layers import Dense
from tensorflow.keras.layers import Flatten
```

2. Load the MNIST data:

```
mnist = tf.keras.datasets.mnist
(train_features,train_labels), (test_features,test_labels) = \
mnist.load_data()
```

train_features has the training images in the form of 28 x 28 pixel values.

train_labels has the training labels. Similarly, **test_features** has the test images in the form of 28 x 28 pixel values. **test_labels** has the test labels.

3. Normalize the data:

```
train_features, test_features = train_features / 255.0, \
                      test_features / 255.0
```

The pixel values of the images range from 0-255. We need to normalize the values by dividing them by 255 so that the range goes from 0 to 1.

4. Build the **sequential** model:

```
model = Sequential()
model.add(Flatten(input_shape=(28,28)))
model.add(Dense(units = 50, activation = 'relu'))
model.add(Dense(units = 20 , activation = 'relu'))
model.add(Dense(units = 10, activation = 'softmax'))
```

There are couple of points to note. The first layer in this case is not actually a layer of neurons but a **Flatten** function. This flattens the 28 x 28 image into a single array of **784**, which is fed to the first hidden layer of **50** neurons. The last layer has **10** neurons corresponding to the 10 classes with a **softmax** activation function.

5. Provide training parameters using the **compile** method:

```
model.compile(optimizer = 'adam', \
              loss = 'sparse_categorical_crossentropy', \
              metrics = ['accuracy'])
```

> **NOTE**
>
> The loss function used here is different from the binary classifier. For a multiclass classifier, the following loss functions are used: **sparse_categorical_crossentropy**, which is used when the labels are not one-hot encoded, as in this case; and, **categorical_crossentropy**, which is used when the labels are one-hot encoded.

6. Inspect the model configuration using the **summary** function:

```
model.summary()
```

The output is as follows:

```
Model: "sequential_3"

_____
Layer (type)                 Output Shape              Param #
=================================================================
flatten_2 (Flatten)          (None, 784)               0
_____
dense_9 (Dense)              (None, 50)                39250
_____
dense_10 (Dense)             (None, 20)                1020
_____
dense_11 (Dense)             (None, 10)                210
=================================================================
Total params: 40,480
Trainable params: 40,480
Non-trainable params: 0
```

Figure 2.33: Deep neural network summary

In the model summary, we can see that there are a total of 40,480 parameters—weights and biases—to learn across the hidden layers to the output layer.

7. Train the model by calling the **fit** method:

```
model.fit(train_features, train_labels, epochs=50)
```

The output will be as follows:

```
Train on 60000 samples
Epoch 1/50
60000/60000 [==============================] - 3s 58us/sample - loss: 0.3271 - accuracy: 0.9064
Epoch 2/50
60000/60000 [==============================] - 3s 49us/sample - loss: 0.1521 - accuracy: 0.9553s - loss: 0.153
Epoch 3/50
60000/60000 [==============================] - 3s 52us/sample - loss: 0.1130 - accuracy: 0.9660
Epoch 4/50
60000/60000 [==============================] - 4s 72us/sample - loss: 0.0925 - accuracy: 0.9717
Epoch 5/50
60000/60000 [==============================] - 3s 52us/sample - loss: 0.0779 - accuracy: 0.9761
Epoch 6/50
60000/60000 [==============================] - 3s 49us/sample - loss: 0.0672 - accuracy: 0.9789
Epoch 7/50
60000/60000 [==============================] - 4s 64us/sample - loss: 0.0589 - accuracy: 0.9814
Epoch 8/50
60000/60000 [==============================] - 3s 50us/sample - loss: 0.0520 - accuracy: 0.9844
Epoch 9/50
60000/60000 [==============================] - 3s 53us/sample - loss: 0.0466 - accuracy: 0.9851
Epoch 10/50
60000/60000 [==============================] - 4s 62us/sample - loss: 0.0408 - accuracy: 0.9868
Epoch 11/50
60000/60000 [==============================] - 3s 47us/sample - loss: 0.0378 - accuracy: 0.9882
```

Figure 2.34: Deep neural network training logs

8. Test the model by calling the **evaluate()** function:

```
model.evaluate(test_features, test_labels)
```

The output will be:

```
10000/10000 [==============================] - 1s 76us/sample - loss:
  0.2072 - accuracy: 0.9718
[0.20719025060918111, 0.9718]
```

Now that the model s trained and tested, in the next few steps, we will run the prediction with some images selected randomly.

9. Load a random image from a test dataset. Let's locate the 200th image:

```
loc = 200
test_image = test_features[loc]
```

10. Let's see the shape of the image using the following command:

```
test_image.shape
```

The output is:

```
(28,28)
```

We can see that the shape of the image is 28 x 28. However, the model expects 3-dimensional input. We need to reshape the image accordingly.

11. Use the following code to reshape the image:

```
test_image = test_image.reshape(1,28,28)
```

12. Let's call the **predict()** method of the model and store the output in a variable called **result**:

```
result = model.predict(test_image)
print(result)
```

result has the output in the form of 10 probability values, as shown here:

```
[[2.9072076e-28 2.1215850e-29 1.7854708e-21
  1.0000000e+00 0.0000000e+00 1.2384960e-15
  1.2660366e-34 1.7712217e-32 1.7461657e-08
  9.6417470e-29]]
```

13. The position of the highest value will be the prediction. Let's use the **argmax** function we learned about in the previous chapter to find out the prediction:

```
result.argmax()
```

In this case, it is **3**:

```
3
```

14. In order to check whether the prediction is correct, we check the label of the corresponding image:

```
test_labels[loc]
```

Again, the value is **3**:

```
3
```

15. We can also visualize the image using **pyplot**:

```
plt.imshow(test_features[loc])
```

The output will be as follows:

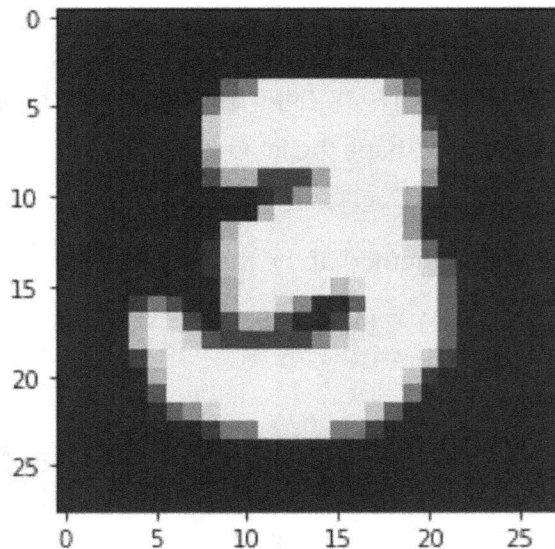

Figure 2.35: Test image visualized

And this shows that the prediction is correct.

> **NOTE**
>
> To access the source code for this specific section, please refer to https://packt.live/2O5KRgd
>
> You can also run this example online at https://packt.live/2O8JHR0. You must execute the entire Notebook in order to get the desired result.

In this exercise, we created a multilayer multiclass neural network model using Keras to classify the MNIST data. With the model we built, we were able to correctly predict a random handwritten digit.

EXPLORING THE OPTIMIZERS AND HYPERPARAMETERS OF NEURAL NETWORKS

Training a neural network to get good predictions requires tweaking a lot of hyperparameters such as optimizers, activation functions, the number of hidden layers, the number of neurons in each layer, the number of epochs, and the learning rate. Let's go through each of them one by one and discuss them in detail.

GRADIENT DESCENT OPTIMIZERS

In an earlier section titled *Perceptron Training Process in TensorFlow*, we briefly touched upon the gradient descent optimizer without going into the details of how it works. This is a good time to explore the gradient descent optimizer in a little more detail. We will provide an intuitive explanation without going into the mathematical details.

The gradient descent optimizer's function is to minimize the loss or error. To understand how gradient descent works, you can think of this analogy: imagine a person at the top of a hill who wants to reach the bottom. At the beginning of the training, the loss is large, like the height of the hill's peak. The functioning of the optimizer is akin to the person descending the hill to the valley at the bottom, or rather, the lowest point of the hill, and not climbing up the hill that is on the other side of the valley.

Remember the learning rate parameter that we used while creating the optimizer? That can be compared to the size of the steps the person takes to climb down the hill. If these steps are large, it is fine at the beginning since the person can climb down faster, but once they near the bottom, if the steps are too large, the person crosses over to the other side of the valley. Then, in order to climb back down to the bottom of the valley, the person will try to move back but will move over to the other side again. This results in going back and forth without reaching the bottom of the valley.

On the other hand, if the person takes very small steps (a very small learning rate), they will take forever to reach the bottom of the valley; in other words, the model will take forever to converge. So, finding a learning rate that is neither too small nor too big is very important. However, unfortunately, there is no rule of thumb to find out in advance what the right value should be—we have to find it by trial and error.

There are two main types of gradient-based optimizers: batch and stochastic gradient descent. Before we jump into them, let's recall that one epoch means a training iteration where the neural network goes through all the training examples:

- In an epoch, when we reduce the loss across all the training examples, it is called **batch gradient descent**. This is also known as **full batch gradient descent**. To put it simply, after going through a full batch, we take a step to adjust the weights and biases of the network to reduce the loss and improve the predictions. There is a similar form of it called mini-batch gradient descent, where we take steps, that is, we adjust weights and biases, after going through a subset of the full dataset.

- In contrast to batch gradient descent, when we take a step at one example per iteration, we have **stochastic gradient descent (SGD)**. The word *stochastic* tells us there is randomness involved here, which, in this case, is the batch that is randomly selected.

Though SGD works relatively well, there are advanced optimizers that can speed up the training process. They include SGD with momentum, Adagrad, and Adam.

THE VANISHING GRADIENT PROBLEM

In the *Training a Perceptron* section, we learned about the forward and backward propagation of neural networks. When a neural network performs forward propagation, the error gradient is calculated with respect to the true label, and backpropagation is performed to see which parameters (the weights and biases) of the neural network have contributed to the error and the extent to which they have done so. The error gradient is propagated from the output layer to the input layer to calculate gradients with respect to each parameter, and in the last step, the gradient descent step is performed to adjust the weights and biases according to the calculated gradient. As the error gradient is propagated backward, the gradients calculated at each parameter become smaller and smaller as it advances to the lower (initial) layers. This decrease in the gradients means that the changes to the weights and biases become smaller and smaller. Hence, our neural network struggles to find the global minimum and does not give good results. This is called the vanishing gradient problem. The problem happens with the use of the sigmoid (logistic) function as an activation function, and hence we use the ReLU activation function to train deep neural network models to avoid gradient complications and improve the results.

HYPERPARAMETER TUNING

Like any other model training process in machine learning, it is possible to perform hyperparameter tuning to improve the performance of the neural network model. One of the parameters is the learning rate. The other parameters are as follows:

- **Number of epochs**: Increasing the number of epochs generally increases the accuracy and lowers the loss

- **Number of layers**: Increasing the number of layers increases the accuracy, as we saw in the exercises with MNIST

- **Number of neurons per layer**: This also increases the accuracy

And once again, there is no way to know in advance what the right number of layers or the right number of neurons per layer is. This has to be figured out by trial and error. It has to be noted that the larger the number of layers and the larger the number of neurons per layer, the greater the computational power required. Therefore, we start with the smallest possible numbers and slowly increase the number of layers and neurons.

OVERFITTING AND DROPOUT

Neural networks with complex architectures and too many parameters tend to fit on all the data points, including noisy labels, leading to the problem of overfitting and neural networks that are not able to generalize well on unseen datasets. To tackle this issue, there is a technique called **dropout**:

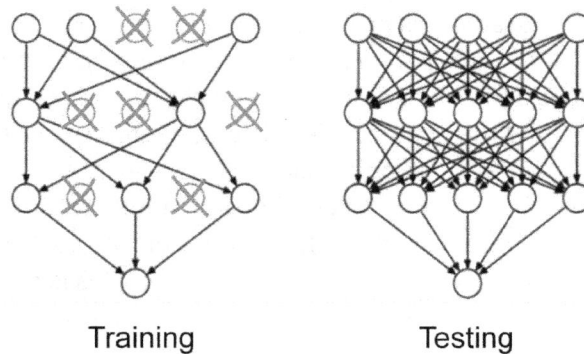

Training   Testing

Figure 2.36: Dropout illustrated

In this technique, a certain number of neurons are deactivated randomly during the training process. The number of neurons to be deactivated is provided as a parameter in the form of a percentage. For example, **Dropout = .2** means 20% of the neurons in that layer will be randomly deactivated during the training process. The same neurons are not deactivated more than once, but a different set of neurons is deactivated in each epoch. During testing, however, all the neurons are activated.

Here is an example of how we can add **Dropout** to a neural network model using Keras:

```
model.add(Dense(units = 300, activation = 'relu')) #Hidden layer1
model.add(Dense(units = 200, activation = 'relu')) #Hidden Layer2
model.add(Dropout(.20))
model.add(Dense(units = 100, activation = 'relu')) #Hidden Layer3
```

In this case, a dropout of 20% is added to **Hidden Layer2**. It is not necessary for the dropout to be added to all layers. As a data scientist, you can experiment and decide what the **dropout** value should be and how many layers need it.

> **NOTE**
>
> A more detailed explanation of dropout can be found in the paper by Nitish Srivastava et al. available here: http://www.jmlr.org/papers/volume15/srivastava14a/srivastava14a.pdf.

As we have come to the end of this chapter, let's test what we have learned so far with the following activity.

ACTIVITY 2.01: BUILD A MULTILAYER NEURAL NETWORK TO CLASSIFY SONAR SIGNALS

In this activity, we will use the Sonar dataset (https://archive.ics.uci.edu/ml/datasets/Connectionist+Bench+(Sonar,+Mines+vs.+Rocks)), which has patterns obtained by bouncing sonar signals off a metal cylinder at various angles and under various conditions. You will build a neural network-based classifier to classify between sonar signals bounced off a metal cylinder (the Mine class), and those bounced off a roughly cylindrical rock (the Rock class). We recommend using the Keras API to make your code more readable and modular, which will allow you to experiment with different parameters easily:

> **NOTE**
>
> You can download the sonar dataset from this link https://packt.live/31Xtm9M.

1. The first step is to understand the data so that you can figure out whether this is a binary classification problem or a multiclass classification problem.

2. Once you understand the data and the type of classification that needs to be done, the next step is network configuration: the number of neurons, the number of hidden layers, which activation function to use, and so on.

   Recall the network configuration steps that we've covered so far. Let's just reiterate a crucial point, the activation function part: for the output (the last) layer, we use sigmoid to do binary classification and Softmax to do multiclass classification.

3. Open the **sonar.csv** file to explore the dataset and see what the target variables are.

4. Separate the input features and the target variables.

5. Preprocess the data to make it neural network-compatible. Hint: one-hot encoding.

6. Define a neural network using Keras and compile it with the right loss function.

7. Print out a model summary to verify the network parameters and considerations.

You are expected to get an accuracy value above 95% by designing a proper multilayer neural network using these steps.

> **NOTE**
>
> The detailed steps for this activity, along with the solutions and additional commentary, are presented on page 390.

SUMMARY

In this chapter, we started off by looking at biological neurons and then moved on to artificial neurons. We saw how neural networks work and took a practical approach to building single-layer and multilayer neural networks to solve supervised learning tasks. We looked at how a perceptron works, which is a single unit of a neural network, all the way to a deep neural network capable of performing multiclass classification. We saw how Keras makes it very easy to create deep neural networks with a minimal amount of code. Lastly, we looked at practical considerations to take into account when building a successful neural network, which involved important concepts such as gradient descent optimizers, overfitting, and dropout.

In the next chapter, we will go to the next level and build a more complicated neural network called a CNN, which is widely used in image recognition.

3

IMAGE CLASSIFICATION WITH CONVOLUTIONAL NEURAL NETWORKS (CNNS)

INTRODUCTION

In this chapter, we will study **convolutional neural networks** (**CNNs**) and image classification. First, we will be introduced to the architecture of CNNs and how to implement them. We will then get hands-on experience of using TensorFlow to develop image classifiers. Finally, we will cover the concepts of transfer learning and fine-tuning and see how we can use state-of-the-art algorithms.

By the end of this chapter, you will have a good understanding of what CNNs are and how programming with TensorFlow works.

INTRODUCTION

In the previous chapters, we learned about traditional neural networks and a number of models, such as the perceptron. We learned how to train such models on structured data for regression or classification purposes. Now, we will learn how we can extend their application to the computer vision field.

Not so long ago, computers were perceived as computing engines that could only process well-defined and logical tasks. Humans, on the other hand, are more complex since we have five basic senses that help us see things, hear noises, feel things, taste foods, and smell odors. Computers were only calculators that could operate large volumes of logical operations, but they couldn't deal with complex data. Compared to the abilities of humans, computers had very clear limitations.

There were some rudimentary attempts to *"give sight"* to computers by processing and analyzing digital images. This field is called computer vision. But it was not until the advent of deep learning that we saw some incredible improvements and results. Nowadays, the field of computer vision has advanced to such an extent that, in some cases, computer vision AI systems are able to process and interpret certain types of images faster and more accurately than humans. You may have heard about the experiment where a group of 15 doctors in China competed against a deep learning system from the company BioMind AI for recognizing brain tumors from X-rays. The AI system took 15 minutes to accurately predict 87% of the 225 input images, while it took 30 minutes for the medical experts to achieve a score of 66% on the same pool of images.

We've all heard about self-driving cars that can automatically make the right decisions depending on traffic conditions or drones that can detect sharks and automatically send alerts to lifeguards. All these amazing applications are possible thanks to the recent development of CNNs.

Computer vision can be split into four different domains:

- **Image classification**, where we need to recognize the main object in an image.

- **Image classification and localization**, where we need to recognize and localize the main object in an image with a bounding box.

- **Object detection**, where we need to recognize multiple objects in an image with bounding boxes.

- **Image segmentation**, where we need to identify the boundaries of objects in an image.

The following figure shows the difference between the four domains:

Figure 3.1: Difference between the four domains of computer vision

In this chapter, we will only look at image classification, which is the most widely used application of CNN. This includes things such as car plate recognition, automatic categorization of the pictures taken with your mobile phone, or creating metadata used by search engines on databases of images.

> **NOTE**
>
> If you're reading the print version of this book, you can download and browse the color versions of some of the images in this chapter by visiting the following link: https://packt.live/2ZUu5G2

DIGITAL IMAGES

Humans can see through their eyes by transforming light into electrical signals that are then processed by the brain. But computers do not have physical eyes to capture light. They can only process information in digital forms composed of bits (0 or 1). So, to be able to "see", computers require a digitized version of an image.

A digital image is formed by a two-dimensional matrix of pixels. For a grayscale image, each of these pixels can take a value between 0 and 255 that represents its intensity or level of gray. A digital image can be composed of one channel for a black and white image or three channels (red, blue, and green) for a color image:

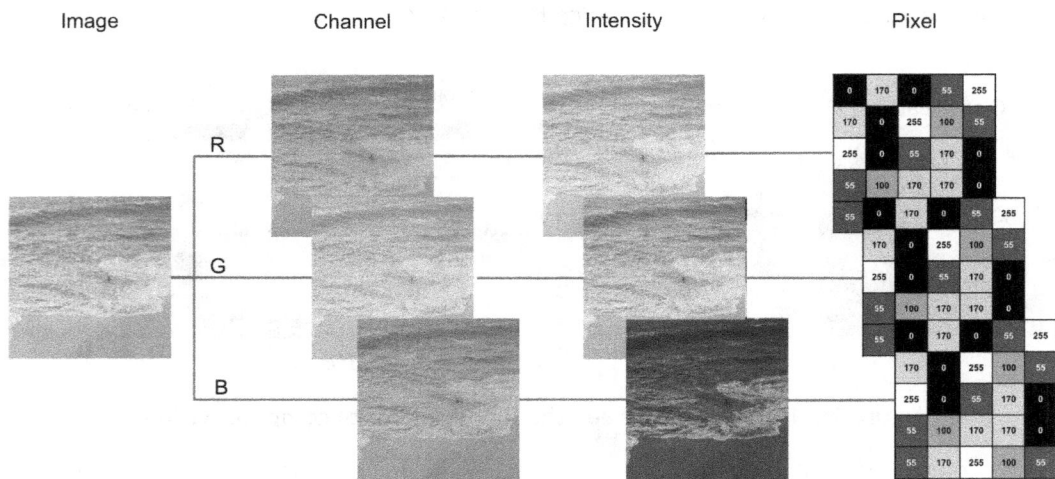

Figure 3.2: Digital representation of an image

A digital image is characterized by its dimensions (height, width, and channel):

- **Height:** How many pixels there are on the vertical axis.

- **Width:** How many pixels there are on the horizontal axis.

- **Channel:** How many channels there are. If there is only one channel, an image will be in grayscale. If there are three channels, the image will be colored.

The following digital image has dimensions (512, 512, 3).

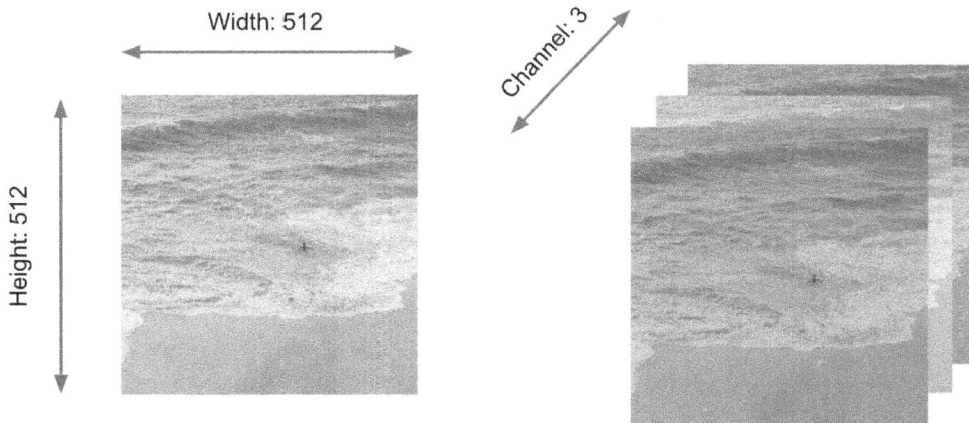

Figure 3.3: Dimensions of a digital image

IMAGE PROCESSING

Now that we know how a digital image is represented, let's discuss how computers can use this information to find patterns that will be used to classify an image or localize objects. So, in order to get any useful or actionable information from an image, a computer has to resolve an image into a recognizable or known pattern. As for any machine learning algorithm, computer vision needs some features in order to learn patterns.

Unlike structured data, where each feature is well defined in advance and stored in separate columns, images don't follow any specific pattern. It is impossible to say, for instance, that the third line will always contain the eye of an animal or that the bottom left corner will always represent a red, round-shaped object. Images can be of anything and don't follow any structure. This is why they are considered to be unstructured data.

However, images do contain features. They contain different shapes (lines, circles, rectangles, and so on), colors (red, blue, orange, yellow, and so on), and specific characteristics related to different types of objects (hair, wheel, leaves, and so on). Our eyes and brain can easily analyze and interpret all these features and identify objects in images. Therefore, we need to simulate the same analytical process for computers. This is where **image filters** (also called kernels) come into play.

Image filters are small matrices specialized in detecting a defined pattern. For instance, we can have a filter for detecting vertical lines only and another one only for horizontal lines. Computer vision systems run such filters in every part of the image and generate a new image with the detected patterns highlighted. These kinds of generated images are called **feature maps**. An example of a feature map where an edge-detection filter is used is shown in the following figure:

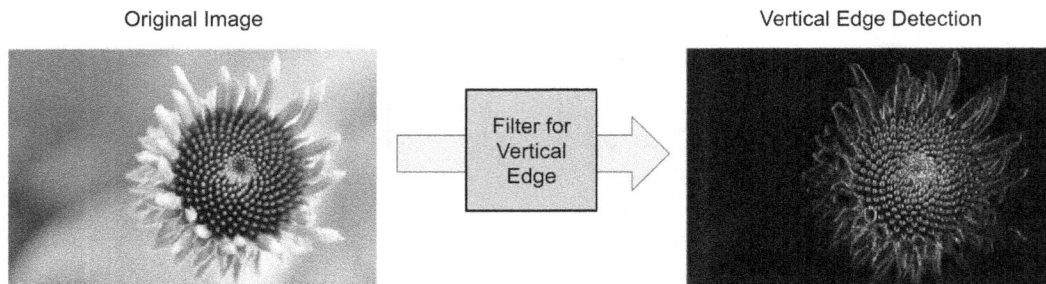

Figure 3.4: Example of a vertical edge feature map

Such filters are widely used in image processing. If you've used Adobe Photoshop before (or any other image processing tool), you will have most likely used filters such as *Gaussian* and *Sharpen*.

CONVOLUTION OPERATIONS

Now that we know the basics of image processing, we can start our journey with CNNs. As we mentioned previously, computer vision relies on applying filters to an image to recognize different patterns or features and generate feature maps. But how are these filters applied to the pixels of an image? You could guess that there is some sort of mathematical operation behind it, and you would be absolutely right. This operation is called convolution.

A convolution operation is composed of two stages:

- An element-wise product of two matrices
- A sum of the elements of a matrix

Let's look at an example of how to convolute two matrices, A and B:

Matrix A

5	10	15
10	20	30
100	150	200

Matrix B

1	0	-1
2	0	-2
1	0	-1

Figure 3.5: Examples of matrices

First, we need to perform an element-wise multiplication with matrices A and B. We will get another matrix, C, as a result, with the following values:

- 1st row, 1st column: $5 \times 1 = 5$
- 1st row, 2nd column: $10 \times 0 = 0$
- 1st row, 3rd column: $15 \times (-1) = -15$
- 2nd row, 1st column: $10 \times 2 = 20$
- 2nd row, 2nd column: $20 \times 0 = 0$
- 2nd row, 3rd column: $30 \times (-2) = -60$
- 3rd row, 1st column: $100 \times 1 = 100$
- 3rd row, 2nd column: $150 \times 0 = 0$
- 3rd row, 3rd column: $200 \times (-1) = -200$

> **NOTE**
>
> An element-wise multiplication is different from a standard matrix multiplication, which operates at the row and column level rather than on each element.

Finally, we just have to perform a sum on all elements of matrix C, which will give us the following:

5+0-15+20+0-60+100+0-200 = -150

The final result of the entire convolution operation on matrices A and B is -150, as shown in the following diagram:

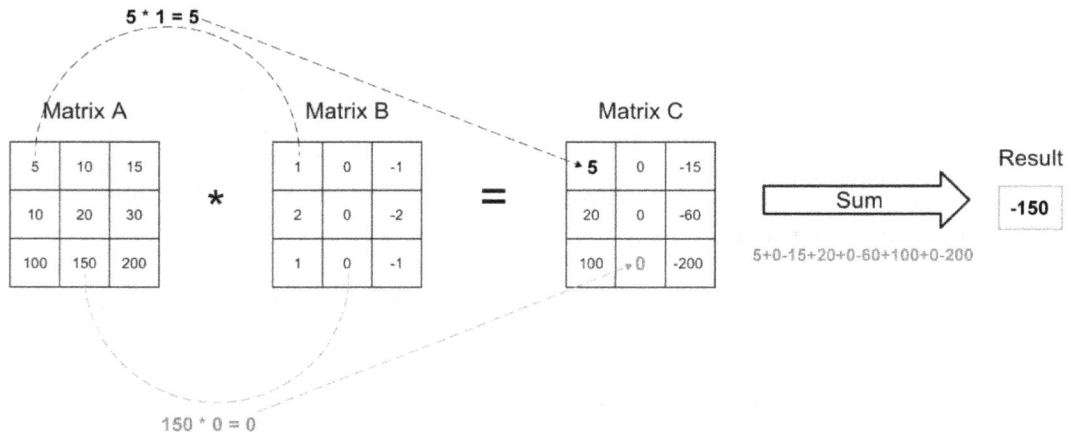

Figure 3.6: Sequence of the convolution operation

In this example, Matrix B is actually a filter (or kernel) called Sobel that is used for detecting vertical lines (there is also a variant for horizontal lines). Matrix A will be a portion of an image with the same dimensions as the filter (this is mandatory in order to perform element-wise multiplication).

> **NOTE**
>
> A filter is, in general, a square matrix such as (3,3) or (5,5).

For a CNN, filters are actually parameters that will be learned (that is, defined) during the training process. So, the values of each filter that will be used will be set by the CNN itself. This is an important concept to go through before we learn how to train a CNN.

EXERCISE 3.01: IMPLEMENTING A CONVOLUTION OPERATION

In this exercise, we will use TensorFlow to implement a convolution operation on two matrices: **[[1,2,3],[4,5,6],[7,8,9]]** and **[[1,0,-1],[1,0,-1],[1,0,-1]]**. Perform the following steps to complete this exercise:

1. Open a new Jupyter Notebook file and name it **Exercise 3.01**.

2. Import the **tensorflow** library:

```
import tensorflow as tf
```

3. Create a tensor called **A** from the first matrix, **([[1,2,3],[4,5,6],[7,8,9]])**. Print its value:

```
A = tf.Variable([[1, 2, 3], [4, 5, 6], [7, 8, 9]])
A
```

The output will be as follows:

```
<tf.Variable 'Variable:0' shape=(3, 3) dtype=int32,
numpy=array([[1, 2, 3],
             [4, 5, 6],
             [7, 8, 9]])>
```

4. Create a tensor called **B** from the first matrix, **([[1,0,-1],[1,0,-1],[1,0,-1]])**. Print its value:

```
B = tf.Variable([[1, 0, -1], [1, 0, -1], [1, 0, -1]])
B
```

The output will be as follows:

```
<tf.Variable 'Variable:0' shape=(3, 3) dtype=int32,
numpy=array([[ 1,  0, -1],
             [ 1,  0, -1],
             [ 1,  0, -1]])>
```

5. Perform an element-wise multiplication on **A** and **B** using **tf.math.multiply()**. Save the result in **mult_out** and print it:

```
mult_out = tf.math.multiply(A, B)
mult_out
```

The expected output will be as follows:

```
<tf.Tensor: id=19, shape=(3, 3), dtype=int32,
numpy=array([[ 1,   0, -3],
            [ 4,   0, -6],
            [ 7,   0, -9]])>
```

6. Perform an element-wise sum on **mult_out** using **tf.math.reduce_sum()**. Save the result in **conv_out** and print it:

```
conv_out = tf.math.reduce_sum(mult_out)
conv_out
```

The expected output will be as follows:

```
<tf.Tensor: id=21, shape=(), dtype=int32, numpy=-6>
```

The result of the convolution operation on the two matrices, **[[1,2,3],[4,5,6],[7,8,9]]** and **[[1,0,-1],[1,0,-1],[1,0,-1]]**, is **-6**.

> **NOTE**
>
> To access the source code for this specific section, please refer to https://packt.live/320pEfC.
>
> You can also run this example online at https://packt.live/2ZdeLFr.
> You must execute the entire Notebook in order to get the desired result.

In this exercise, we used the built-in functions of TensorFlow to perform a convolution operation on two matrices.

STRIDE

So far, we have learned how to perform a single convolution operation. We learned that a convolution operation uses a filter of a specific size, say, (3, 3), that is, 3 × 3, and applies it on a portion of the image of a similar size. If we have a large image, let's say of size (512, 512), then we can just look at a very tiny part of the image.

Taking tiny parts of the image at a time, we need to perform the same convolution operation on the entire space of a given image. To do so, we will apply a technique called sliding. As the name implies, sliding is where we apply the filter to an adjacent area of the previous convolution operation: we just slide the filter and apply convolution.

If we start from the top-left corner of an image, we can slide the filter by one pixel at a time to the right. Once we get to the right edge, we can slide down the filter by one pixel. We repeat this sliding operation until we've applied convolution to the entire space of the image:

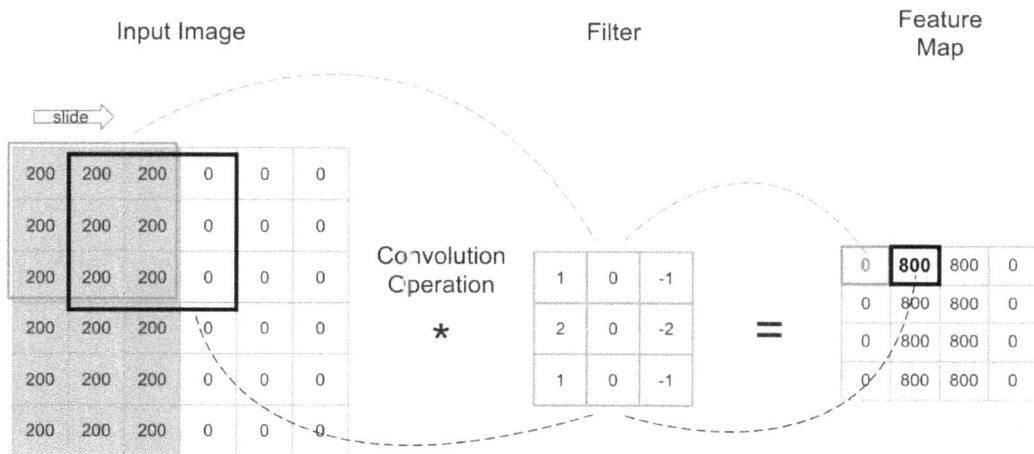

Figure 3.7: Example of stride

Rather than sliding by 1 pixel only, we can choose a bigger sliding window, such as 2 or 3 pixels. The parameter defining the value of this sliding window is called **stride**. With a bigger stride value, there will be fewer overlapping pixels, but the resulting feature map will have smaller dimensions, so you will be losing a bit of information.

In the preceding example, we applied a Sobel filter on an image that has been split horizontally with dark values on the left-hand side and white ones on the right-hand side. The resulting feature map has high values (800) in the middle, which indicates that the Sobel filter found a vertical line in that area. This is how sliding convolution helps to detect specific patterns in an image.

PADDING

In the previous section, we learned how a filter can go through all the pixels of an image with pixel sliding. Combined with the convolution operation, this process helps to detect patterns (that is, extract features) in an image.

Applying a convolution to an image will result in a feature map that has smaller dimensions than the input image. A technique called padding can be used in order to get the exact same dimensions for the feature map as for the input image. It consists of adding a layer of pixels with a value of **0** to the edge:

Input Image

Padded Image

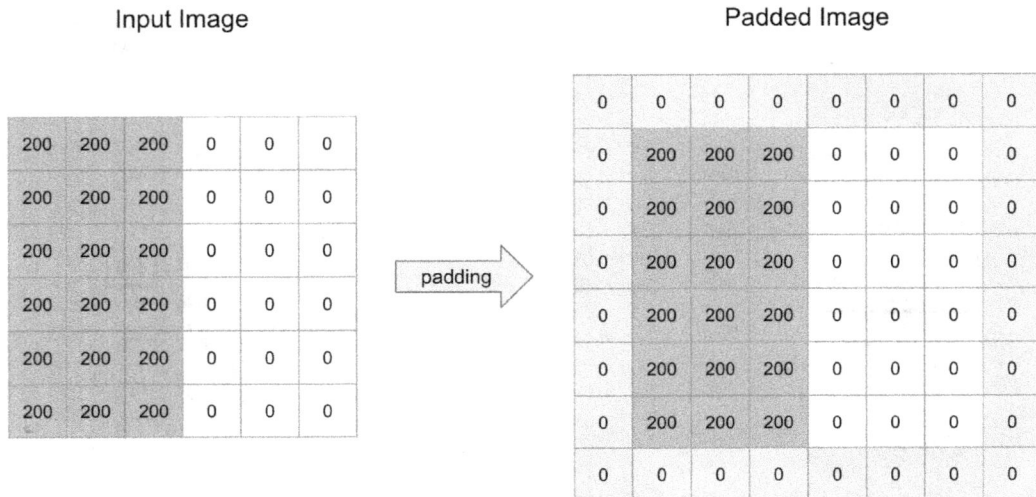

Figure 3.8: Example of padding

In the preceding example, the input image has the dimensions (6,6). Once padded, its dimensions increased to (8,8). Now, we can apply convolution on it with a filter of size (3,3):

Input Image

Filter

Feature Map

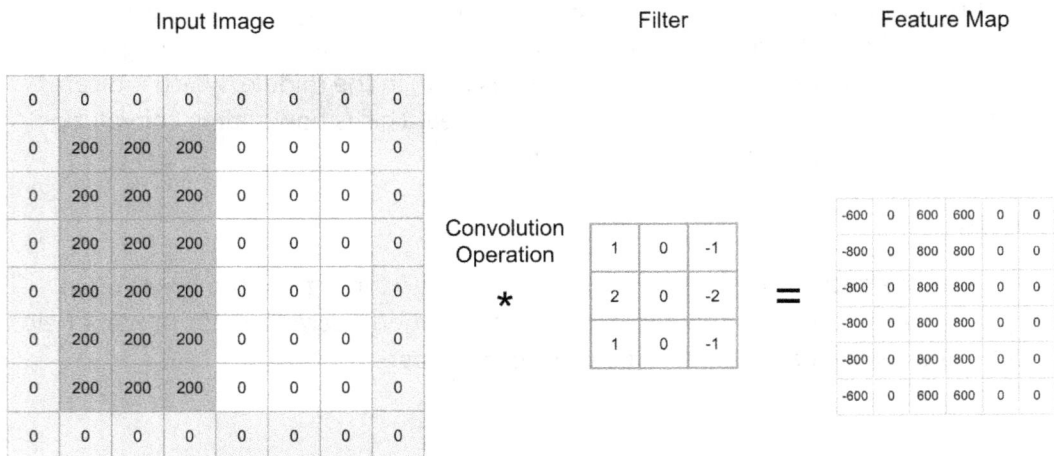

Figure 3.9: Example of padded convolution

The resulting image after convoluting the padded image is (6,6) in terms of its dimensions, which is the exact same dimensions as for the original input image. The resulting feature map has high values in the middle of the image, just like the previous example without padding. So, the filter can still find the same pattern in the image. But you may notice now that we have very low values (-800) on the left edge. This is actually fine as lower values mean the filter hasn't found any pattern in this area.

The following formulas can be used for calculating the output dimensions of a feature map after a convolution:

$$output\ width = \frac{(w + 2 * p - f)}{s} + 1$$

$$output\ height = \frac{(h + 2 * p - f)}{s} + 1$$

Figure 3.10: Formulas for calculating the output dimensions of a feature map

Here, we have the following:

- **w**: Width of the input image
- **h**: Height of the input image
- **p**: Number of pixels used on each side for padding
- **f**: Filter size
- **s**: Number of pixels in the stride

Let's apply this formula to the preceding example:

- **w** = 6
- **h** = 6
- **p** = 1
- **f** = 3
- **s** = 1

Then, calculate the output dimensions as follows:

$$output\ width\ =\ \frac{(6\ +\ 2\ *\ 1\ -\ 3)}{1}\ +1=\frac{5}{1}+1$$

$$output\ height\ =\ \frac{(6\ +\ 2\ *\ 1\ -\ 3)}{1}\ +1=\frac{5}{1}+1$$

Figure 3.11: Output – dimensions of the feature map

So, the dimensions of the resulting feature map are (6,6).

CONVOLUTIONAL NEURAL NETWORKS

In *Chapter 2, Neural Networks*, you learned about traditional neural networks, such as perceptrons, that are composed of fully connected layers (also called dense layers). Each layer is composed of neurons that perform matrix multiplication, followed by a non-linear transformation with an activation function.

CNNs are actually very similar to traditional neural networks, but instead of using fully connected layers, they use convolutional layers. Each convolution layer will have a defined number of filters (or kernels) that will apply the convolution operation with a given stride on an input image with or without padding and can be followed by an activation function.

CNNs are widely used for image classification, where the network will have to predict the right class for a given input. This is exactly the same as classification problems for traditional machine learning algorithms. If the output can only be from two different classes, it will be a **binary classification**, such as recognizing dogs versus cats. If the output can be more than two classes, it will be a **multi-class classification** exercise, such as recognizing 20 different sorts of fruits.

In order to make such predictions the last layer of a CNN model needs to be a fully connected layer with the relevant activation function according to the type of prediction problem. You can use the following list of activation functions as a rule of thumb:

Problem Type	Activation Function for Last Layer	Output
Regression	None (or identity function)	A numerical number that can take any value.
Binary classification	sigmoid	A numerical number ranging from 0 to 1 corresponding to the probability of the observation.
Multi-class classification	softmax	Multiple numerical numbers (depending on the number of classes) ranging from 0 to 1 corresponding to the probability of each class.

Figure 3.12: List of activation functions

To gain a better perspective of its structure, here's what a simple CNN model looks like:

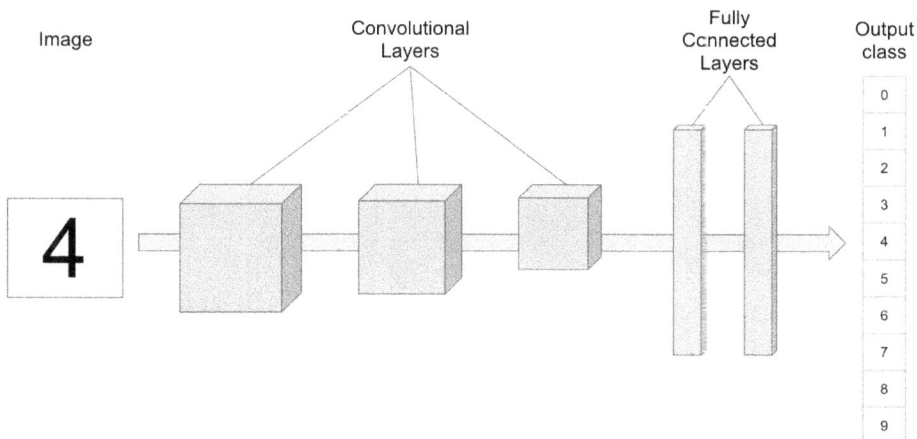

Figure 3.13: Structure of a simple CNN model

We have learned a lot about CNNs already. There is one more concept we need to go through in order to reduce the training time of a CNN before jumping into our first exercise: pooling layers.

POOLING LAYERS

Pooling layers are used to reduce the dimensions of the feature maps of convolution layers. But why do we need to perform such downsampling? One of the main reasons is to reduce the number of calculations that are performed in the networks. Adding multiple layers of convolution with different filters can have a significant impact on the training time. Also, reducing the dimensions of feature maps can eliminate some of the noise in the feature map and help us focus only on the detected pattern. It is quite typical to add a pooling layer after each convolutional layer in order to reduce the size of the feature maps.

A pooling operation acts very similarly to a filter, but rather than performing a convolution operation, it uses an aggregation function such as average or max (max is the most widely used function in the current CNN architecture). For instance, **max pooling** will look at a specific area of the feature map and find the maximum values of its pixels. Then, it will perform a stride and find the maximum value among the neighbor pixels. It will repeat this process until it processes the entire image:

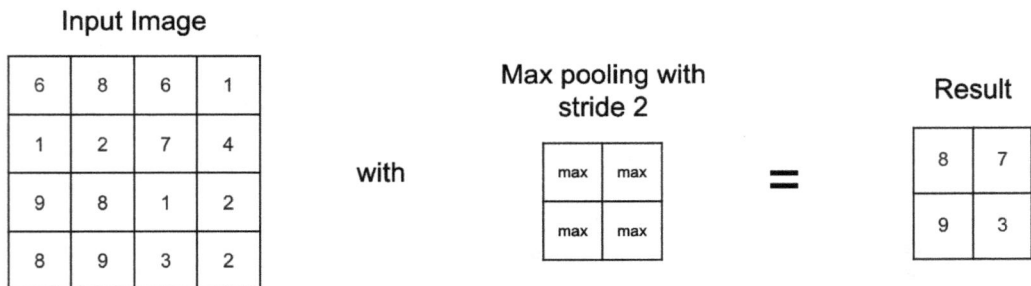

Figure 3.14: Max pooling with stride 2 on an input image

In the preceding example, we used a max pooling (which is the most widely used function for pooling) of size (2, 2) and a stride of 2. We looked at the top-left corner of the feature map and found the maximum value among the pixels, 6, 8, 1, and 2, and got a result of 8. Then, we slid the max pooling by a stride of 2 and performed the same operation on the group of pixels, that is, 6, 1, 7, and 4. We repeated the same operation on the bottom groups and got a new feature map of size (2,2).

A CNN model with max pooling will look like this:

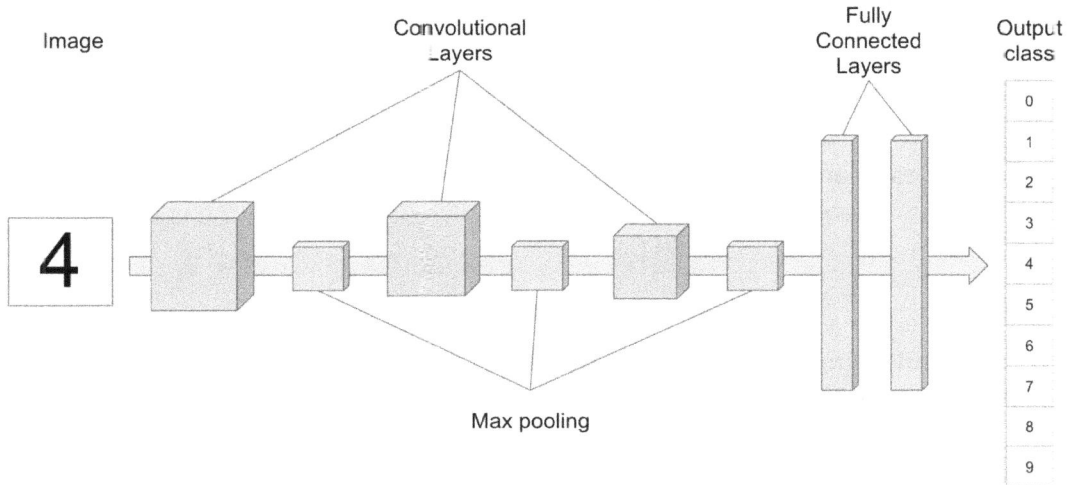

Figure 3.15: Example of the CNN architecture with max pooling

For instance, the preceding model can be used for recognizing handwritten digits (from 0 to 9). There are three convolution layers in this model, followed by a max pooling layer. The final layers are fully connected and are responsible for making the predictions of the digit that's been detected.

The overhead of adding pooling layers is much less than computing convolution. This is why they will speed up the training time.

CNNS WITH TENSORFLOW AND KERAS

So far, you've learned a lot about how CNN works under the hood. Now, it is finally time to see how we can implement what we have learned. We will be using the Keras API from TensorFlow 2.0.

The Keras API provides a high-level API for building your own CNN architecture. Let's look at the main classes we will be using for CNN.

First, to create a convolution layer, we will need to instantiate a **Conv2D ()** class and specify the number of kernels, their size, the stride, padding, and activation function:

```
from tensorflow.keras import layers

layers.Conv2D(64, kernel_size=(3, 3), stride=(2,2), \
              padding="same", activation="relu")
```

In the preceding example, we have created a convolution layer with **64** kernels that are **(3, 3)** in dimension with a stride of **2**, a padding to get the same output dimension as the input (**padding='same'**), and ReLU as the activation function.

> **NOTE**
>
> You can learn more about this class by going to TensorFlow's documentation website: https://www.tensorflow.org/api_docs/python/tf/keras/layers/Conv2D

In order to add a max pooling layer, you will have to use the **MaxPool2D()** class and specify its dimensions and stride, as shown in the following code snippet:

```
from tensorflow.keras import layers

layers.MaxPool2D(pool_size=(3, 3), strides=1)
```

In the preceding code snippet, we have instantiated a max pooling layer of size **(3,3)** with a stride of **1**.

> **NOTE**
>
> You can learn more about this class by going to TensorFlow's documentation website: https://www.tensorflow.org/api_docs/python/tf/keras/layers/MaxPool2D

For a fully connected layer, we will use the **Dense()** class and specify the number of units and the activation function:

```
from tensorflow.keras import layers

layers.Dense(units=1, activation='sigmoid')
```

The preceding code shows us how to create a fully connected layer that has **1** output unit and uses **sigmoid** as the activation function.

Finally, while manipulating input data, we may have to change its dimensions before feeding it to a CNN model. If we are using NumPy arrays, we can use the **reshape** method (as seen in *Chapter 1, Building Blocks of Deep Learning*), as follows:

```
features_train.reshape(60000, 28, 28, 1)
```

Here, we have transformed the dimension of **features_train** to **(60000, 28, 28, 1)**, which corresponds to the format (number of observations, height, width, channel). This is needed when working with grayscale images to add the channel dimension. In this example, the dimensions of a grayscale image, **(28,28)**, will be reshaped to **(28,28,1)** and there will be **60000** images in total.

In TensorFlow, you can use the **reshape** method as follows:

```
from tensorflow.keras import layers
layers.Reshape((60000, 28, 28, 1))
```

Now that we have learned how to design a CNN in TensorFlow, it's time to put this all into practice on the famous MNIST dataset.

> **NOTE**
>
> You can learn more about Reshape by going to TensorFlow's documentation website: https://www.tensorflow.org/api_docs/python/tf/keras/layers/Reshape

EXERCISE 3.02: RECOGNIZING HANDWRITTEN DIGITS (MNIST) WITH CNN USING KERAS

In this exercise, we will be working on the MNIST dataset (which we worked on in *Chapter 2, Neural Networks*), which contains images of handwritten digits. However, this time, we will be using a CNN model. This dataset was originally shared by Yann Lecun, one of the most renowned deep learning researchers. We will build a CNN model and then train it to recognize handwritten digits. The CNN will be composed of two layers of convolution with 64 kernels each, followed by two fully connected layers that have 128 and 10 units, respectively.

TensorFlow provides this dataset directly from its API. Perform the following steps to complete this exercise:

> **NOTE**
>
> You can read more about this dataset on TensorFlow's website:
> https://www.tensorflow.org/datasets/catalog/mnist

1. Open a new Jupyter Notebook file and name it **Exercise 3.02**.

2. Import **tensorflow.keras.datasets.mnist** as **mnist**:

   ```
   import tensorflow.keras.datasets.mnist as mnist
   ```

3. Load the **mnist** dataset using **mnist.load_data()** and save the results into **(features_train, label_train), (features_test, label_test)**:

   ```
   (features_train, label_train), (features_test, label_test) = \
   mnist.load_data()
   ```

4. Print the content of **label_train**:

   ```
   label_train
   ```

   The expected output will be as follows:

   ```
   array([5, 0, 4, ..., 5, 6, 8], dtype=uint8)
   ```

   The label column contains numeric values that correspond to the 10 handwritten digits: **0** to **9**.

5. Print the shape of the training set:

   ```
   features_train.shape
   ```

   The expected output will be as follows:

   ```
   (60000, 28, 28)
   ```

   The training set is composed of **60000** observations of shape **28** by **28**.

6. Print the **shape** of the testing set:

   ```
   features_test.shape
   ```

The expected output will be as follows:

```
(10000, 28, 28)
```

The testing set is composed of **10000** observations of shape **28** by **28**.

7. Reshape the training and testing sets with the dimensions **(number_observations, 28, 28, 1)**:

```
features_train = features_train.reshape(60000, 28, 28, 1)
features_test = features_test.reshape(10000, 28, 28, 1)
```

8. Standardize **features_train** and **features_test** by dividing them by **255**:

```
features_train = features_train / 255.0
features_test = features_test / 255.0
```

9. Import **numpy** as **np**, **tensorflow** as **tf**, and **layers** from **tensorflow.keras**:

```
import numpy as np
import tensorflow as tf
from tensorflow.keras import layers
```

10. Set **8** as the seed for **numpy** and **tensorflow** using **np.random_seed()** and **tf.random.set_seed()**, respectively:

```
np.random.seed(8)
tf.random.set_seed(8)
```

> **NOTE**
>
> The results may still differ slightly after setting the seeds.

11. Instantiate a **tf.keras.Sequential()** class and save it to a variable called **model**:

```
model = tf.keras.Sequential()
```

12. Instantiate a **layers.Conv2D()** class with **64** kernels of shape **(3,3)**, **activation='relu'**, and **input_shape=(28,28,1)**, and save it to a variable called **conv_layer1**:

```
conv_layer1 = layers.Conv2D(64, (3,3), activation='relu', \
                            input_shape=(28, 28, 1))
```

13. Instantiate a **layers.Conv2D()** class with **64** kernels of shape **(3,3)** and **activation='relu'** and save it to a variable called **conv_layer2**:

> **NOTE**
>
> It is only required to specify the **input_shape** parameter for the first layer. For the following layers, CNN would infer it automatically.

```
conv_layer2 = layers.Conv2D(64, (3,3), activation='relu')
```

14. Instantiate a **layers.Flatten()** class with **128** neurons, **activation='relu'**, and save it to a variable called **fc_layer1**:

```
fc_layer1 = layers.Dense(128, activation='relu')
```

15. Instantiate a **layers.Flatten()** class with **10** neurons, **activation='softmax'**, and save it to a variable called **fc_layer2**:

```
fc_layer2 = layers.Dense(10, activation='softmax')
```

16. Add the four layers you just defined to the model using **.add()**, add a **MaxPooling2D()** layer of size **(2,2)** in between each of the convolution layers, and add a **Flatten()** layer before the first fully connected layer to flatten the feature maps:

```
model.add(conv_layer1)
model.add(layers.MaxPooling2D(2, 2))
model.add(conv_layer2)
model.add(layers.MaxPooling2D(2, 2))
model.add(layers.Flatten())
model.add(fc_layer1)
model.add(fc_layer2)
```

17. Instantiate a **tf.keras.optimizers.Adam()** class with **0.001** as the learning rate and save it to a variable called **optimizer**:

```
optimizer = tf.keras.optimizers.Adam(0.001)
```

18. Compile the neural network using **.compile()** with **loss='sparse_categorical_crossentropy'**, **optimizer=optimizer**, **metrics=['accuracy']**:

```
model.compile(loss='sparse_categorical_crossentropy', \
              optimizer=optimizer, metrics=['accuracy'])
```

19. Print the summary of the model:

```
model.summary()
```

The expected output will be as follows:

```
Model: "sequential"

Layer (type)                  Output Shape           Param #
=================================================================
conv2d (Conv2D)               (None, 26, 26, 64)     640

max_pooling2d (MaxPooling2D)  (None, 13, 13, 64)     0

conv2d_1 (Conv2D)             (None, 11, 11, 64)     36928

max_pooling2d_1 (MaxPooling2  (None, 5, 5, 64)       0

flatten (Flatten)             (None, 1600)           0

dense (Dense)                 (None, 128)            204928

dense_1 (Dense)               (None, 10)             1290
=================================================================
Total params: 243,786
Trainable params: 243,786
Non-trainable params: 0
```

Figure 3.16: Summary of the model

The preceding summary shows us that there are more than 240,000 parameters to be optimized with this model.

20. Fit the neural networks with the training set and specify **epochs=5**, **validation_split=0.2**, and **verbose=2**:

```
model.fit(features_train, label_train, epochs=5,\
          validation_split = 0.2, verbose=2)
```

The expected output will be as follows:

```
Train on 48000 samples, validate on 12000 samples
Epoch 1/5
48000/48000 - 32s - loss: 0.1345 - accuracy: 0.9585 - val_loss: 0.0545 - val_accuracy: 0.9843
Epoch 2/5
48000/48000 - 30s - loss: 0.0435 - accuracy: 0.9866 - val_loss: 0.0496 - val_accuracy: 0.9841
Epoch 3/5
48000/48000 - 30s - loss: 0.0296 - accuracy: 0.9905 - val_loss: 0.0453 - val_accuracy: 0.9868
Epoch 4/5
48000/48000 - 31s - loss: 0.0217 - accuracy: 0.9931 - val_loss: 0.0488 - val_accuracy: 0.9866
Epoch 5/5
48000/48000 - 36s - loss: 0.0153 - accuracy: 0.9951 - val_loss: 0.0417 - val_accuracy: 0.9886

<tensorflow.python.keras.callbacks.History at 0x16b6b6350>
```

Figure 3.17: Training output

We trained our CNN on 48,000 samples, and we used 12,000 samples as the validation set. After training for five epochs, we achieved an accuracy score of **0.9951** for the training set and **0.9886** for the validation set. Our model is overfitting a bit.

21. Let's evaluate the performance of the model on the testing set:

```
model.evaluate(features_test, label_test)
```

The expected output will be as follows:

```
10000/10000 [==============================] - 1s 86us/sample -
loss: 0.0312 - accuracy: 0.9903 [0.03115778577708088, 0.9903]
```

With this, we've achieved an accuracy score of **0.9903** on the testing set.

> **NOTE**
>
> To access the source code for this specific section, please refer to https://packt.live/2W2VLYI.
>
> You can also run this example online at https://packt.live/3iKAVGZ. You must execute the entire Notebook in order to get the desired result.

In this exercise, we designed and trained a CNN architecture to recognize the images of handwritten digit images from the MNIST dataset and achieved an almost perfect score.

DATA GENERATOR

In the previous exercise, we built our first multi-class CNN classifier on the MNIST dataset. We loaded the entire dataset into the model as it wasn't very big. But for bigger datasets, we will not be able to do this. Thankfully, Keras provides an API called **data generator** that we can use to load and transform data in batches.

Data generators are also very useful for image classification. Sometimes, an image dataset comes in the form of a folder with predefined structures for the training and testing sets and for the different classes (all images that belong to a class will be stored in the same folder). The data generator API will be able to understand this structure and feed the CNN model properly with the relevant images and corresponding information. This will save you a lot of time as you won't need to build a custom pipeline to load images from the different folders.

On top of this, data generators can divide the images into batches of images and feed them sequentially to the model. You don't have to load the entire dataset into memory in order to perform training. Let's see how they work.

First, we need to import the **ImageDataGenerator** class from **tensorflow. keras.preprocessing**:

```
from tensorflow.keras.preprocessing.image \
import ImageDataGenerator
```

Then, we can instantiate it by providing all the image transformations we want it to perform. In the following example, we will just normalize all the images from the training set by dividing them by **255** so that all the pixels will have a value between **0** and **1**:

```
train_imggen = ImageDataGenerator(rescale=1./255)
```

After this step, we will create a data generator by using the **.flow_from_ directory()** method and will specify the path to the training directory, **batch_size**, the **target_size** of the image, the shuffle, and the type of class:

```
train_datagen = train_imggen.\
                flow_from_directory(batch_size=32, \
                                    directory=train_dir, \
                                    shuffle=True, \
                                    target_size=(100, 100), \
                                    class_mode='binary')
```

> **NOTE**
>
> You need to create a separate data generator for the validation set.

Finally, we can train our model using the **.fit_generator()** method by providing the data generators for the training and validation sets, the number of epochs, and the number of steps per epoch, which corresponds to the number of images divided by the batch size (as integer):

```
model.fit_generator(train_data_gen, \
                    steps_per_epoch=total_train // batch_size, \
                    epochs=5, validation_data=val_data_gen, \
                    validation_steps=total_val // batch_size)
```

This method is very similar to the `.fit()` method you saw earlier, but rather than training the CNN on the entire dataset in one go, it will train by batches of images using the data generator we defined. The number of steps defines how many batches will be required to process the entire dataset.

Data generators are quite useful for loading data from folders and feeding the model in batches of images. But they can also perform some data processing, as shown in the following section.

EXERCISE 3.03: CLASSIFYING CATS VERSUS DOGS WITH DATA GENERATORS

In this exercise, we will be working on the cats versus dogs dataset, which contains images of dogs and cats. We will build two data generators for the training and validation sets and a CNN model to recognize images of dogs or cats. Perform the following steps to complete this exercise:

> **NOTE**
>
> The dataset we'll be using is a modified version from the Kaggle cats versus dogs dataset: https://www.kaggle.com/c/dogs-vs-cats/data. The modified version, which only uses a subset of 25,000 images, has been provided by Google at https://storage.googleapis.com/mledu-datasets/cats_and_dogs_filtered.zip.

1. Open a new Jupyter Notebook file and name it **Exercise 3.03**.

2. Import the **tensorflow** library:

```
import tensorflow as tf
```

3. Create a variable called **file_url** containing the link to the dataset:

```
file_url = 'https://github.com/PacktWorkshops'\
          '/The-Deep-Learning-Workshop/raw/master'\
          '/Chapter03/Datasets/Exercise3.03'\
          '/cats_and_dogs_filtered.zip'
```

NOTE

In the aforementioned step, we are using the dataset stored at
https://packt.live/3jZKRNw. If you have stored the dataset at any other URL,
please change the highlighted path accordingly. Watch out for the slashes in
the string below. Remember that the backslashes (\ ) are used to split the
code across multiple lines, while the forward slashes (/) are part of
the URL.

4. Download the dataset using **tf.keras.get_file** with **'cats_and_dogs.
    zip', origin=file_url, extract=True** as parameters and save the
    result to a variable called **zip_dir**:

    ```
    zip_dir = tf.keras.utils.get_file('cats_and_dogs.zip', \
                            origin=file_url, extract=True)
    ```

5. Import the **pathlib** library:

    ```
    import pathlib
    ```

6. Create a variable called **path** containing the full path to the **cats_and_dogs_
    filtered** directory using **pathlib.Path(zip_dir).parent**:

    ```
    path = pathlib.Path(zip_dir).parent / 'cats_and_dogs_filtered'
    ```

7. Create two variables called **train_dir** and **validation_dir** that take the
    full paths to the train and validation folders, respectively:

    ```
    train_dir = path / 'train'
    validation_dir = path / 'validation'
    ```

8. Create four variables called **train_cats_dir**, **train_dogs_dir**,
    validation_cats_dir, and **validation_dogs_dir** that take the full
    paths to the cats and dogs folders for the train and validation sets, respectively:

    ```
    train_cats_dir = train_dir / 'cats'
    train_dogs_dir = train_dir /'dogs'
    validation_cats_dir = validation_dir / 'cats'
    validation_dogs_dir = validation_dir / 'dogs'
    ```

9. Import the **os** package. We will need this in the next step in order to count the
    number of images from a folder:

    ```
    import os
    ```

10. Create two variables called **total_train** and **total_val** that will get the number of images for the training and validation sets:

```
total_train = len(os.listdir(train_cats_dir)) \
                + len(os.listdir(train_dogs_dir))
total_val = len(os.listdir(validation_cats_dir)) \
                + len(os.listdir(validation_dogs_dir))
```

11. Import **ImageDataGenerator** from **tensorflow.keras. preprocessing**:

```
from tensorflow.keras.preprocessing.image\
import ImageDataGenerator
```

12. Instantiate two **ImageDataGenerator** classes and call them **train_image_ generator** and **validation_image_generator**. These will rescale the images by dividing them by **255**:

```
train_image_generator = ImageDataGenerator(rescale=1./255)
validation_image_generator = ImageDataGenerator(rescale=1./255)
```

13. Create three variables called **batch_size**, **img_height**, and **img_width** that take the values **16**, **100**, and **100**, respectively:

```
batch_size = 16
img_height = 100
img_width = 100
```

14. Create a data generator called **train_data_gen** using **.flow_from_ directory()** and specify the batch size, the path to the training folder, **shuffle=True**, the target size as **(img_height, img_width)**, and the class mode as **binary**:

```
train_data_gen = train_image_generator.flow_from_directory\
                (batch_size=batch_size, directory=train_dir, \
                shuffle=True, \
                target_size=(img_height, img_width), \
                class_mode='binary')
```

15. Create a data generator called **val_data_gen** using **.flow_from_ directory()** and specify the batch size, paths to the validation folder, **shuffle=True**, the target size as **(img_height, img_width)**, and the class mode as **binary**:

```
val_data_gen = validation_image_generator.flow_from_directory\
            (batch_size=batch_size, \
             directory=validation_dir, \
             target_size=(img_height, img_width), \
             class_mode='binary')
```

16. Import **numpy** as **np**, **tensorflow** as **tf**, and **layers** from **tensorflow. keras**:

```
import numpy as np
import tensorflow as tf
from tensorflow.keras import layers
```

17. Set **8** (this is totally arbitrary) as the **seed** for **numpy** and **tensorflow** using **np.random_seed()** and **tf.random.set_seed()**, respectively:

```
np.random.seed(8)
tf.random.set_seed(8)
```

18. Instantiate a **tf.keras.Sequential()** class into a variable called **model** with the following layers: A convolution layer with **64** kernels of shape **3**, **ReLU** as the activation function, and the required input dimensions; a max pooling layer; a convolution layer with **128** kernels of shape **3** and **ReLU** as the activation function; a max pooling layer; a flatten layer; a fully connected layer with **128** units and **ReLU** as the activation function; a fully connected layer with **1** unit and **sigmoid** as the activation function.

    The code will look as follows:

```
model = tf.keras.Sequential([
    layers.Conv2D(64, 3, activation='relu', \
                  input_shape=(img_height, img_width ,3)),\
    layers.MaxPooling2D() \
    layers.Conv2D(128, 3, activation='relu'),\
    layers.MaxPooling2D() \
    layers.Flatten(),\
    layers.Dense(128, activation='relu'),\
    layers.Dense(1, activation='sigmoid')])
```

19. Instantiate a **tf.keras.optimizers.Adam()** class with **0.001** as the learning rate and save it to a variable called **optimizer**:

```
optimizer = tf.keras.optimizers.Adam(0.001)
```

20. Compile the neural network using `.compile()` with **loss='binary_crossentropy', optimizer=optimizer, metrics=['accuracy']**:

```
model.compile(loss='binary_crossentropy', \
              optimizer=optimizer, metrics=['accuracy'])
```

21. Print a summary of the model using `.summary()`:

```
model.summary()
```

The expected output will be as follows:

```
Model: "sequential_8"

_____
Layer (type)                    Output Shape              Param #
=================================================================
conv2d_28 (Conv2D)              (None, 98, 98, 64)        1792
_____
max_pooling2d_18 (MaxPooling    (None, 49, 49, 64)        0
_____
conv2d_29 (Conv2D)              (None, 47, 47, 128)       73856
_____
max_pooling2d_19 (MaxPooling    (None, 23, 23, 128)       0
_____
flatten_6 (Flatten)             (None, 67712)             0
_____
dense_18 (Dense)                (None, 128)               8667264
_____
dense_19 (Dense)                (None, 1)                 129
=================================================================
Total params: 8,743,041
Trainable params: 8,743,041
Non-trainable params: 0
_____
```

Figure 3.18: Summary of the model

The preceding summary shows us that there are more than **8,700,000** parameters to be optimized with this model.

22. Fit the neural networks with **fit_generator()** and provide the train and validation data generators, **epochs=5**, the steps per epoch, and the validation steps:

```
model.fit_generator(train_data_gen, \
                    steps_per_epoch=total_train // batch_size, \
                    epochs=5, \
                    validation_data=val_data_gen,\
                    validation_steps=total_val // batch_size)
```

The expected output will be as follows:

```
Train for 125 steps, validate for 62 steps
Epoch 1/5
125/125 [==============================] - 25s 202ms/step - loss: 0.7185 - accuracy: 0.5350 - val_loss: 0.6891 - val_
accuracy: 0.5071
Epoch 2/5
125/125 [==============================] - 25s 199ms/step - loss: 0.6677 - accuracy: 0.5990 - val_loss: 0.6666 - val_
accuracy: 0.5907
Epoch 3/5
125/125 [==============================] - 26s 204ms/step - loss: 0.5761 - accuracy: 0.6990 - val_loss: 0.6473 - val_
accuracy: 0.6492
Epoch 4/5
125/125 [==============================] - 25s 197ms/step - loss: 0.4822 - accuracy: 0.7670 - val_loss: 0.6766 - val_
accuracy: 0.6724
Epoch 5/5
125/125 [==============================] - 25s 200ms/step - loss: 0.3913 - accuracy: 0.8150 - val_loss: 0.7113 - val_
accuracy: 0.6512

<tensorflow.python.keras.callbacks.History at 0x15428dcd0>
```

Figure 3.19: Training output

> **NOTE**
>
> The expected output will be close to the one shown. You may have slightly different accuracy values due to some randomness in weights initialization.

We've trained our CNN for five epochs and achieved an accuracy score of **0.85** for the training set, and **0.7113** for the validation set. Our model is overfitting quite a lot. You may want to try the training with different architectures to see whether you can improve this score and reduce overfitting. You can also try feeding this model with some images of cats or dogs of your choice and see the output predictions.

> **NOTE**
>
> To access the source code for this specific section, please refer to https://packt.live/31XQmp9.
>
> You can also run this example online at https://packt.live/2ZW10tW. You must execute the entire Notebook in order to get the desired result.

DATA AUGMENTATION

In the previous section, you were introduced to data generators that can do a lot of the heavy lifting, such as feeding the model from folders rather than columnar data for you regarding data processing for neural networks. So far, we have seen how to create them, load data from a structured folder, and feed the model by batch. We only performed one image transformation with it: rescaling. However, data generators can perform many more image transformations.

But why do we need to perform data augmentation? The answer is quite simple: to prevent overfitting. By performing data augmentation, we are increasing the number of images in a dataset. For one image, we can generate, for instance, 10 different variants of the same image. So, the size of your dataset will be multiplied by 10.

Also, with data augmentation, we have a set of images with a broader range of visuals. For example, selfie pictures can be taken from different angles, but if your dataset only contains selfie pictures that are straight in terms of their orientation, your CNN model will not be able to interpret other images with different angles correctly. By performing data augmentation, you are helping your model generalize better to different types of images. However, as you may have guessed, there is one drawback: data augmentation will also increase the training time as you have to perform additional data transformations.

Let's take a quick look at some of the different types of data argumentation that we can do.

HORIZONTAL FLIPPING

Horizontal flipping returns an image that is flipped horizontally:

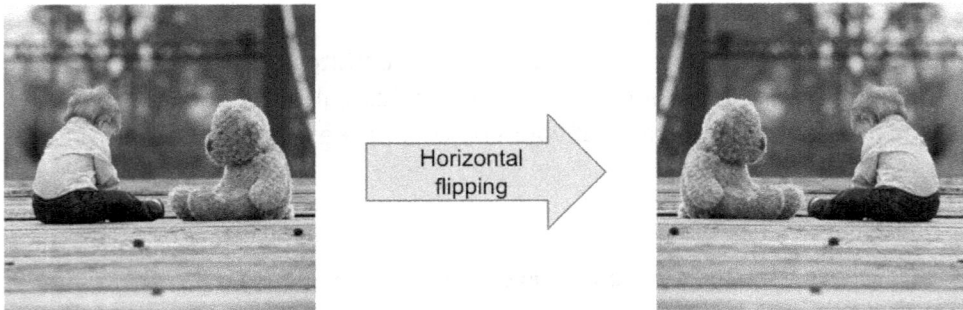

Figure 3.20: Example of horizontal flipping

VERTICAL FLIPPING

Vertical flipping will flip an image vertically:

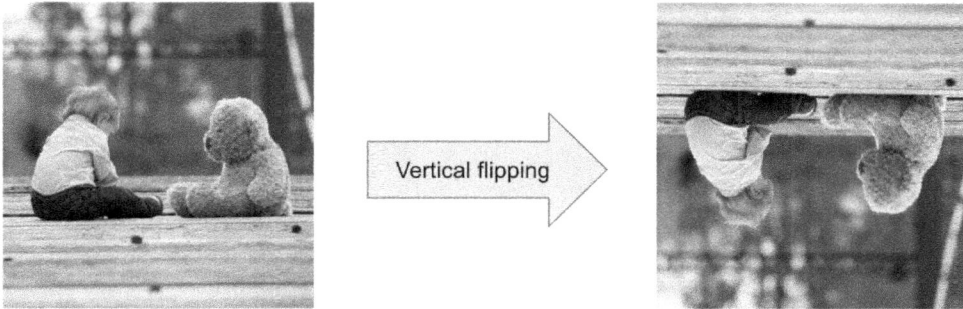

Figure 3.21: Example of vertical flipping

ZOOMING

An image can be zoomed in and provide different sizes of objects in the image:

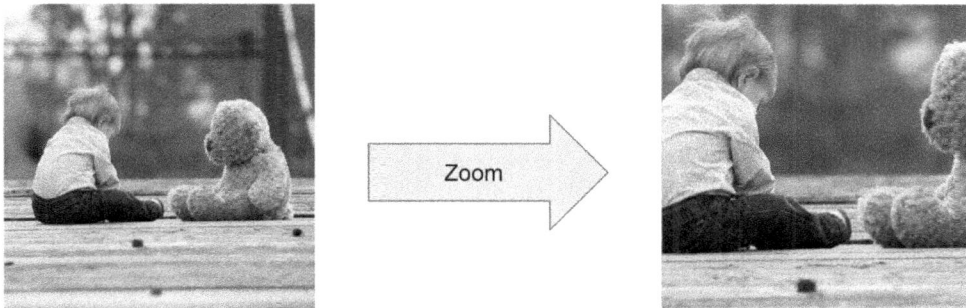

Figure 3.22: Example of zooming

HORIZONTAL SHIFTING

Horizontal shifting, as its name implies, will shift the image along the horizontal axis but keep it the same size. With this transformation, the image may be cropped, and new pixels need to be generated to fill the void. A common technique is to copy the neighboring pixels or to fill that space with black pixels:

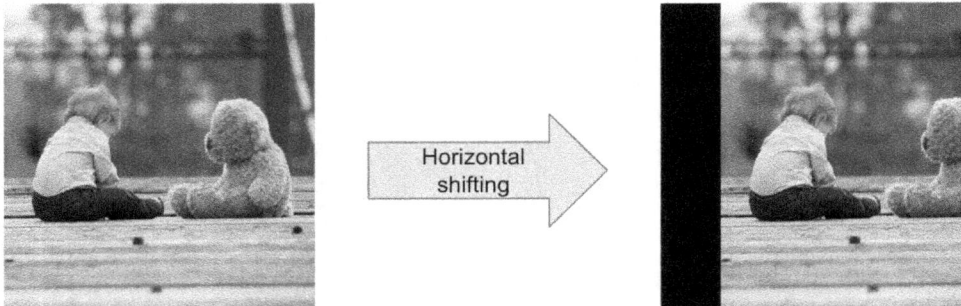

Figure 3.23: Example of horizontal shifting

VERTICAL SHIFTING

Vertical shifting is similar to horizontal shifting, but along the vertical axis:

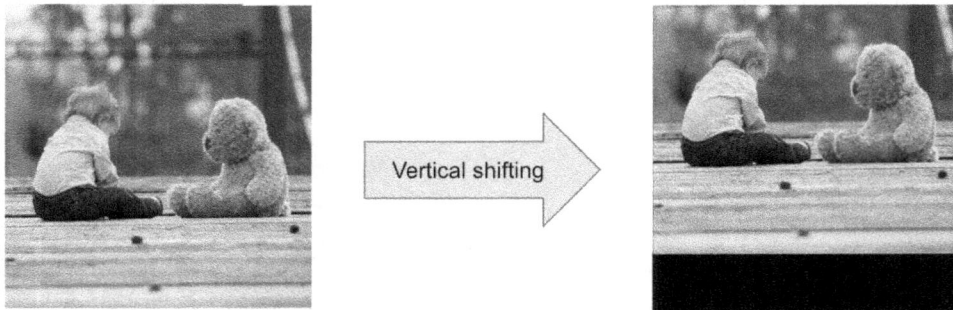

Figure 3.24: Example of vertical shifting

ROTATING

A rotation with a particular angle can be performed on an image like so:

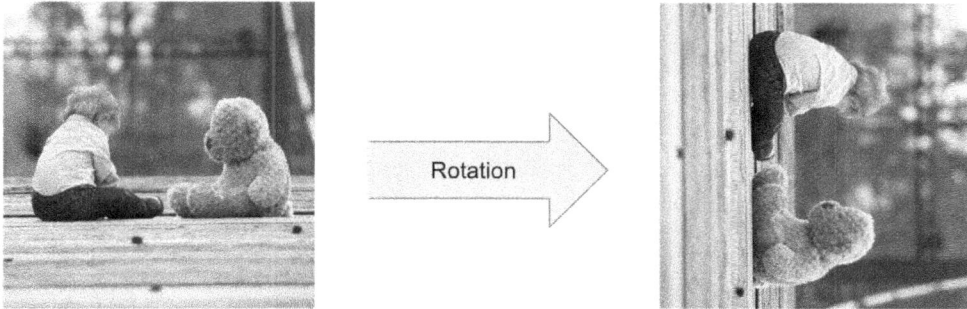

Figure 3.25: Example of rotating

SHEARING

Shearing transforms the image by moving one of the edges along the axis of the edge. After doing this, the image distorts from a rectangle to a parallelogram:

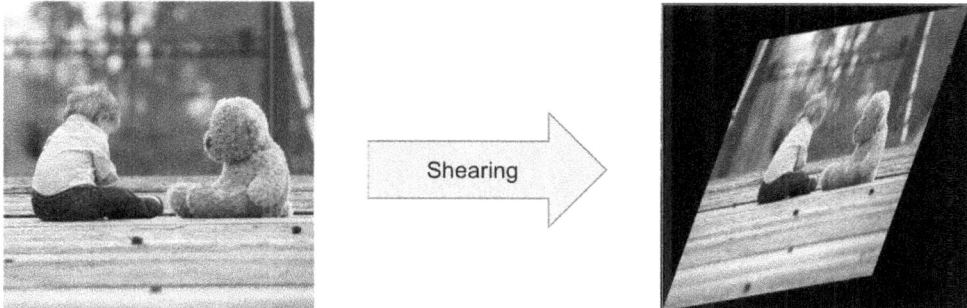

Figure 3.26: Example of shearing

With **Keras**, all these data transformation techniques can be added to **ImageDataGenerator**:

```
from tensorflow.keras.preprocessing.image import ImageDataGenerator
ImageDataGenerator(rescale=1./255, \
                   horizontal_flip=True, zoom_range=0.2, \
                   width_shift_range=0.2, \
                   height_shift_range=0.2, \
                   shear_range=0.2, rotation_range=40, \
                   fill_mode='nearest')
```

Now that we have a general understanding of data argumentation, let's look at how to implement it in our models in the following exercise.

EXERCISE 3.04: IMAGE CLASSIFICATION (CIFAR-10) WITH DATA AUGMENTATION

In this exercise, we will be working on the CIFAR-10 dataset (Canadian Institute for Advanced Research), which is composed of 60,000 images of 10 different classes: airplanes, cars, birds, cats, deer, dogs, frogs, horses, ships, and trucks. We will build a CNN model and use data augmentation to recognize these categories. Perform the following steps to complete this exercise:

> **NOTE**
>
> You can read more about this dataset on TensorFlow's website: https://www.tensorflow.org/api_docs/python/tf/keras/datasets/cifar10.

1. Open a new Jupyter Notebook file and name it **Exercise 3.04**.

2. Import **tensorflow.keras.datasets.cifar10**:

   ```
   from tensorflow.keras.datasets import cifar10
   ```

3. Load the CIFAR-10 dataset using **cifar10.load_data()** and save the results to **(features_train, label_train), (features_test, label_test)**:

   ```
   (features_train, label_train), (features_test, label_test) = \
   cifar10.load_data()
   ```

4. Print the shape of **features_train**:

```
features_train.shape
```

The expected output will be as follows:

```
(50000, 32, 32, 3
```

The training set is composed of **50000** images that have the dimensions **(32,32,3)**.

5. Create three variables called **batch_size**, **img_height**, and **img_width** that take the values **16**, **32**, and **32**, respectively:

```
batch_size = 16
img_height = 32
img_width = 32
```

6. Import **ImageDataGenerator** from **tensorflow.keras. preprocessing**:

```
from tensorflow.keras.preprocessing.image import ImageDataGenerator
```

7. Create an **ImageDataGenerator** instance called **train_img_gen** with data augmentation: rescaling (by dividing by 255), **width_shift_range=0.1**, **height_shift_range=0.1**, and horizontal flipping:

```
train_img_gen = ImageDataGenerator\
                (rescale=1./255, width_shift_range=0.1, \
                 height_shift_range=0.1, horizontal_flip=True)
```

8. Create an **ImageDataGenerator** instance called **val_img_gen** with rescaling (by dividing by 255):

```
val_img_gen = ImageDataGenerator(rescale=1./255)
```

9. Create a data generator called **train_data_gen** using the **.flow()** method and specify the batch size, features, and labels from the training set:

```
train_data_gen = train_img_gen.flow\
                 (features_train, label_train, \
                  batch_size=batch_size)
```

10. Create a data generator called **val_data_gen** using the **.flow()** method and specify the batch size, features, and labels from the testing set:

```
val_data_gen = train_img_gen.flow\
               (features_test, label_test, \
                batch_size=batch_size)
```

11. Import **numpy** as **np**, **tensorflow** as **tf**, and **layers** from **tensorflow. keras**:

```
import numpy as np
import tensorflow as tf
from tensorflow.keras import layers
```

12. Set **8** as the seed for **numpy** and **tensorflow** using **np.random_seed()** and **tf.random.set_seed()**:

```
np.random.seed(8)
tf.random.set_seed(8)
```

13. Instantiate a **tf.keras.Sequential()** class into a variable called **model** with the following layers: a convolution layer with **64** kernels of shape **3**, ReLU as the activation function, and the necessary input dimensions; a max pooling layer; a convolution layer with **128** kernels of shape **3** and ReLU as the activation function; a max pooling layer; a flatten layer; a fully connected layer with **128** units and ReLU as the activation function; a fully connected layer with **10** units and Softmax as the activation function.

The code will be as follows:

```
model = tf.keras.Sequential([
        layers.Conv2D(64, 3, activation='relu', \
                      input_shape=(img_height, img_width ,3)), \
        layers.MaxPooling2D(), \
        layers.Conv2D(128, 3, activation='relu'), \
        layers.MaxPooling2D(), \
        layers.Flatten(), \
        layers.Dense(128, activation='relu'), \
        layers.Dense(10, activation='softmax')])
```

14. Instantiate a **tf.keras.optimizers.Adam()** class with **0.001** as the learning rate and save it to a variable called **optimizer**:

```
optimizer = tf.keras.optimizers.Adam(0.001)
```

15. Compile the neural network using `.compile()` with **loss='sparse_categorical_crossentropy', optimizer=optimizer, metrics=['accuracy']**:

```
model.compile(loss='sparse_categorical_crossentropy', \
              optimizer=optimizer, metrics=['accuracy'])
```

16. Fit the neural networks with **fit_generator()** and provide the train and validation data generators, **epochs=5**, the steps per epoch, and the validation steps:

```
model.fit_generator(train_data_gen, \
                    steps_per_epoch=len(features_train) \
                                    // batch_size, \
                    epochs=5, \
                    validation_data=val_data_gen, \
                    validation_steps=len(features_test) \
                                     // batch_size)
```

The expected output will be as follows:

```
Train for 3125 steps, validate for 625 steps
Epoch 1/5
3125/3125 [==============================] - 54s 17ms/step - loss: 1.3866 - accuracy: 0.5010 - val_loss: 1.2076 - val
_accuracy: 0.5745
Epoch 2/5
3125/3125 [==============================] - 51s 16ms/step - loss: 1.1451 - accuracy: 0.5971 - val_loss: 1.1110 - val
_accuracy: 0.6096
Epoch 3/5
3125/3125 [==============================] - 52s 17ms/step - loss: 1.0448 - accuracy: 0.6314 - val_loss: 1.0195 - val
_accuracy: 0.6500
Epoch 4/5
3125/3125 [==============================] - 58s 18ms/step - loss: 0.9860 - accuracy: 0.6523 - val_loss: 1.0126 - val
_accuracy: 0.6434
Epoch 5/5
3125/3125 [==============================] - 52s 16ms/step - loss: 0.9388 - accuracy: 0.6713 - val_loss: 0.9722 - val
_accuracy: 0.6582

<tensorflow.python.keras.callbacks.History at 0x25cb9be90>
```

Figure 3.27: Training logs for the model

NOTE

To access the source code for this specific section, please refer to https://packt.live/31ZLyQk.

You can also run this example online at https://packt.live/2OcmahS. You must execute the entire Notebook in order to get the desired result.

In this exercise, we trained our CNN on five epochs, and we achieved an accuracy score of **0.6713** on the training set and **0.6582** on the validation set. Our model is overfitting slightly, but its accuracy score is quite low. You may wish to try this on different architectures to see whether you can improve this score by, for instance, adding more convolution layers.

> **NOTE**
>
> The expected output for the preceding exercise will be close to the one shown (Figure 3.27). You may have slightly different accuracy values due to some randomness in weights initialization.

ACTIVITY 3.01: BUILDING A MULTICLASS CLASSIFIER BASED ON THE FASHION MNIST DATASET

In this activity, you will train a CNN to recognize images of clothing that belong to 10 different classes. You will apply some data augmentation techniques to reduce the risk of overfitting. You will be using the Fashion MNIST dataset provided by TensorFlow. Perform the following steps to complete this activity:

> **NOTE**
>
> The original dataset was shared by Han Xiao. You can read more about this dataset on TensorFlow's website here: https://www.tensorflow.org/datasets/catalog/mnist

1. Import the Fashion MNIST dataset from TensorFlow.

2. Reshape the training and testing sets.

3. Create a data generator with the following data augmentation:

```
rescale=1./255,
rotation_range=40,
width_shift_range=0.1,
height_shift_range=0.1,
shear_range=0.2,
zoom_range=0.2,
```

```
horizontal_flip=True,
fill_mode='nearest'
```

4. Create the neural network architecture with the following layers: A convolutional layer with **Conv2D(64, (3,3), activation='relu')** followed by **MaxPooling2D(2,2)**; a convolutional layer with **Conv2D(64, (3,3), activation='relu')** followed by **MaxPooling2D(2,2)**; a flatten layer; a fully connected layer with **Dense(128, activation=relu)**; a fully connected layer with **Dense(10, activation='softmax')**.

5. Specify an Adam optimizer with a learning rate of **0.001**.

6. Train the model.

7. Evaluate the model on the testing set.

The expected output will be as follows:

```
Train for 3750 steps, validate for 625 steps
Epoch 1/5
3750/3750 [==============================] - 55s 15ms/step - loss: 0.8060 - accuracy: 0.6972 - val_loss: 0.6738 - val
_accuracy: 0.7481
Epoch 2/5
3750/3750 [==============================] - 52s 14ms/step - loss: 0.5892 - accuracy: 0.7788 - val_loss: 0.5588 - val
_accuracy: 0.7911
Epoch 3/5
3750/3750 [==============================] - 54s 14ms/step - loss: 0.5211 - accuracy: 0.8041 - val_loss: 0.4928 - val
_accuracy: 0.8170
Epoch 4/5
3750/3750 [==============================] - 53s 14ms/step - loss: 0.4864 - accuracy: 0.8162 - val_loss: 0.4896 - val
_accuracy: 0.8186
Epoch 5/5
3750/3750 [==============================] - 7022s 2s/step - loss: 0.4590 - accuracy: 0.8271 - val_loss: 0.4696 - val
_accuracy: 0.8334

<tensorflow.python.keras.callbacks.History at 0x1469b82d0>
```

Figure 3.28: Training logs for the model

The expected accuracy scores should be around **0.82** for the training and validation sets.

> **NOTE**
>
> The detailed steps for this activity, along with the solutions and additional commentary, are presented on page 394.

SAVING AND RESTORING MODELS

In the previous section, we learned how we can use data augmentation to generate different variants of an image. This will increase the size of the dataset but will also help the model train on a wider variety of images and help it generalize better.

Once you've trained your model, you will most likely want to deploy it in production and use it to make live predictions. To do so, you will need to save your model as a file. This file can then be loaded by your prediction service so that it can be used as an API or data science tool.

There are different components of a model that can be saved:

- The model's architecture with all the network and layers used

- The model's trained weights

- The training configuration with the loss function, optimizer, and metrics

In TensorFlow, you can save the entire model or each of these components separately. Let's learn how to do this.

SAVING THE ENTIRE MODEL

To save all the components into a single artifact, use the following code:

```
model.save_model(filepath='path_to_model/cnn_model')
```

To load the saved model, use the following code:

```
loaded_model = tf.keras.models.load_model\
            (filepath='path_to_model/cnn_model')
```

SAVING THE ARCHITECTURE ONLY

You can save just the architecture of the model as a **json** object. Then, you will need to use the **json** package to save it to a file, as shown in the following code snippet:

```
import json

config_json = model.to_json()
with open('config.json', 'w') as outfile:
    json.dump(config_json, outfile)
```

Then, you will load it back using the **json** package:

```
import json

with open('config.json') as json_file:
    config_data = json.load(json_file)
loaded_model = tf.keras.models.model_from_json(config_data)
```

SAVING THE WEIGHTS ONLY

You can save just the weights of the model as follows:

```
model.save_weights('path_to_weights/weights.h5')
```

Then, you will load them back after instantiating the architecture of your new model:

```
new_model.load_weights('path_to_weights/weights.h5')
```

This is particularly useful f you want to train your model even more later. You will load the saved weights and keep training your model and updating its weights further.

> **NOTE**
>
> .h5 is the file extension used by default by TensorFlow.

TRANSFER LEARNING

So far, we've learned a lot about designing and training our own CNN models. But as you may have noticed, some of our models are not performing very well. This can be due to multiple reasons, such as the dataset being too small or our model requiring more training.

But training a CNN takes a lot of time. It would be great if we could reuse an existing architecture that has already been trained. Luckily for us, such an option does exist, and it is called transfer learning. TensorFlow provides different implementations of state-of-the-art models that have been trained on the ImageNet dataset (over 14 million images).

> **NOTE**
>
> You can find the list of available pretrained models in the TensorFlow documentation: https://www.tensorflow.org/api_docs/python/tf/keras/applications

To use a pretrained model, we need to import its implemented class. Here, we will be importing a **VGG16** model:

```
import tensorflow as tf
from tensorflow.keras.applications import VGG16
```

Next, we will define the input dimensions of the images from our dataset. Let's say we have images of **(100, 100, 3)**:

```
img_dim = (100, 100, 3)
```

Then, we will instantiate a **VGG16** model:

```
base_model = VGG16(input_shape=img_dim, \
                   weights='imagenet', include_top=True)
```

Now, we have a **VGG16** model trained on the **ImageNet** dataset. The **include_top=True** parameter is used to specify that we will be using the same last layers to predict ImageNet's 20,000 categories of images.

Now, we can use this pretrained model to make predictions:

```
base_model.predict(input_img)
```

But what if we want to use this pretrained model to predict different classes other than the ones from ImageNet? In this situation, we will need to replace the last fully connected layers of the pretrained models that are used for prediction and train them on the new classes. These last few layers are referred to as the top (or head) of the model. We can do this by specifying **include_top=False**:

```
base_model = VGG16(input_shape=img_dim, \
                   weights='imagenet', include_top=False)
```

After this, we will need to freeze this model so that it can't be trained (that is, its weights will not be updated):

```
base_model.trainable = False
```

Then, we will create a new fully connected layer with the parameter of our choice. In this example, we will add a **Dense** layer with **20** units and a **softmax** activation function:

```
prediction_layer = tf.keras.layers.Dense(20, activation='softmax')
```

We will then add the new fully connected layer to our base model:

```
new_model = tf.keras.Sequential([base_model, prediction_layer])
```

Finally, we will train this model, but only the weights for the last layer will be updated:

```
optimizer = tf.keras.optimizers.Adam(0.001)
new_model.compile(loss='sparse_categorical_crossentropy', \
                  optimizer=optimizer, metrics=['accuracy'])
new_model.fit(features_train, label_train, epochs=5, \
              validation_split = 0.2, verbose=2)
```

We just created a new model from a pretrained model and adapted it in order to make predictions for our own dataset. We achieved this by replacing the last layers according to the predictions we want to make. Then, we trained only these new layers to make the right predictions. Using transfer learning, you leveraged the existing weights of the **VGG16** model, which were trained on ImageNet. This has saved you a lot of training time and can significantly increase the performance of your model.

FINE-TUNING

In the previous section, we learned how to apply transfer learning and use pretrained models to make predictions on our own dataset. With this approach, we froze the entire network and trained only the last few layers that were responsible for making the predictions. The convolutional layers stay the same, so all the filters are set in advance and you are just reusing them.

But if the dataset you are using is very different from ImageNet, these pretrained filters may not be relevant. In this case, even using transfer learning will not help your model accurately predict the right outcomes. There is a solution for this, which is to only freeze a portion of the network and train the rest of the model rather than just the top layers, just like we do with **transfer learning**.

In the early layers of the networks, the filters tend to be quite generic. For instance, you may find filters that detect horizontal or vertical lines at that stage. The filters closer to the end of the network (close to the top or head) are usually more specific to the dataset you are training on. So, these are the ones we want to retrain. Let's learn how we can achieve this in TensorFlow.

First, let's instantiate a pretrained **VGG16** model:

```
base_model = VGG16(input_shape=img_dim, \
                   weights='imagenet', include_top=False)
```

We will need to set the threshold for the layers so that they're frozen. In this example, we will freeze the first 10 layers:

```
frozen_layers = 10
```

Then, we will iterate through these layers and freeze them individually:

```
for layer in base_model.layers[:frozen_layers]:
  layer.trainable = False
```

Then, we will add our custom fully connected layer to our base model:

```
prediction_layer = tf.keras.layers.Dense(20, activation='softmax')
new_model = tf.keras.Sequential([base_model, prediction_layer])
```

Finally, we will train this model:

```
optimizer = tf.keras.optimizers.Adam(0.001)
new_model.compile(loss='sparse_categorical_crossentropy', \
                  optimizer=optimizer, metrics=['accuracy'])
new_model.fit(features_train, label_train, epochs=5, \
              validation_split = 0.2, verbose=2)
```

In this case, our model will train and update all the weights from the threshold layer we defined. They will use the pretrained weights as the initialized values for the first iteration.

With this technique, called fine-tuning, you can still leverage pretrained models by partially training them to fit your dataset.

ACTIVITY 3.02: FRUIT CLASSIFICATION WITH TRANSFER LEARNING

In this activity, we will train a CNN to recognize images of fruits that belong to 120 different classes. We will use transfer learning and data augmentation to do so. We will be using the Fruits 360 dataset (https://arxiv.org/abs/1712.00580), which was originally shared by Horea Muresan, Mihai Oltean, *Fruit recognition from images using deep learning, Acta Univ. Sapientiae, Informatica Vol. 10, Issue 1, pp. 26-42, 2018.*

It contains more than 82,000 images of 120 different types of fruits. We will be using a subset of this dataset with more than 16,000 images. Perform the following steps to complete this activity:

> **NOTE**
>
> The dataset can be found here: https://packt.live/3gEjHsX

1. Import the dataset and unzip the file using TensorFlow.

2. Create a data generator with the following data augmentation:

```
rescale=1./255,
rotation_range=40,
width_shift_range=0.1,
height_shift_range=0.1,
shear_range=0.2,
zoom_range=0.2,
horizontal_flip=True,
fill_mode='nearest'
```

3. Load a pretrained **VGG16** model from TensorFlow.

4. Add two fully connected layers on top of **VGG16**: A fully connected layer with **Dense(1000, activation='relu')** and a fully connected layer with **Dense(120, activation='softmax')**.

5. Specify an Adam optimizer with a learning rate of **0.001**.

6. Train the model.

7. Evaluate the model on the testing set.

The expected accuracy scores should be around **0.89** to **0.91** for the training and validation sets. The output will be similar to this:

```
Epoch 1/5
3750/3750 [==============================] - 55s 15ms/step - loss: 0.8060 - accuracy: 0.6972 - val_loss: 0.6738 - val_accuracy: 0.7481
Epoch 2/5
3750/3750 [==============================] - 52s 14ms/step - loss: 0.5892 - accuracy: 0.7788 - val_loss: 0.5588 - val_accuracy: 0.7911
Epoch 3/5
3750/3750 [==============================] - 54s 14ms/step - loss: 0.5211 - accuracy: 0.8041 - val_loss: 0.4928 - val_accuracy: 0.8170
Epoch 4/5
3750/3750 [==============================] - 53s 14ms/step - loss: 0.4864 - accuracy: 0.8162 - val_loss: 0.4896 - val_accuracy: 0.8186
Epoch 5/5
3750/3750 [==============================] - 7022s 2s/step - loss: 0.4590 - accuracy: 0.8271 - val_loss: 0.4696 - val_accuracy: 0.8334
```

Figure 3.29: Expected output of the activity

> **NOTE**
>
> The detailed steps for this activity, along with the solutions and additional commentary, are presented on page 398.

SUMMARY

We started our journey in this chapter with an introduction to computer vision and image processing, where we learned the different applications of such technology, how digital images are represented, and analyzed this with filters.

Then, we dived into the basic elements of CNN. We learned what a convolution operation is, how filters work in detecting patterns, and what stride and padding are used for. After understanding these building blocks, we learned how to use TensorFlow to design CNN models. We built our own CNN architecture to recognize handwritten digits.

After this, we went through data generators and learned how they can feed our model with batches of images rather than loading the entire dataset. We also learned how they can perform data augmentation transformations to expand the variety of images and help the model generalize better.

Finally, we learned about saving a model and its configuration, but also about how to apply transfer learning and fine-tuning. These techniques are very useful for reusing pretrained models and adapting them to your own projects and datasets. This will save you a lot of time as you won't have to train the model from scratch.

In the next chapter, you will learn about another very interesting topic that is used for natural language processing: embeddings.

4

DEEP LEARNING FOR TEXT — EMBEDDINGS

OVERVIEW

In this chapter, we will begin our foray into **Natural Language Processing** for text. We will start by using the **Natural Language Toolkit** to perform text preprocessing on raw text data, where we will tokenize the raw text and remove punctuations and stop words. As we progress through this chapter, we will implement classical approaches to text representation, such as one-hot encoding and the **TF-IDF** approach. This chapter demonstrates the power of word embeddings and explains the popular deep learning-based approaches for embeddings. We will use the **Skip-gram** and **Continuous Bag of Words** algorithms to generate our own word embeddings. We will explore the properties of the embeddings, the different parameters of the algorithms, and generate vectors for phrases. By the end of this chapter, you will be able to handle text data and start using word embeddings by using pre-trained models, as well as your own embeddings.

INTRODUCTION

How does Siri know exactly what to do when you ask her to "*play a mellow song from the 80s*"? How does Google find the most relevant results for even your ill-formed search queries in a fraction of a second? How does your translation app translate text from German to English almost instantly? How does your email client protect you and automatically identify all those malicious spam/phishing emails? The answer to all these questions, and what powers many more amazing applications, is using **Natural Language Processing** (**NLP**).

So far, we've dealt with structured, numeric data – images that were also numeric matrices. In this chapter, we'll begin our discussion by talking about handling text data and unlock the skills needed to harness this goldmine of unstructured information. We will discuss a key idea in this chapter – representation, particularly using embeddings. We will discuss the considerations and implement the approaches for representation. We will begin with the simplest approaches and end with word embeddings – an amazingly powerful approach for representing text data. Word embeddings will help you get state-of-the-art results in NLP tasks when coupled with deep learning approaches.

NLP is a field concerned with helping machines make sense of natural (human) language. As shown in the following figure, NLP resides at the intersection of linguistics, computer science, and artificial intelligence:

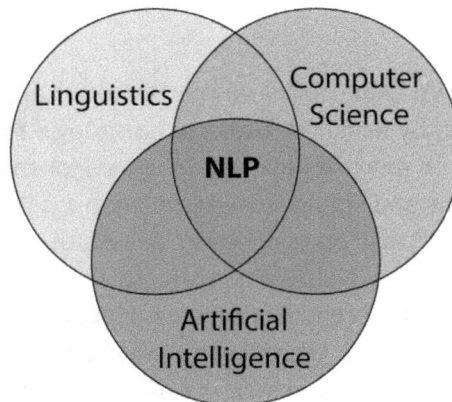

Figure 4.1: Where NLP fits

It is a vast field – think of all the places language (spoken and written) is used. NLP enables and powers the kind of applications listed in the prececing figure, including the following:

- Classification of documents into categories (text classification)

- Translation between languages, say, German to English (sequence-to-sequence learning)

- Automatically classifying the sentiment of a tweet or a movie review (sentiment analysis)

- Chatbots that reply to your query instantly, 24/7

Before we go any further, we need to acknowledge and appreciate that NLP isn't easy. Consider the following sentence: '*The boy saw a man with a telescope.*"

Who had the telescope? Did the boy use a telescope to see the man through it? Cr was the man carrying a telescope with him? There is an ambiguity that we can't resolve with this sentence alone. Maybe some more context will help us figure this out.

Let's consider this sentence, then: "*Rahim convinced Mohan to buy a television for himself.*" Who was the TV bought for – Rahim or Mohan? This is another case of ambiguity that we may be able to resolve with more context, but again, it may be very difficult for a machine/program.

Let's consider another example: "*Rahim has quit skydiving.*" This sentence implies that Rahim did a fair amount of skydiving. There is a presupposition in this sentence, which is hard for a machine to infer.

Language is a complex system that uses symbols (words/terms) and combines them in many ways to communicate ideas. Making sense of language is not always very easy, and there are many reasons for this. Ambiguity is by far the biggest reason: words can have different meanings in different contexts. Add to that subtext, different perspectives, and so on. We can never be sure if the same words are understood the same way by different people. A poem can be interpreted in many ways by those who read it, where each reader brings their unique perspective and understanding of the world and employs them to make sense of the poem in their own way.

DEEP LEARNING FOR NATURAL LANGUAGE PROCESSING

The emergence of deep learning has had a strong positive impact on many fields, and NLP is no exception. By now, you can appreciate that deep learning approaches have given us accuracies like never before, and this has helped us improve in many areas. There are several tasks in NLP that have gained tremendously from deep learning approaches. Applications that use sentiment prediction, machine translation, and chatbots previously required a lot of manual intervention. With deep learning and NLP, these tasks are completely automated and bring with them impressive performance. The simple, high-level view shown in *Figure 4.2* shows how deep learning can be used for processing natural language. Deep learning provides us with not only great representations of natural language that machines can understand but also very powerful modeling approaches well suited for tasks in NLP.

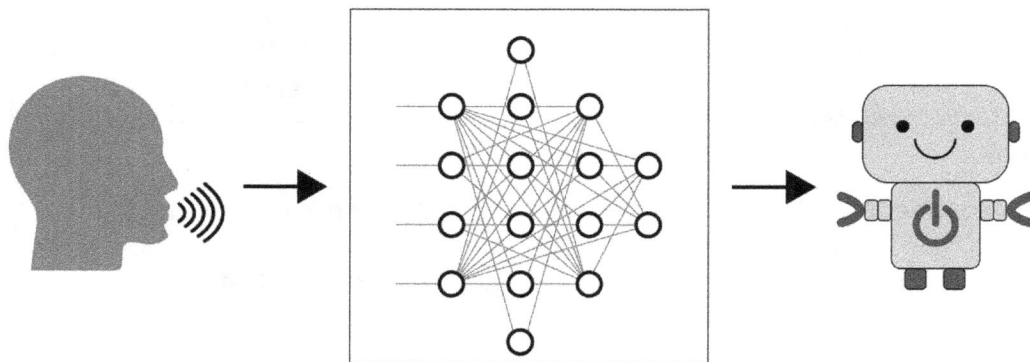

Figure 4.2: Deep learning for NLP

That being said, we need to be cautious to avoid underestimating the difficulty of getting machines to perform tasks involving human language and the field of NLP. Deep learning hasn't solved all of the challenges in NLP, but it has indeed caused a paradigm shift in the way several tasks in NLP are approached and has helped advance some applications in this field, making otherwise difficult tasks accessible and easy for anyone and everyone. We will perform some of these in *Chapter 5, Deep Learning for Sequences*.

One such key task is text data representation – which is, in simple terms, converting raw text into something a model would understand. Word embeddings constitute a deep learning-based approach that has changed the game and gives a very powerful representation of text. We'll discuss embeddings in detail and create our own embeddings later in this chapter. First, let's get our hands dirty by working with some text and performing some very important data preparation.

GETTING STARTED WITH TEXT DATA HANDLING

Let's get some test data into Python to begin. First, we'll create some toy data of our own and get familiar with the tools. Then, we'll use Lewis Carrol's classic work, "*Alice's Adventures in Wonderland*", which is available through Project Gutenberg (gutenberg. org). Conveniently enough, we have this easily accessible through the **Natural Language ToolKit** (**NLTK**), a great library for performing NLP from scratch.

> **NOTE**
>
> The code implementations for this chapter can be found at: https://packt.live/3gEgkSP. All the code in this chapter must be run in a single Jupyter Notebook.

NLTK should come with the Anaconda distribution. If not, you can install NLTK by using the following command in the command line:

```
pip install nltk
```

This should work on Windows. For macOS and Linux, you can use the following command:

```
$ sudo pip install -U nltk
```

Our dummy data can be created using the following command (we're using Jupyter Notebooks here; feel free to use any interface):

```
raw_txt = """Welcome to the world of Deep Learning for NLP! \
             We're in this together, and we'll learn together. \
             NLP is amazing, \
             and Deep Learning makes it even more fun. \
             Let's learn!"""
```

We have the text in **raw_txt**, which is a string variable, so now, we're ready to start processing it.

TEXT PREPROCESSING

Text preprocessing refers to the process of getting the text data ready for your primary analysis/model. Regardless of your end goal – which could be sentiment analysis, classification, clustering, or any of the many others – you need to get your raw text data cleaned up and ready for analysis. This is the first part of any application involving NLP

What do we mean by **clean up**, and when is the text data ready? We know that the text data we encounter in our day-to-day lives can be very messy (think about social media, product reviews, service reviews, and so on) and has various imperfections. Depending on the task at hand and the kind of data you're dealing with, the imperfections you care about will vary, and **cleaning up** can mean very different things. As an example, in some applications, preprocessing could just mean "dividing the sentences into individual terms." The steps you take here can and will have an impact on the final outcome of your analysis. Let's discuss this in more detail.

TOKENIZATION

The first step in preprocessing is inevitably **tokenization** – splitting the raw input text sequence into **tokens**. In simple terms, it is breaking the raw text into constituent elements that you want to work on. This token can be a paragraph, sentence, word, or even a character. If you want to separate a paragraph into sentences, then you would tokenize the paragraph into sentences. If you want to separate the words in a sentence, then you would tokenize the sentence into words.

For our raw text, first, we want to separate the sentences. To do so, we have multiple options in Python – here, we'll use the tokenize API in NLTK.

> **NOTE**
>
> We'll be using Jupyter Notebooks throughout this book, which is something that we recommend. However, feel free to use any IDE you wish.

Before we can use the API, we have to **import nltk** and download the **punkt** sentence tokenizer. Then, we need to import the **tokenize** library. All this can be done using the following commands:

```
import nltk
nltk.download('punkt')
from nltk import tokenize
```

The tokenize API has utilities to extract different levels of tokens (sentences, words, or characters) for different types of data (a very handy tweet tokenizer, too). We'll use the **sent_tokenize()** method here. The **sent_tokenize()** method breaks input text into constituent sentences. Let's see it in action:

```
tokenize.sent_tokenize(raw_txt)
```

This should give us the following individual sentences:

```
['Welcome to the world of Deep Learning for NLP!',
 "We're in this together, and we'll learn together.",
 'NLP is amazing, and Deep Learning makes it even more fun.',
 "Let's learn!"]
```

Looking at the output, it seems like **sent_tokenize()** is doing a pretty good job. It has correctly identified the sentence boundaries and given us the four sentences, as expected. Let's assign the result to a variable for ease of handling and let's check the data type of the result and its constituents:

```
txt_sents = tokenize.sent_tokenize(raw_txt)
type(txt_sents), len(txt_sents)
```

The following is the output of the preceding code:

```
(list, 4)
```

As we can see, it's a list with four elements, where each element contains the sentence as a string.

We can try breaking sentences into individual words using the **word_tokenize()** method. This method breaks a given sentence into its constituent words. It uses smart rules to figure out word boundaries. Let's use list comprehension (comprehensions in Python are a concise approach to constructing new sequences) for a bit for convenience:

```
txt_words = [tokenize.word_tokenize(sent) for sent in txt_sents]
type(txt_words), type txt_words[0])
```

The preceding command gives us the following output:

```
(list, list)
```

The output is as expected – the elements of the resulting list are lists themselves, containing the words that form the sentence. Let's also print out the first two elements of the result:

```
print(txt_words[:2])
```

The output would be as follows:

```
[['Welcome', 'to', 'the', 'world', 'of',
  'Deep', 'Learning', 'for', 'NLP', '!'],
 ['We', "'re", 'in', 'this', 'together',
  ',', 'and', 'we', "'ll", 'learn', 'together', '.']]
```

The sentences have been broken into individual words. We can also see that contractions like "we'll" have been broken into constituents, that is, "we" and "'ll". All punctuation (commas, periods, exclamation marks, and so on) are separate tokens. This is very convenient for us if we wish to remove them, which we will do later.

NORMALIZING CASE

Another common step is to normalize case – we usually don't want "car", "CAR", "Car", and "caR" to be treated as separate entities. To do so, we typically convert all text into lowercase (we could also convert it into uppercase if we wanted).

All strings in Python have a **lower()** method to them, so converting a string variable (**strvar**) into lowercase is as simple as **strvar.lower()**.

> **NOTE**
>
> We could have used this right in the beginning, before tokenization, and it would have been as simple as **raw_txt = raw_txt.lower()**.

We will normalize the case of our data using the **lower()** method after tokenizing into individual sentences. We'll accomplish this with the following commands:

```
txt_sents  = [sent.lower() for sent in txt_sents]
txt_words = [tokenize.word_tokenize(sent) for sent in txt_sents]
```

Let's print out a couple of sentences to see what the result looks like:

```
print(txt_words[:2])
```

The output will be as follows:

```
[['welcome', 'to', 'the', 'world', 'of',
  'deep', 'learning', 'for', 'nlp', '!'],
 ['we', "'re", 'in', 'this', 'together',
  ',', 'and', 'we', "'ll", 'learn', 'together', '.']]
```

We can see that the output has all the terms in lowercase this time. We've taken the raw text, broken it into sentences, normalized the case, and then broken that down into words. Now, we have all the tokens that we need, but we still seem to have a lot of punctuation marks as tokens that we need to get rid of. Let's go ahead and perform more "cleanup".

REMOVING PUNCTUATION

We can see that the data currently has all punctuation as separate tokens. Again, bear in mind that there could be tasks where punctuations could be important. As an example, when performing sentiment analysis, that is, predicting if the sentiment in the text is positive or negative, an exclamation can add value. For our task, let's remove these since we're only interested in representing the terms of language. To do so, we need to have a list of all the punctuation marks we want to remove. Luckily, we have such a list in the string base library in Python, which we can simply import and assign to a list variable:

```
from string import punctuation
list_punct = list(punctuation)
print(list_punct)
```

You should get the following output:

```
['!', '"', '#', '$', '%', '&', "'", '(', ')', '*', '+', ',',
 '-', '.', '/', ':', ';', '<', '=', '>', '?', '@',
 '[', '\\', ']', '^', '_', '`', '{', '|', '}', '~']
```

All the usual punctuation marks are available here. If there are any additional punctuation marks you want to remove, you can simply add them to the **list_ punct** variable.

We can define a function to remove punctuation from a given list of tokens. This function will expect a list of tokens, from which it will drop the tokens that are available in the **list_punct** variable:

```
def drop_punct(input_tokens):
    return [token for token in input_tokens \
            if token not in list_punct]
```

We can test this out on some dummy tokens using the following command:

```
drop_punct(["let",".","us",".","go","!"])
```

We get the following result:

```
['let', 'us', 'go']
```

The function works as intended. Now, we need to pass the **txt_words** variable we modified in the previous section to the **drop_punct** function we just created. We will store our result in a new variable called **txt_words_nopunct**:

```
txt_words_nopunct = [drop_punct(sent) for sent in txt_words]
print(txt_words_nopunct)
```

We will get the following output:

```
[['welcome', 'to', 'the', 'world', 'of',
  'deep', 'learning', 'for', 'nlp'],
 ['we', "'re", 'in', 'this', 'together', 'and',
  'we', "'ll", 'learn', 'together'],
 ['nlp', 'is', 'amazing', 'and',
  'deep', 'learning', 'makes', 'it', 'even', 'more', 'fun'],
 ['let', "'s", 'learn']]
```

As you can see from the preceding output, the function we created has removed all the punctuation marks from our raw text. Now, the data looks much cleaner without the punctuation, but we still need to get rid of non-informative terms. We'll discuss that in the next section.

REMOVING STOP WORDS

In day-to-day language, we have a lot of terms that don't add a lot of information/value*. These are typically referred to as "stop words". We can think of these as belonging to two broad categories:

1. **General/functional**: These are filler words in the language that don't provide a lot of information but help stitch together other informative words to form meaningful sentences, such as "the", "an", "of", and so on.

2. **Contextual**: These aren't general functional terms, but given the context, don't add a lot of value. If you're working with reviews of a mobile phone, where all reviews are talking about the phone, the term "phone" itself may not add a lot of information.

> **NOTE**
>
> *The notion of "value" changes with each task. Functional words such as "the" and "and" may not be important for, say, automatic document categorization into subjects, but can be very important for other applications, such as part-of-speech tagging (identifying verbs, adjectives, nouns, pronouns, and so on).

Functional stop words are conveniently built into NLTK. We just need to import them and then we can store them in a variable. Once stored, they can be accessed just like any Python list. Let's import them and see how many of these words we have:

```
import nltk
nltk.download("stopwords")
from nltk.corpus import stopwords
list_stop = stopwords.words("english")
len(list_stop)
```

We will see the following output:

```
179
```

We can see that we have 179 built-in stop words. Let's also print some of them:

```
print(list_stop[:50])
```

The output will be as follows:

```
['i', 'me', 'my', 'myself', 'we', 'our', 'ours', 'ourselves',
 'you', "you're", "you've", "you'll", "you'd", 'your',
 'yours', 'yourself', 'yourselves', 'he', 'him', 'his',
 'himself', 'she', "she's", 'her', 'hers', 'herself',
 'it', "it's", 'its', 'itself', 'they', 'them',
 'their', 'theirs', 'themselves', 'what', 'which', 'who', 'whom',
 'this', 'that', "that'll", 'these', 'those', 'am', 'is', 'are',
 'was', 'were', 'be']
```

We can see that most of these terms are very commonly used "filler" terms that have a "functional" role in the language, and don't add a lot of information.

Now, removing stop words can be done the same way we removed punctuation.

EXERCISE 4.01: TOKENIZING, CASE NORMALIZATION, PUNCTUATION, AND STOP WORD REMOVAL

In this exercise, we will remove stop words from the data, and also apply everything we have learned so far. We'll start by performing tokenization (sentences and words); then, we'll perform case normalization, followed by punctuation and stop word removal.

> **NOTE**
>
> Before commencing this exercise, ensure that you are using a Jupyter Notebook where you have downloaded both the **punkt** sentence tokenizer and the **stopwords** corpus, as demonstrated in the *Text Preprocessing* section.

We'll keep the code concise this time. We'll be defining and manipulating the **raw_txt** variable. Let's get started:

1. Run the following commands to import **nltk** and the **tokenize** module from it:

```
import nltk
from nltk import tokenize
```

2. Define the **raw_txt** variable so that it contains the text "**Welcome to the world of deep learning for NLP! We're in this together, and we'll learn together. NLP is amazing, and deep learning makes it even more fun. Let's learn!**":

```
raw_txt = """Welcome to the world of deep learning for NLP! \
            We're in this together, and we'll learn together. \
            NLP is amazing, \
            and deep learning makes it even more fun. \
            Let's learn."""
```

3. Use the **sent_tokenize()** method to separate the raw text into individual sentences and store the result in a variable. Use the **lower()** method to convert the string into lowercase before tokenizing:

```
txt_sents = tokenize.sent_tokenize(raw_txt.lower())
```

> **NOTE**
>
> The **txt_sents** variable we've just created will be used later on in the chapter as well.

4. Using list comprehension, apply the **word_tokenize()** method to separate each sentence into its constituent words:

```
txt_words = [tokenize.word_tokenize(sent) for sent in txt_sents]
```

5. Import **punctuation** from the **string** module and convert it into a list:

```
from string import punctuation
stop_punct = list(punctuation)
```

6. Import the built-in stop words for English from NLTK and save them in a variable:

```
from nltk.corpus import stopwords
stop_nltk = stopwords.words("english")
```

7. Create a combined list that contains the punctuations as well as the NLTK stop words. Note that we can remove them together in one go:

```
stop_final = stop_punct + stop_nltk
```

8. Define a function that will remove stop words and punctuation from the input sentence, provided as a collection of tokens:

```
def drop_stop(input_tokens):
    return [token for token in input_tokens \
               if token not in stop_final]
```

9. Remove redundant tokens by applying the function to the tokenized sentences and store the result in a variable:

```
txt_words_nostop = [drop_stop(sent) for sent in txt_words]
```

10. Print the first cleaned-up sentence from the data:

```
print(txt_words_nostop[0])
```

With the stop words removed, the result will look like this:

```
['welcome', 'world', 'deep', 'learning', 'nlp']
```

> **NOTE**
>
> To access the source code for this specific section, please refer to https://packt.live/2VVNEgf.
>
> You can also run this example online at https://packt.live/38Gr54r.
> You must execute the entire Notebook in order to get the desired result.

In this exercise, we performed all the cleanup steps we've learned about so far. This time around, we combined certain steps and made the code more concise. These are some very common steps that we should apply when dealing with text data. You could try to further optimize and modularize by defining a function that returns the result after all the processing steps. We encourage you to try it out.

So far, the steps in the cleanup process were steps that got rid of tokens that weren't very useful in our assessment. But there are a few more things we could do to make our data even better – we can try using our understanding of the language to combine tokens, identify tokens that have practically the same meaning, and remove further redundancy. A couple of popular approaches are stemming and lemmatization.

> **NOTE**
>
> The variables we have created in this exercise will be used in later sections of the chapter as well. Ensure that you're completing this exercise first before moving to the upcoming exercises and activities.

STEMMING AND LEMMATIZATION

"Eat", "eats", "eating", "ate" – aren't they all just variations of the same word, all referring to the same action? In most text and spoken language, in general, we have multiple forms of the same word. Typically, we don't want these to be considered as separate tokens. A search engine would need to return similar results if the query is "red shoes" or "red shoe"– it would be a terrible search experience otherwise. We acknowledge that such cases are very common and that we need a strategy to handle such cases. But what should we do with the variants of a word? A reasonable approach is to map them all to a common token so that they are all treated the same.

Stemming is a rule-based approach to achieve normalization by reducing a word to its "stem". The stem is the root of the word before any affixes (an element added to make a variant) are added. This approach is rather simple – chop off the suffix to get the stem. A popular algorithm is the **Porter stemming** algorithm, which applies a series of such rules:

Rule	Term	Stem
s->	cats	cat
ies->i	trophies	trophi
es->e	drives	drive
ing->e	driving	drive

Figure 4.3: Examples of the Porter stemming algorithm's rule-based approach

> **NOTE**
>
> The full set of Porter stemming algorithm rules can be found at
> http://snowball.tartarus.org/algorithms/porter/stemmer.html.

Let's look at the Porter stemming algorithm in action. Let's import the **PorterStemmer** function from the **'stem'** module in NLTK and create an instance of it:

```
from nltk.stem import PorterStemmer
stemmer_p = PorterStemmer()
```

Note that the stemmer works on individual tokens, not sentences as a whole. Let's see how the stemmer stems the word "**driving**":

```
print(stemmer_p.stem("driving"))
```

The output will be as follows:

```
drive
```

Let's see how we can apply this to a whole sentence. Note that we will have to tokenize the sentence:

```
txt = "I mustered all my drive, drove to the driving school!"
```

The following code is used for tokenizing the sentence and applying the stemmer to each term:

```
tokens = tokenize.word_tokenize(txt)
print([stemmer_p.stem(word) for word in tokens])
```

The output is as follows:

```
['I', 'muster', 'all', 'my', 'drive', ',', 'drove', 'to',
 'the', 'drive', 'school', '!']
```

We can see that the stemmer has correctly reduced "mustered" to "muster" and "driving" to "drive", while "drove" is untouched. Also, note that the result of a stemmer need not be a valid English word.

Lemmatization is a more sophisticated approach that refers to a dictionary and finds a valid root form (the lemma) of the word. Lemmatization works best when the part of speech of the word is also provided – it considers the role the term is playing and returns the appropriate form. The output from a lemmatization step is always a valid English word. However, lemmatization is computationally very expensive, and for it to work well, it needs the part-of-speech tag, which typically isn't available in the data. Let's have a brief look at it. First, let's import **WordNetLemmatizer** from **nltk.stem** and instantiate it:

```
nltk.download('wordnet')
from nltk.stem import WordNetLemmatizer
lemmatizer = WordNetLemmatizer()
```

Let's apply the **lemmatizer** on the term **ponies**:

```
lemmatizer.lemmatize("ponies")
```

The following is the output:

```
'pony'
```

For our discussions, stemming is sufficient. The result from stemming may not always be a valid word. For example, **poni** is the stem for **ponies** but isn't a valid English word. Also, there may be some inaccuracies, but for the objective of mapping to a common word, this crude method works just fine.

EXERCISE 4.02: STEMMING OUR DATA

In this exercise, we will continue with data preprocessing. We removed the stop words and punctuation in the previous exercise. Now, we will use the Porter stemming algorithm to stem the tokens. Since we'll be using the **txt_words_nostop** variable we created previously, let's continue with the same Jupyter Notebook we created in *Exercise 4.01, Tokenizing, Case Normalization, Punctuation, and Stop Word Removal*. The variable, at this point, will contain the following text:

```
[['welcome', 'world', 'deep', 'learning', 'nlp'],
 ["'re", 'together', "'ll", 'learn', 'together'],
 ['nlp', 'amazing', 'deep', 'learning', 'makes', 'even', 'fun'],
 ['let', "'s", 'learn']]
```

The following are the steps to complete this exercise:

1. Import **PorterStemmer** from NLTK using the following command:

    ```
    from nltk.stem import PorterStemmer
    ```

2. Instantiate the stemmer:

    ```
    stemmer_p = PorterStemmer()
    ```

3. Apply the stemmer to the first sentence in **txt_words_nostop**:

    ```
    print([stemmer_p.stem(token) for token in txt_words_nostop[0]])
    ```

    When we print the result, we get the following output:

    ```
    ['welcom', 'world', 'deep', 'learn', 'nlp']
    ```

    We can see that **welcome** has been changed to **welcom** and **learning** to **learn**. This is consistent with the rules of the Porter stemming algorithm.

4. Apply the stemmer to all the sentences in the data. You could use loops, or a nested list comprehension:

    ```
    txt_words_stem = [[stemmer_p.stem(token) for token in sent] \
                        for sent in txt_words_nostop]
    ```

5. Print the output using the following command:

    ```
    txt_words_stem
    ```

The output will be as follows:

```
[['welcom', 'world', 'deep', 'learn', 'nlp'],
 ["'re", 'togeth', "'ll", 'learn', 'togeth'],
 ['nlp', 'amaz', 'deep', 'learn', 'make', 'even', 'fun'],
 ['let', "'s", 'learn']]
```

It looks like plenty of modifications have been made by the stemmer. Many of the words aren't valid anymore but are still recognizable, and that's okay.

> **NOTE**
>
> To access the source code for this specific section, please refer to https://packt.live/2VVNEgf
>
> You can also run this example online at https://packt.live/38Gr54r. You must execute the entire Notebook in order to get the desired result.

In this exercise, we used the Porter stemming algorithm to stem the terms of our tokenized data. Stemming works on individual terms, so it needs to be applied after tokenizing into terms. Stemming reduced some terms to their base form, which weren't necessarily valid English words.

BEYOND STEMMING AND LEMMATIZATION

Beyond stemming and lemmatization, there are many specific approaches to handle word variations. We have techniques such as phonetic hashing to identify spelling variations of a word induced by pronunciations. Then, there is spelling correction to identify and rectify errors in spelling. Another potential step is abbreviation handling so that *television* and *TV* are treated the same. The result from these steps can be further augmented by performing domain-specific term handling. You get the drift... there are a lot of steps possible, and, depending on your data and the criticality of your application, you may include some of these in your processing.

In general, though, the steps we performed together are largely sufficient – case normalization, tokenization, stop word, and punctuation removal, followed by stemming/lemmatization. These are some common steps that most NLP applications include.

DOWNLOADING TEXT CORPORA USING NLTK

So far, we've performed these steps on dummy data that we created. Now, it's time to try out our newly acquired skills on a larger and more authentic text. First, let's acquire that text – Lewis Carrol's classic work, "*Alice's Adventures in Wonderland*", which is available through Project Gutenberg and accessible through NLTK.

You may need to download the **'gutenberg'** corpus through NLTK. First, import NLTK using the following command:

```
import nltk
```

Then, use the **nltk.download()** command to open up an app, that is, the **NLTK Downloader** interface (shown in the following screenshot):

```
nltk.download()
```

We can see that the app has multiple tabs. Click the **Corpora** tab:

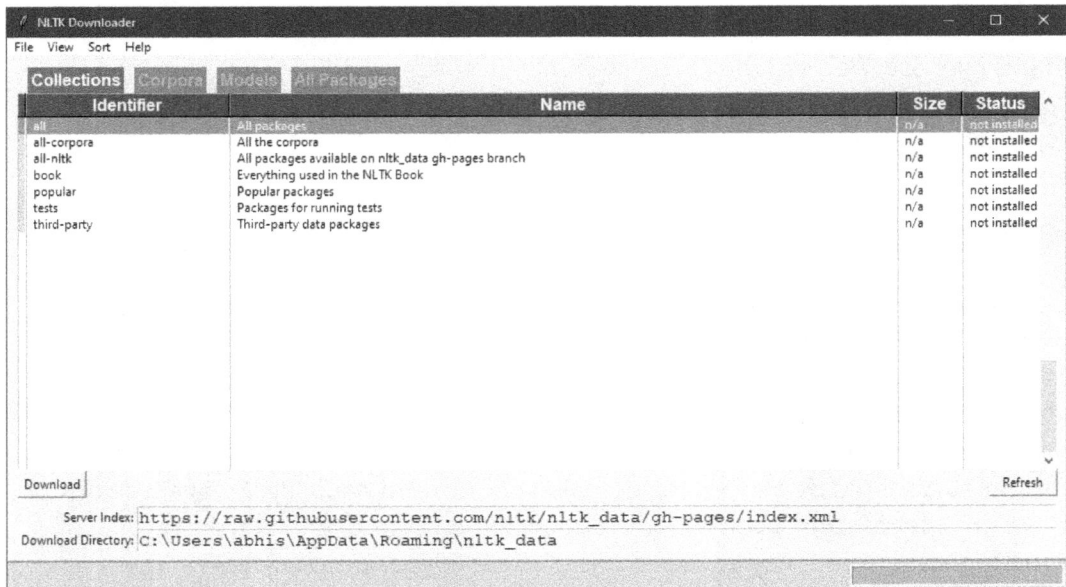

Figure 4.4: NLTK Downloader

In the **Corpora** tab, scroll down until you reach **gutenberg**. If the status is **not installed**, go ahead and click the **Download** button in the lower-left corner. That should install the **gutenberg** corpus:

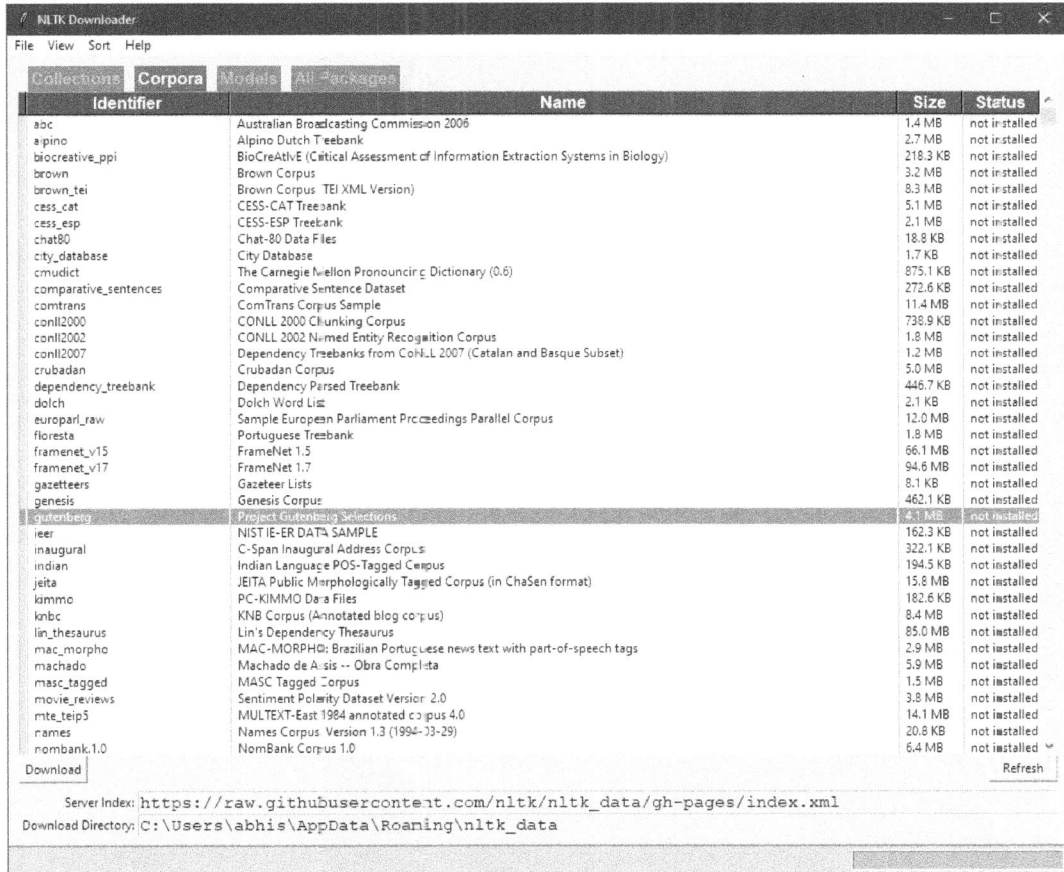

Figure 4.5: NLTK Downloader's Corpora tab

Close the interface. Now, you can access some classic texts right from NLTK. We'll read in the text and store it in a variable:

```
alice_raw = nltk.corpus.gutenberg.raw('carroll-alice.txt')
```

The text is stored in **alice_raw**, which is one big character string. Let's have a look at the first few characters of this string:

```
alice_raw[:800]
```

The output will be as follows:

```
"[Alice's Adventures in Wonderland by Lewis Carroll 1865]
 \n\nCHAPTER I. Down the Rabbit-Hole\n\nAlice was beginning
 to get very tired of sitting by her sister on the\nbank,
 and of having nothing to do: once or twice she had peeped
 into the\nbook her sister was reading, but it had no pictures
 or conversations in\nit, 'and what is the use of a book,'
 thought Alice 'without pictures or\nconversation?'
 \n\nSo she was considering in her own mind
 (as well as she could, for the\nhot day made her feel
 very sleepy and stupid), whether the pleasure\nof making
 a daisy-chain would be worth the trouble of getting up
 and\npicking the daisies, when suddenly a White Rabbit
 with pink eyes ran\nclose by her.\n\nThere was nothing
 so VERY remarkable in that; nor did Alice think
 it so\nVERY much out of the way to hear the Rabbit"
```

We can see the raw text in the output, which contains the usual imperfections that we expect – varying case, stop words, punctuation, and so on.

We're ready. Let's test out our skills through an activity.

ACTIVITY 4.01: TEXT PREPROCESSING OF THE 'ALICE IN WONDERLAND' TEXT

In this activity, you will apply all the preprocessing steps you've learned about so far to a much larger, real text. We'll work with the text for Alice in Wonderland that we stored in the **alice_raw** variable:

```
alice_raw[:800]
```

The text currently looks like this:

```
"[Alice's Adventures in Wonderland by Lewis Carroll 1865]
 \n\nCHAPTER I. Down the Rabbit-Hole\n\nAlice was beginning
 to get very tired of sitting by her sister on the\nbank,
 and of having nothing to do: once or twice she had peeped
 into the\nbook her sister was reading, but it had no pictures
 or conversations in\nit, 'and what is the use of a book,'
 thought Alice 'without pictures or\nconversation?
 '\n\nSo she was considering in her own mind
 (as well as she could, for the\nhot day made her feel
 very sleepy and stupid), whether the pleasure\nof making
```

```
a daisy-chain would be worth the trouble of getting up
and\npicking the daisies, when suddenly a White Rabbit
with pink eyes ran\nclose by her.\n\nThere was nothing
so VERY remarkable in that: nor did Alice think
it so\nVERY much out of the way to hear the Rabbit"
```

By the end of this activity, you will have cleaned and tokenized the data, removed a lot of imperfections, removed stop words and punctuation, and have applied stemming on the data.

> **NOTE**
>
> Before beginning this activity, make sure you have the **gutenberg** corpus installed and the **alice_raw** variable created, as shown in the previous section titled *Downloading Text Corpora Using NLTK*.

The following are the steps you need to perform:

1. Continuing in the same Jupyter Notebook, use the raw text in the **'alice_raw'** variable. Change the raw text to lowercase.

2. Tokenize the sentences.

3. Import punctuation from the **string** module and the stop words from NLTK.

4. Create a variable holding the contextual stop words, that is, **--** and **said**.

5. Create a master list for stop words to remove that contain terms from punctuation, NLTK stop words and contextual stop words.

6. Define a function to drop these tokens from any input sentence (tokenized).

7. Use the **PorterStemmer** algorithm from NLTK to perform stemming on the result.

8. Print out the first five sentences from the result.

> **NOTE**
>
> The detailed steps for this activity, along with the solutions and additional commentary, are presented on page 405.

The expected output looks like this:

```
[['alic', "'s", 'adventur', 'wonderland', 'lewi', 'carrol',
  '1865', 'chapter', 'i.', 'rabbit-hol', 'alic', 'begin',
  'get', 'tire', 'sit', 'sister', 'bank', 'noth', 'twice',
  'peep', 'book', 'sister', 'read', 'pictur', 'convers',
  "'and", 'use', 'book', 'thought', 'alic', "'without",
  'pictur', 'convers'],
 ['consid', 'mind', 'well', 'could', 'hot', 'day', 'made',
  'feel', 'sleepi', 'stupid', 'whether', 'pleasur', 'make',
  'daisy-chain', 'would', 'worth', 'troubl', 'get', 'pick',
  'daisi', 'suddenli', 'white', 'rabbit',
  'pink', 'eye', 'ran', 'close'],
 ['noth', 'remark', 'alic', 'think', 'much', 'way', 'hear',
  'rabbit', 'say', "'oh", 'dear'],
 ['oh', 'dear'],
 ['shall', 'late']]
```

Let's take a look at what we have achieved so far and what lies ahead.

So far, we've learned how to perform text preprocessing – the process of getting the text data ready for our primary analysis/model. We started with raw text data that has, potentially, many imperfections. We learned how to handle many of these imperfections and are now at a juncture where we are comfortable with handling text data and getting it ready for further analysis. This is an important first part in any NLP application. So, we took raw text data and got clean data in return. What's next?

The next section is a very important one since it has a very strong bearing on the quality of your analysis. It is known as representation. Let's discuss it.

TEXT REPRESENTATION CONSIDERATIONS

We have processed our raw input data into cleaned text. Now, we need to transform this cleaned text into something a predictive model understands. But what does a predictive model understand? Does it understand the different words? Does it read a word as we do? Can it work with the text that we supply to it?

By now, you understand that models work on numbers. The input to a model is a stream of numbers. It doesn't understand images, but it can work with matrices and numbers representing those images. For handling images, the key idea is to convert them into numbers and generate features out of them. The idea is the same for text: we need to convert the text into numbers, which will act as features for the model.

Representation is all about converting the text into numbers/features that the model understands. Doesn't sound like there is much to it, right? If you think that, then here's something for you to consider: input features are very important for any modeling exercise, and representation is the process of creating those features. It has a very significant effect on the outcome of your model and is a process that you should pay a great deal of attention to.

How do you go about text representation, then? What's the "best" way to represent text, if there is such a thing at all? Let's discuss a few approaches.

CLASSICAL APPROACHES TO TEXT REPRESENTATION

Text representation approaches have evolved significantly over the years, and the advent of neural networks and deep neural networks has made a significant impact on the way we now represent text (more on that later). We have come a long way indeed: from handcrafting features to marking if a certain word is present in the text, to creating powerful representations such as word embeddings. While there are a lot of approaches, some more suitable for the task than the others, we will discuss a few major classical approaches and work with all of them in Python.

ONE-HOT ENCODING

One-hot encoding is, perhaps, one of the most intuitive approaches toward text representation. A one-hot encoded feature for a word is a binary indicator of the term being present in the text. It's a simple approach that is easy to interpret – the presence or absence of a word. To understand this better, let's consider our sample text before stemming, and let's see how one-hot encoding works for a particular term of interest, say, **nlp**.

Let's see what the text currently looks like using the following command:

```
txt_words_nostop
```

We can see that the text looks like this:

```
[['welcome', 'world', 'deep', 'learning', 'nlp'],
 ["'re", 'together', "'ll", 'learn', 'together'],
 ['nlp', 'amazing', 'deep', 'learning', 'makes', 'even', 'fun'],
 ['let', "'s", 'learn']]
```

Our word of interest is **nlp**. Here's what the one-hot encoded feature for it would look like:

Input text	"nlp"
['welcome', 'world', 'deep', 'learning', 'nlp']	1
["'re", 'together', "'ll", 'learn', 'together']	0
['nlp', 'amazing', 'deep', 'learning', 'makes', 'even', 'fun']	1
['let', "'s", 'learn']	0

Figure 4.6: One-hot encoded feature for 'nlp'

We can see that the feature is **1**, but only for the sentences where the term **nlp** is present and is **0** otherwise. We can make such indicator variables for each word that we're interested in. So, if we're interested in three terms, we make three such features:

Input text	One hot encoding for		
	"nlp"	"deep"	"learn"
['welcome', 'world', 'deep', 'learning', 'nlp']	1	1	0
["'re", 'together', "'ll", 'learn', 'together']	0	0	1
['nlp', 'amazing', 'deep', 'learning', 'makes', 'even', 'fun']	1	1	0
['let', "'s", 'learn']	0	0	1

Figure 4.7: One-hot encoded features for 'nlp', 'deep', and 'learn'

Let's recreate this using Python in an exercise.

EXERCISE 4.03: CREATING ONE-HOT ENCODING FOR OUR DATA

In this exercise, we will replicate the preceding example. The target terms are **nlp**, **deep**, and **learn**. We will create a one-hot encoded feature for these terms using our own function and store the result in a **numpy** array.

Again, we'll be using the **txt_words_nostop** variable we created in *Exercise 4.01, Tokenizing, Case Normalization, Punctuation, and Stop Word Removal*. So, you will need to continue this exercise in the same Jupyter Notebook. Follow these steps to complete this exercise:

1. Print out the **txt_words_nostop** variable to see what we're working with:

```
print(txt_words_nostop)
```

The output will be as follows:

```
[['welcome', 'world', 'deep', 'learning', 'nlp'],
 ["'re", 'together', "'ll", 'learn', 'together'],
 ['nlp', 'amazing', 'deep', 'learning', 'makes', 'even', 'fun'],
 ['let', "'s", 'learn']]
```

2. Define a list with the target terms, that is, **"nlp"**, **"deep"**, **"learn"**:

```
target_terms = ["nlp","deep","learn"]
```

3. Define a function that takes in a single tokenized sentence and returns a **0** or **1** for each target term, depending on its presence in the text. Note that the length of the output is fixed at **3**:

```
def get_onehot(sent):
    return [1 if term in sent else 0 for term in target_terms]
```

We're iterating over the target terms and checking if they're available in the input sentence.

4. Apply the function to each sentence in our text and store the result in a variable

```
one_hot_mat = [get_onehot(sent) for sent in txt_words_nostop]
```

5. Import **numpy**, create a **numpy array** from the result, and print it:

```
import numpy as np
np.array(one_hot_mat)
```

The array's output is as follows:

```
array([[1, 1, 0],
       [0, 0, 1],
       [1, 1, 0],
       [0, 0, 1]])
```

We can see that the output contains four rows, one for each sentence. Each of the columns in the array contains the one-hot encoding for a target term. The values for "learn" are 0, 1, 0, 1, which is consistent with our expectations.

> **NOTE**
>
> To access the source code for this specific section, please refer to https://packt.live/2VVNEgf.
>
> You can also run this example online at https://packt.live/38Gr54r. You must execute the entire Notebook in order to get the desired result.

In this exercise, we saw how we can generate features from text using one-hot encoding. The example used a list of target terms. This may work when you have a very specific objective in mind where we know exactly which terms are useful. Indeed, this was the method that was heavily employed until a few years ago, where people handcrafted features from text. In many situations, this is not feasible – since we don't know exactly which terms are important, we use one-hot encoding for a large number of terms (5,000, 10,000, or even more).

The other aspect is whether the presence/absence of the term enough for most situations. Do we not want to include more information? Maybe the frequency of the term instead of just its presence, or maybe even some other smarter measure? Let's see how this works.

TERM FREQUENCIES

We discussed that one-hot encoding merely indicates the presence or absence of a term. A reasonable argument here is that the frequency of terms is also important. It may be that a term that's present more times in a document is more important for the document. Maybe representing the term by its frequency is a better approach than simply the indicator. The frequency approach is straightforward – for each term, count the number of times it appears in a particular text. If a term is absent from the document/text, it gets a 0. We do this for all the terms in our vocabulary. Therefore, we have as many features as the number of words in our vocabulary (something we can choose; this can be thought of as a hyperparameter). We should note that after the preprocessing steps, the "*terms*" that we're working with are tokens that may not be valid words in the language:

> **NOTE:**
>
> The *vocabulary* is the superset of all the terms that we'll use in the final model. Vocabulary size refers to the number of unique terms in the vocabulary. You could have 20,000 unique terms in the raw text but choose to work with the most frequent 10,000 terms; this would be the effective vocabulary size.

Consider the following image; if we had *N* documents and had V(**t1, t2, t3 … t$_v$**) words in our working vocabulary, the representation for the data would be a matrix of dimensions *N* × *V*.

Document Index	t1	t2	t3	t4	t5	t6	t7	t8	.	t$_{v-1}$	t$_v$
1	0	0	2	7	3	9	0	10	4	0	0
2	4	0	0	2	9	0	2	0	0	9	0
3	9	3	0	0	0	6	5	0	0	8	0
4	9	0	10	0	0	0	2	3	0	0	0
5	0	8	5	0	4	9	0	0	0	0	6
6	7	0	8	0	0	0	8	0	0	4	0
7	10	0	0	0	7	0	0	0	2	0	0
8	0	5	0	0	0	0	0	0	0	3	0
9	10	0	10	0	0	0	0	9	0	2	0
.	0	7	0	0	0	0	7	0	3	0	0
.	10	0	0	0	0	1	0	0	6	5	5
.	5	0	0	5	0	0	0	0	1	8	10
N	0	10	2	0	5	0	6	0	0	1	0

Figure 4.8: Document-term matrix

This matrix is our **Document-Term Matrix** (**DTM**) – where each row represents a document, and each column represents a term. The values in the cells can represent some measure (count, or any other measure). We'll work with term frequencies in this section.

We could create our own function again, but we have a very handy utility called **'CountVectorizer'** for this in **scikit-learn** that we'll use instead. Let's familiarize ourselves with it, beginning by importing the utility:

```
from sklearn.feature_extraction.text import CountVectorizer
```

The vectorizer can work with raw text, as well as tokenized data (as in our case). To work on the raw text, we would use the following code, where we will create a DTM with term frequencies from our raw text (**txt_sents**).

Before we begin, let's take a quick look at the contents of this variable:

```
txt_sents
```

The output should be as follows:

```
['welcome to the world of deep learning for nlp!',
 "we're in this together, and we'll learn together.",
 'nlp is amazing, and deep learning makes it even more fun.',
 "let's learn!"]
```

> **NOTE**
>
> If the contents of the **txt_sents** variable have been overwritten while working on *Activity 4.01, Text Preprocessing of the 'Alice in Wonderland' Text*, you can revisit *Step 3* of *Exercise 4.01, Tokenizing, Case Normalization, Punctuation, and Stop Word Removal* and redefine the variable so that its contents match the preceding output.

Now, let's instantiate the vectorizer. Note that we need to provide the vocabulary size. This picks the top *n* terms from the data for creating the matrix:

```
vectorizer = CountVectorizer(max_features = 5)
```

We chose five terms here; the result will contain five columns in the matrix. Let's train (**'fit'**) the vectorizer on the data:

```
vectorizer.fit(txt_sents)
```

The vectorizer has now learned a vocabulary – the top five terms – and has created an index for each term in the vocabulary. Let's have a look at the vocabulary:

```
vectorizer.vocabulary_
```

The preceding attribute gives us the following output:

```
{'deep': 1, 'we': 4, 'together': 3, 'and': 0, 'learn': 2}
```

We can see which terms have been picked (the top five).

Now, let's apply the vectorizer to the data to create the DTM. A minor detail: the result from a vectorizer is a sparse matrix. To view it, we'll convert it into an array:

```
txt_dtm = vectorizer.fit_transform(txt_sents)
txt_dtm.toarray()
```

Have a look at the output:

```
array([[0, 1, 0, 0, 0],
       [1, 0, 1, 2, 2],
       [1, 1, 0, 0, 0],
       [0, 0, 1, 0, 0]], dtype=int64)
```

The second document (the second row) has a frequency of **2** for the last two terms. What are those terms? Well, indices 3 and 4 are the terms **'together'** and **'we'**, respectively. Let's print out the original text to see if the output is as expected:

```
txt_sents
```

The output will be as follows:

```
['welcome to the world of deep learning for nlp!',
 "we're in this together, and we'll learn together.",
 'nlp is amazing, and deep learning makes it even more fun.',
 "let's learn!"]
```

This is just as we expected, and it looks like the count vectorizer works just fine.

Notice that the vectorizer tokenizes the sentence as well. If you don't want that and want to use preprocessed tokens instead (**txt_words_stem**), you simply need to pass a dummy tokenizer and preprocessor to **CountVectorizer**. Let's see how that works. First, we create a function that does nothing and simply returns the tokenized sentence/document:

```
def do_nothing(doc):
    return doc
```

Now, we'll instantiate the vectorizer to use this function as the preprocessor and tokenizer:

```
vectorizer = CountVectorizer(max_features=5,
                             preprocessor=do_nothing,
                             tokenizer=do_nothing)
```

Here, we're fitting and transforming the data in one step using the **fit_transform()** method from the tokenizer, and then we view the result. The method identifies the unique terms as the *vocabulary* when fitting on the data, then counts and returns the occurrence of each term for each document when transforming. Let's see it in action:

```
txt_dtm = vectorizer.fit_transform(txt_words_stem)
txt_dtm.toarray()
```

The output array will be as follows:

```
array([[0, 1, 1, 1, 0],
       [1, 0, 1, 0, 2],
       [0, 1, 1, 1, 0],
       [0, 0, 1, 0, 0]], dtype=int64)
```

We can see that the output is different from that of the previous result. Is this difference expected? To understand, let's look at the vocabulary of the vectorizer:

```
vectorizer.vocabulary_
```

The output will be as follows:

```
{'deep': 1, 'learn': 2, 'nlp': 3, 'togeth': 4, "'ll": 0}
```

We're working with preprocessed data, remember? We have already removed stop words and stemmed. Let's try printing out the input data just to be sure:

```
txt_words_stem
```

The output will be as follows:

```
[['welcom', 'world', 'deep', 'learn', 'nlp'],
 ["'re", 'togeth', "'ll", 'learn', 'togeth'],
 ['nlp', 'amaz', 'deep', 'learn', 'make', 'even', 'fun'],
 ['let', "'s", 'learn']]
```

We can see that the DTM is working according to the new vocabulary and the frequencies that were obtained after preprocessing.

So, this was the second approach of generating features from text data, that is, using the frequencies of the terms. In the next section, we will look at another very popular method.

THE TF-IDF METHOD

Does the high frequency of a term in a document mean that the word is very important for the document? Not really. What if that term is very common in all the documents? A common assumption that's employed in text data handling is that if a term is present in all documents, it may not be very differentiating or important for this particular document at hand. Seems like a reasonable assumption. Once more, let's consider the example of the term "*mobile*" when we're working with mobile phone reviews. The term is likely to be present in a very high proportion of reviews. But if your task is identifying the sentiment in the reviews, the term may not add a lot of information.

We can bump up the importance of terms that are present in the document but rare in the entire data and decrease the importance of terms that are present in most of the documents.

The TF-IDF method, which stands for *Term Frequency – Inverse Document Frequency*, defines **Inverse Document Frequency (IDF)** as follows:

$$idf(t) = \log[n / df(t)] + 1$$

Figure 4.9: Equation for TF-IDF

n is the total number of documents, while *df(t)* is the number of documents where the term *t* occurs. This is used as a factor to adjust the term frequency. You can see that it works just as we want it to – it increases the importance for rare terms, and decreases it for common terms. Note that there are variations to this formula, but we'll stick with what **scikit-learn** uses. Like **CountVectorizer**, the TF-IDF vectorizer tokenizes the sentence and learns the vocabulary, but instead of returning the counts for a term in a document, it returns the adjusted (IDF-multiplied) counts.

Now, let's apply this interesting new approach to our data.

EXERCISE 4.04: DOCUMENT-TERM MATRIX WITH TF-IDF

In this exercise, we'll implement the third approach to feature generation from text – TF-IDF. We will use scikit-learn's **TfidfVectorizer** utility and create the DTM for our raw text data. Since we're using the **txt_sents** variable we created earlier in this chapter, we'll need to use the same Jupyter Notebook. The text contained in the variable currently looks like this:

```
['welcome to the world of deep learning for nlp!',
 "we're in this together, and we'll learn together.",
 'nlp is amazing, and deep learning makes it even more fun.',
 "let's learn!"]
```

> **NOTE**
>
> If the contents of the **txt_sents** variable have been overwritten while working on *Activity 4.01, Text Preprocessing of the 'Alice in Wonderland' Text*, you can revisit *Step 3* of *Exercise 4.01, Tokenizing, Case Normalization, Punctuation, and Stop Word Removal* and redefine the variable so that its contents match the preceding output.

The following are the steps to perform:

1. Import the **TfidfVectorizer** utility from **scikit learn**:

```
from sklearn.feature_extraction.text import TfidfVectorizer
```

2. Instantiate the **vectorizer** with a vocabulary size of **5**:

```
vectorizer_tfidf = TfidfVectorizer(max_features=5)
```

3. Fit the **vectorizer** on the raw data of **txt_sents**:

```
vectorizer_tfidf.fit(txt_sents)
```

4. Print out the vocabulary learned by the **vectorizer**:

```
vectorizer_tfidf.vocabulary_
```

The trained vocabulary will look as follows:

```
{'deep': 1, 'we': 4, 'together': 3, 'and': 0, 'learn': 2}
```

Notice that the vocabulary is the same as that of the count vectorizer. This is expected. We're not changing the vocabulary; we're adjusting its importance for the documents.

5. Transform the data using the trained vectorizer:

```
txt_tfidf = vectorizer_tfidf.transform(txt_sents)
```

6. Print out the resulting DTM:

```
txt_tfidf.toarray()
```

The output will be as follows:

```
array([[0.         , 1.         ,
        0.         , 0.         , 0.         ],
       [0.25932364, 0.         , 0.25932364,
        0.65783832, 0.65783832],
       [0.70710678, 0.70710678, 0.         ,
        0.         , 0.         ],
       [0.         , 0.         , 1.         ,
        0.         , 0.         ]])
```

We can clearly see that the output values are different from the frequencies and that the values less than 1 indicate that many values have been lowered after multiplication with IDF.

7. We also need to see the IDF for each of the terms in the vocabulary to check if the factor is indeed working as we expect it to. Print out the IDF values for the terms using the **idf_** attribute:

```
vectorizer_tfidf.idf_
```

The output will be as follows:

```
array([1.51082562, 1.51082562, 1.51082562,
       1.91629073, 1.91629073])
```

The terms **'and'**, **'deep'**, and **'learn'** have a lower IDF, while the terms **'together'** and **'we'** have a higher IDF. This is just as we expect it to be – the terms **'together'** and **'we'** appear only in one document, while the others appear in two. So, the TF-IDF scheme is indeed giving more importance to rarer words.

> **NOTE**
>
> To access the source code for this specific section, please refer to https://packt.live/2VVNEgf.
>
> You can also run this example online at https://packt.live/38Gr54r. You must execute the entire Notebook in order to get the desired result.

In this exercise, we saw how we can represent text using the TF-IDF approach. We also saw how the approach downweighs more frequent terms by noticing that the IDF values were lower for higher-frequency terms. We ended up with a DTM containing the TF-IDF values for the terms.

SUMMARIZING THE CLASSICAL APPROACHES

We've just looked at three approaches to the classical way of text representation. We began with one-hot encoding, where the feature for a term was simply marking its presence in a document. The count/frequency-based approach attempted to add the importance of the term by using its frequency in a document. The TF-IDF approach attempted to use a "normalized" importance value of the term, factoring in how common the term is across the documents.

All three approaches that we've discussed so far fall under the "*Bag of Words*" approach to representation. So, why are they called "*Bag of Words*"? For a couple of reasons. The first reason is that they don't retain the order of the tokens – once in the bag, the position of the terms/tokens doesn't matter. The second reason is that this approach retains features for individual terms. So, for each document, you have, in a way, a "mixed bag of tokens", or a "*bag of words*", for simplicity.

The result from all three approaches had a dimensionality of $N \times V$, where N is the number of documents and V is the vocabulary size. Note that all three representations are very sparse – a typical sentence is very short (maybe 20 words), but the vocabulary size is typically in the thousands, resulting in most of the cells of the DTM being 0. This doesn't seem ideal. Well, there's this and a few more shortcomings of such representations, which we'll see shortly, that have led to the success of deep learning-based methods for representation. Let's discuss these ideas next.

DISTRIBUTED REPRESENTATION FOR TEXT

Why are word embeddings so popular? Why are we claiming they are amazingly powerful? What makes them so special? To understand and appreciate word embeddings, we need to acknowledge the shortcomings of the representations so far.

The terms "*footpath*" and "*sidewalk*" are synonyms. Do you think the approaches we've discussed so far will be able to capture this information? Well, you could manually go in and replace "*sidewalk*" with "*footpath*" so that both have the same token eventually, but can you do this for all possible synonyms in the language?

The terms "*hot*" and "*cold*" are antonyms. Do the previous Bag-of-Words representations capture this? What about "*dog*" being a type of "*animal*"? "*Cockpit*" being a part of a "*plane*"? Differentiating between a dog's bark and a tree's bark? Can you handle all these cases manually?

All the preceding are examples of "**semantic associations**" between the terms – simply put, their meanings are linked in some way or another. Bag-of-words representations can't capture these. This is where the notion of distributional semantics comes in. The key idea of distributional semantics is that terms with similar distributions have similar meanings.

A quick, fun quiz for you: Guess the meaning of the term *furbaby* from the following text:

"*I adopted a young Persian furbaby a month back. Like all furbabys, it loves to scratch its back and hates water, but unlike other furbabys, it miserably fails at catching a mouse.*"

You may have guessed it right: *furbaby* is referring to a cat. This was easy, wasn't it?

But how did you do that? Nowhere has the term cat been used. You looked at the context (the terms surrounding it) for "*furbaby*" and, based on your understanding of language and the world, you figured that these terms are generally associated with cats. You intuitively used this notion: words with similar meaning appear in similar contexts. If "*furbaby*" and "*cat*" appeared in similar contexts, their meaning must be similar.

"*You shall know a word by the company it keeps.*"

This famous, and now overused, quote by John Firth captures this idea very well. It's overused for all the right reasons. Let's see how this notion is employed in word embeddings.

WORD EMBEDDINGS AND WORD VECTORS

Word embeddings are representations of each term as a vector with low dimensionality. The one-hot encoded representation for a term was also a vector, but with dimensionality in the several thousands. Word embeddings/word vectors have much lower dimensionality and result from distributional semantics-based approaches – essentially, the representation captures the notion that words with similar meanings appear in similar contexts.

Word vectors attempt to capture the meanings of terms. This idea makes them very powerful, assuming, of course, they have been created correctly. With word vectors, vector operations such as adding/subtracting vectors and dot products are possible and have some very interesting meanings. There is also this great property that items with similar meanings are spatially closer. All of this leads to some amazing results.

A very interesting result is that word vectors can perform well on analogy tasks. Analogy tasks are defined as tasks of the format – "*a* is to *b* as *x* is to ?" – that is, find an entity that has the same relation to *x* as *b* has to *a*. As an example, if you ask "man is to uncle as a woman is to ?", the result would be "**aunt**" (more on this later). You can also find out semantic regularities between terms – the relationships between terms and sets of terms. Let's look at the following figure, which is based on word vectors/embeddings, to understand this better:

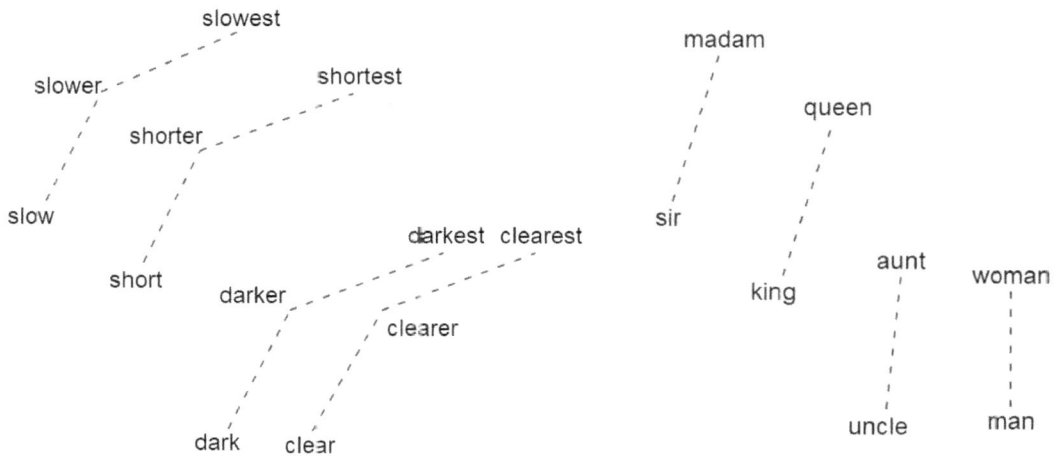

Figure 4.10: Semantic relationships between terms

The preceding figure shows some examples. The vectors can have high dimensionality (up to 300 or even more), so dimensionality reduction to two dimensions is performed to visualize it. The dotted connection between the two terms represents the relation between the terms. The direction of this connection is the important bit. On the left panel, we can see that the segment connecting **slow** and **slower** is parallel to the segment connecting **short** and **shorter**. What does this mean? This means that the word embeddings learned that the relationship between **short** and **shorter** is the same as the relationship between **slow** and **slower**. Likewise, the embeddings learned that the relationship between **clearer** and **clearest** is the same as that between **darker** and **darkest**. Pretty neat, right?

Similarly, the right-hand side of *Figure 4.10* shows that the embeddings learned that the relationship between **sir** and **madam** is the same as that between **king** and **queen**. Embeddings have also captured other kinds of semantic associations between terms, which we discussed in the previous section. Isn't that amazing?

This would not be possible with the approaches we discussed earlier. Word embeddings truly are working around the "meaning" of terms. We hope you can already appreciate the utility and power of word vectors. If you're not convinced yet, we'll soon be working with them and will see this for ourselves.

To generate word embeddings, there are several algorithms we can use. We will discuss two major approaches and apprise you of some other popular approaches. We will see how the distributional semantics approach is leveraged to derive these word embeddings.

WORD2VEC

Back in school, to test if we understood the meaning of certain terms, our language teachers used a very popular technique: "*fill in the blanks*". Based on the words around it, we needed to identify the word that would best fill that blank. If you understood the meaning well, you would do well. Think about this – isn't this distributional semantics?

In the *'furbaby'* example, you could predict the term `'cat'` because you understood the contexts and terms it occurs with. The exercise was effectively a "fill-in-the-blank" exercise. You could fill the blank only because you understood the meaning of 'cat'.

If you can predict a term given some context around it, you understand the meaning of the term.

This simple idea is exactly the formulation behind the **word2vec** algorithm. The **word2vec** algorithm/process is a prediction exercise, a massive "fill-in-the-blank" exercise, in a way. In short, this is what the algorithm does:

Given the contextual words, predict the missing target word.

That's all there is to it. The **word2vec** algorithm predicts the target word given the context. Let's understand how these are defined.

Consider the sentence, "*The Persian cat eats fish and hates bathing.*" We define context as some fixed number of the terms to the left and right of the target word, which is in the center. For our example, let `'cat'` be the target word, and let's take two words on either side of the target as our context:

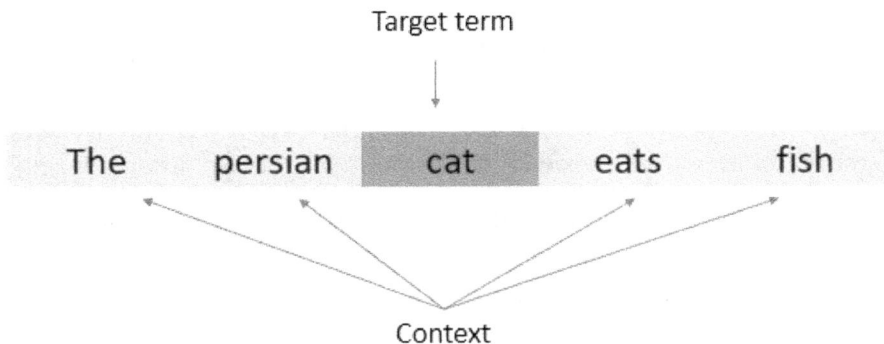

Figure 4.11: "cat" as the target term

The five terms together form a **'window'**, which has the target term at the center and the context terms around it. In this example, since we are considering two terms on either side, the window size is 2 (more on these parameters later). The window is a sliding one and moves over the terms in the sentence. The next window would have

C1	The	persian	cat	eats	fish	
C2		persian	cat	eats	fish	and
C3			cat	eats	fish	and	hates
C4				eats	fish	and	hates	bathing

Figure 4.12: Windows for the target term

C1, **C2**, **C3**, and **C4** denote the contexts for each window. In **C3**, "fish" is the target word, which is predicted using the terms "cat", "eats", "and", and "hates". The formulation is clear, but how does the model learn the representations? Let's discuss that next:

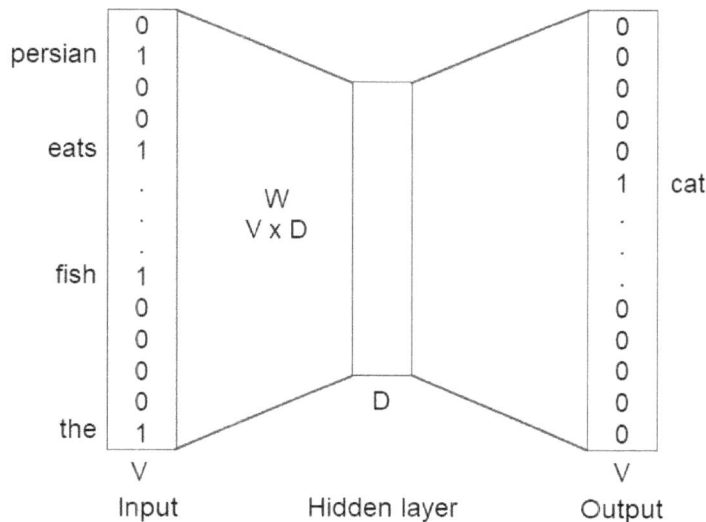

The Persian **cat** eats fish

Figure 4.13: The CBOW architecture with an example

The model shown in the preceding figure uses a neural network with a single hidden layer. The output layer is for the target term and is one-hot encoded with *V* outputs, one for each term – the predicted term, which is `'cat'`, is the term that gets `'hot'` in the output, of course. The input layer for the context terms is also size *V*, but fires for all the terms in the context. The hidden layer is of dimensionality *V* x *D* (where *D* is the dimension of the vectors). This hidden layer is where these magical representations of the terms are learned. Note that there is just one input layer, as the weights matrix *W* suggests.

While the network trains, predicting the target word better with each epoch, the parameters of the hidden layer are also getting updates. These parameters are effectively D-length vectors for each term. This D-length vector for a term is our word embedding for that term. After the iterations complete, we would have learned our word embeddings for all the terms in the vocabulary. Pretty neat, isn't it?

The approach we just discussed is the CBOW approach to training word vectors. The context is a simple bag of words (as we discussed in the previous section on classical approaches; order doesn't matter, remember), hence the name. There is another popular approach, the Skip-gram approach, which inverts the approach of the CBOW method – it predicts the context words based on the center word. This approach may seem a little less intuitive initially but works well. We'll discuss the differences between the results from CBOW and Skip-gram later in this chapter:

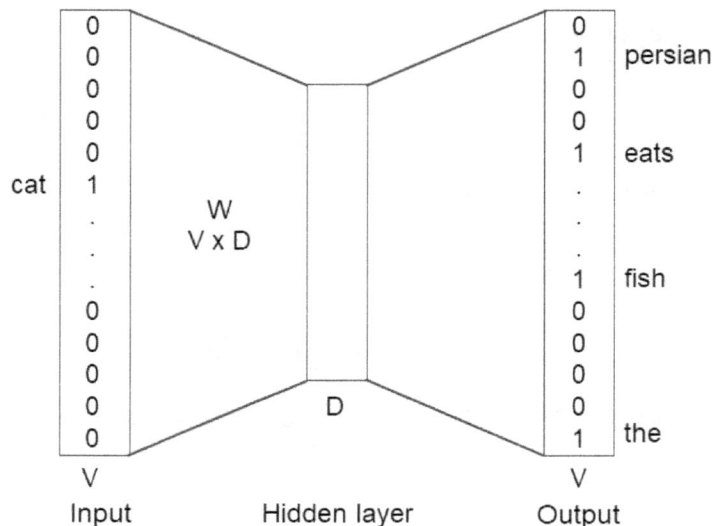

The Persian cat **eats fish**

Figure 4.14: The Skip-gram architecture

Let's see the CBOW approach in action in Python. We'll create our own word embeddings and assess if we can indeed get the amazing results we have been claiming so far.

TRAINING OUR OWN WORD EMBEDDINGS

There are many implementations of the **word2vec** algorithm available in different packages. We will use the implementation in **Gensim**, which is a great package for many NLP tasks. The implementation of word2vec in Gensim is close to the original paper by *Mikolov et al.* in 2013 (https://arxiv.org/pdf/1301.3781.pdf). Gensim also supports other algorithms for word embeddings; more on this later.

If you don't have Gensim installed, you can install it by typing the following command into a Jupyter Notebook:

```
!pip install gensim
```

The dataset we'll use is the **text8** corpus (http://mattmahoney.net/dc/textdata.html), which is the first billion characters from Wikipedia. It should, therefore, cover data from a variety of topics, not specific to one domain. Conveniently, Gensim has a utility (the **downloader** API) to read in the data. Let's read in the data after importing the **downloader** utility from Gensim:

```
import gensim.downloader as api
dataset = api.load("text8")
```

This step downloads the **text8** data and can take a while, depending on your internet connectivity. Alternatively, the data is available here (https://packt.live/3gKXU2D) to be downloaded and read using the **Text8Corpus** utility in Gensim, as shown in the following code:

```
from gensim.models import word2vec
dataset = word2vec.Text8Corpus("text8")
```

The **text8** data is now available as an iterable, which can simply be passed to the **word2vec** algorithm.

Before we train the embeddings, to make the results reproducible, let's set the seed as **1** for random number generation using NumPy:

```
np.random.seed(1)
```

> **NOTE**
>
> Although we have set the seed, there are more causes for variation of results. Some of this is because of an internal hash seed that the Python version on your system may use. Using multiple cores can also cause the results to vary. In any case, while the values you see may be different, and there could be some changes in the order of the results, the output you see should largely agree with ours. Note that this applies to all the practical elements pertaining to word vectors in this chapter.

Now, let's train our first word embedding by using the **word2Vec** method:

```
model = word2vec.Word2Vec(dataset)
```

This may take a minute or two, or less, depending on your system. Once complete, we will have our trained word vectors in the model and have access to multiple handy utilities to work with these word vectors. Let's access the word vector/embedding for a term:

```
print(model.wv["animal"])
```

The output will be as follows:

```
[-1.354541    0.21255197 -1.2803084    1.5077848    0.778622    -0.3017604
 -0.76313347  1.2620598    0.33203518 -0.12264135 -0.8749983  -0.49952352
 -3.4714737   0.23336382 -1.5944424  -0.77313447  1.6004859    0.39750227
 -0.21382928  0.79679984  2.230377   -1.2451155  -0.23759073 -0.6826417
  0.06757552  1.1354308  -0.9048583  -1.0423294  -0.6095223  -0.8687781
 -1.5961674   0.94031745  0.82677937 -1.435147   -0.967001   -0.9331538
  1.2132032  -1.4953172  -1.0877444   1.2012008    0.16687249 -1.3708639
 -0.3737326  -2.015771    2.219466    0.43797886 -1.4105217    0.13743232
```

Figure 4.15: The embedding for "animal"

You have a series of numbers – the vector for the term. Let's find the length of the vector:

```
len(model.wv["animal"])
```

The length of the vector is as follows:

```
100
```

The representation for each term is now a vector of length 100 (the length is a hyperparameter we can change; we used the default setting to get started). The vector for any term can be accessed as we did previously. Among the other handy utilities is the **most_similar()** method, which helps us find the terms that are the most similar to a target term. Let's see it in action:

```
model.wv.most_similar("animal")
```

The output will be as follows:

```
[('insect', 0.75981861352920053),
 ('animals', 0.729228138923645),
 ('aquatic', 0.66794979572299614),
 ('insects', 0.65222656726683716),
 ('organism', 0.64866477251005286),
 ('mammal', 0.6478426456451416),
 ('eating', 0.6435647010803223),
 ('ants', 0.6415578722953796),
 ('humans', 0.6414449214935303),
 ('feces', 0.6313734650611877)]
```

The output is a list of tuples, with each tuple containing the term and its similarity score with the term "animal".

We can see **insect, animals, insects**, and **mammal** in the top-most similar terms to "animal". This seems like a very good result, right? But how is the similarity being calculated? Words are being represented by vectors, and the vectors are trying to capture meaning – the similarity between terms is the similarity between their corresponding vectors. The **most_similar()** method uses **cosine similarity** between the vectors and returns the terms with the highest values. The value corresponding to each term in the result is the cosine similarity with the target word's vector.

Cosine similarity measures are suitable here as we expect terms that are similar in meaning to be spatially together. Cosine similarity is the cosine of the angle between the vectors. Terms with similar meaning and representation will have an angle closer to 0 and a similarity score closer to 1, whereas terms with completely unrelated meanings will have an angle closer to 90, and a cosine similarity closer to 0. Let's see what the model has learned as top terms related to "happiness":

```
model.wv.most_similar("happiness")
```

The most similar items turn out to be the following (the most similar ones are at the top):

```
[('humanity', 0.7819231748580933),
 ('perfection', 0.7699881792068481),
 ('pleasure', 0.7422512769699097),
 ('righteousness', 0.7402842044830322),
 ('desires', 0.7374188899993896),
 ('dignity', 0.7189303040504456),
 ('goodness', 0.7103697657585144),
 ('fear', 0.7047020196914673),
 ('mankind', 0.7046756744384766),
 ('salvation', 0.6990150213241577)]
```

Humanity, mankind, goodness, righteousness, and compassion -- we have some life lessons here. It seems to have learned what many people seemingly can't figure out in their entire lifetime. Remember, it is just a series of matrix multiplications.

SEMANTIC REGULARITIES IN WORD EMBEDDINGS

We mentioned earlier that these representations capture regularities in language and are good at solving simple analogy tasks. The offsets between vector embeddings seem to capture the analogical relationship between words. So, for example, *"king"* - *"man"* + *"woman"* is expected to result in *"queen"*. Let's see if the model that we trained on the **text8** corpus also understands some regularities.

We'll use the **most_similar()** method here, which allows us to add and subtract vectors from each other. We'll provide **'king'** and **'woman'** as vectors to add to each other, use **'man'** to subtract from the result, and then check out the five terms that are the most similar to the resulting vector:

```
model.wv.most_similar(positive=['woman', 'king'], \
                      negative=['man'], topn=5)
```

The output will be as follows:

```
[('queen', 0.6803990602493286),
 ('empress', 0.6331825852394104),
 ('princess', 0.6145625114440918),
 ('throne', 0.6131302714347839),
 ('emperor', 0.6064509153366089)]
```

The top result is **'queen'**. Looks ike the model is capturing these regularities. Let's try out another example. "Man" is to "uncle" as "woman" is to ? Or in an arithmetic form, what is the vector closest to *uncle - man + woman = ?*

```
model.wv.most_similar(positive=['uncle', 'woman'], \
                      negative=['man'], topn=5)
```

The following is the output of the preceding code:

```
[('aunt', 0.8145735263824463),
 ('grandmother', 0.8067640066146851),
 ('niece', 0.7993890643119812),
 ('wife', 0.7965766787528992),
 ('widow', 0.7914236187934875)]
```

This seems to be working great. Notice that all the top five results are for the feminine gender. So, we took **uncle**, removed the masculine elements, added feminine elements, and now we have some really good results.

Let's look at some other examples of vector arithmetic. We can take vectors for two different terms and average them to arrive at vectors for a phrase as well. Let's try it for ourselves.

> **NOTE**
>
> Taking the average of individual vectors is just one of the many ways of arriving at phrase vectors. Variations range from weighted averages to more complex mathematical functions.

EXERCISE 4.05: VECTORS FOR PHRASES

In this exercise, we will begin to create vectors for two different phrases, **get happy** and **make merry**, by taking the average of the individual vectors. We will find a similarity between the representations for the phrases. You will need to continue this exercise in the same Jupyter Notebook we have been using throughout this chapter. Follow these steps to complete this exercise:

1. Extract the vector for the term "*get*" and store it in a variable:

    ```
    v1 = model.wv['get']
    ```

2. Extract the vector for the term "*happy*" and store it in a variable:

    ```
    v2 = model.wv['happy']
    ```

3. Create a vector as the element-wise average of the two vectors, **(v1 + v2)/2**. This is our vector for the entire phrase "get happy":

    ```
    res1 = (v1+v2)/2
    ```

4. Similarly, extract vectors for the terms "*make*" and "*merry*":

    ```
    v1 = model.wv['make']
    v2 = model.wv['merry']
    ```

5. Create a vector for the phrase by averaging the individual vectors:

    ```
    res2 = (v1+v2)/2
    ```

6. Using the **cosine_similarities()** method in the model, find the cosine similarity between the two:

    ```
    model.wv.cosine_similarities(res1, [res2])
    ```

    The cosine similarity comes out as follows:

    ```
    array([0.5798107], dtype=float32)
    ```

The result is a cosine similarity of about **0.58**, which is positive and much higher than **0**. This means that the model thinks the phrases "get happy" and "make merry" are similar in meaning. Not bad, right? Instead of a simple average, we could use weighted averages, or come up with more sophisticated methods of combining individual vectors.

> **NOTE**
>
> To access the source code for this specific section, please refer to https://packt.live/2VVNEgf.
>
> You can also run this example online at https://packt.live/38Gr54r. You must execute the entire Notebook in order to get the desired result.

In this exercise, we saw how we could use vector arithmetic to represent phrases, instead of individual terms, and we saw that meaning is still captured. This brings us to a very important lesson – *vector arithmetic on word embeddings has meaning.*

These vector arithmetic operations work on the meaning of terms, resulting in some very interesting results.

We hope you now appreciate the power of word embeddings. We realize that these results come from just some matrix multiplication and take a minute to train on our dataset. Word embeddings are almost magical, and it is pleasantly surprising how such a simple prediction formulation results in such a powerful representation.

When we created the word vectors previously, we didn't pay much attention to the controls/parameters. There are many, but only some have a significant impact on the quality of the representations. We will now come to understand the different parameters of the **word2vec** algorithm and see the effect of changing these for ourselves.

EFFECT OF PARAMETERS — "SIZE" OF THE VECTOR

The `size` parameter of the **word2vec** algorithm is the length of the vector for each term. By default, as we saw earlier, this is 100. We will try reducing this parameter and assess the differences, if any, in the results. Let's retrain the word embeddings, with `size` as 30 this time:

```
model = word2vec.Word2Vec(dataset, size=30)
```

Now, let's check the analogy task from earlier, that is, **king - man + woman**:

```
model.wv.most_similar(positive=['woman', 'king'], \
                      negative=['man'], topn=5)
```

This should give us the following output:

```
[('emperor', 0.8314059972763062),
 ('empress', 0.8250986933708191),
 ('son', 0.8157491683959961),
 ('prince', 0.8060941696166992),
 ('archbishop', 0.8003251552581787)]
```

We can see that **queen** isn't present in the top five results. It looks like by using a very low dimensionality, we aren't capturing enough information in the representation for a term.

EFFECT OF PARAMETERS – "WINDOW SIZE"

The **window size** parameter defines the context; concretely, the window size is the number of terms to the left and to the right of the target term while building the context. The effect of this parameter is not very obvious. The general observation is that when you use a higher window size (say, 20), the top similar terms seem to be terms that are used along with the target term, not necessarily having a similar meaning. On the other hand, reducing the window size (to, say, 2), returns the top terms that are very similar in meaning, and are synonyms in many cases.

SKIP-GRAM VERSUS CBOW

Choosing between Skip-gram and CBOW as the learning algorithm is exercised by setting **sg = 1** for Skip-gram (the default is **sg = 0**, that is, CBOW). Recall that the Skip-gram approach predicts the context words based on the central target word. This flips the formulation of CBOW, where the context words are used to predict the target word. But how do we choose between the two? What are the benefits of one over the other? To see for ourselves, let's train embeddings using Skip-gram and compare some results with what we had for CBOW. To begin, let's take a particular example for CBOW. First, we'll recreate the CBOW word vectors with the default vector size by not specifying the size parameter. Oeuvre is a term for the body of work of an artist/performer. We'll see the most similar terms for the uncommon term, **oeuvre**:

```
model = word2vec.Word2Vec(dataset)
model.wv.most_similar("oeuvre", topn=5)
```

The following terms come out as the most similar terms:

```
[('baglione', 0.7203884124755859),
 ('chateaubriand', 0.7119786143302917),
 ('kurosawa', 0.6956337690353394),
 ('swinburne', 0.6926312446594238),
 ('poetess', 0.6910216808319092)]
```

We can see that most results are the names of artists (**swinburne**, **kurosawa**, and **baglione**) or food dishes (chateaubriand). None of the top five results are close in meaning to the target term. Now, let's retrain our vectors using the Skip-gram method and see the result on the same task:

```
model_sg = word2vec.Word2Vec(dataset, sg=1)
model_sg.wv.most_similar("oeuvre", topn=5)
```

This gives us the following output:

```
[('masterful', 0.834753334522473),
 ('orchestration', 0.814994163606262),
 ('mussorgsky', 0.811679601693115),
 ('showcasing', 0.808014631213623),
 ('lithographs', 0.805435299673352)]
```

We can see that the top terms are much closer in meaning (**masterful**, **orchestration**, **showcasing**). So, the Skip-gram method seems to work better for rare words.

Why is this so? The CBOW method smooths over a lot of the distributional statistics by effectively averaging overall context words (remember, all the context terms together go as an input), while Skip-gram does not. When you have a small dataset, the smoothing that's done by CBOW is desirable. If you have a small/moderately sized dataset, and if you are concerned about the representation of rare terms, then Skip-gram is a good option.

EFFECT OF TRAINING DATA

A very important decision while training your word vectors is the underlying data. The patterns and similarities will be learned from the data you supply to the algorithm, and we expect the model to learn differently from data from different domains, different kinds of settings. and so on. To appreciate this, we load different corpora from different contexts and see how the embeddings vary.

The Brown corpus is a collection of general text, collected from 15 different topics to make it general (from politics to religion, books to music, and many other themes). It contains 500 text samples and about 1 million words. The "movie" corpus contains movie-review data from IMDb. Both of these are available in NLTK.

EXERCISE 4.06: TRAINING WORD VECTORS ON DIFFERENT DATASETS

In this exercise, we will train our own word vectors on the Brown corpus and the IMDb movie reviews corpus. We will assess the differences in the representations learned and the effect of the underlying training data. Follow these steps to complete this exercise:

1. Import the Brown and IMDb movie reviews corpus from NLTK:

```
nltk.download('brown')
nltk.download('movie_reviews')
from nltk.corpus import brown, movie_reviews
```

2. The corpora have a convenient method, **sent()**, to extract the individual sentences and words (tokenized sentences, which can be directly passed to the **word2vec** algorithm). Since both the corpora are rather small, use the Skip-gram method to create the embeddings:

```
model_brown = word2vec.Word2Vec(brown.sents(), sg=1)
model_movie = word2vec.Word2Vec(movie_reviews.sents(), sg=1)
```

We now have two embeddings that have been learned on different contexts for the same term. Let's see the most similar terms for **money** from the model on the Brown corpus.

3. Print out the *top five terms* most similar to **money** from the model that were learned on the Brown corpus:

```
model_brown.wv.most_similar('money', topn=5)
```

The following is the output of the preceding code:

```
[('job', 0.8477444648742676),
 ('care', 0.8424298763275146),
 ('friendship', 0.8394286632537842),
 ('risk', 0.8268661499023438),
 ('permission', 0.8243911862373352)]
```

We can see that the top term is **'job'**; fair enough. Let's see what the model learned regarding movie reviews.

4. Print out the top five terms most similar to **money** from the model that learned from the movie corpus:

```
model_movie.wv.most_similar('money', topn=5)
```

The following are the top terms:

```
[('cash', 0.7299771904945374),
 ('ransom', 0.7130625247955322),
 ('record', 0.7028014063835144),
 ('risk', 0.6977001428604126),
 ('paid', 0.6940697431564331)]
```

The top terms are **cash** and **ransom**. Considering the language being used in movies, and thus in movie reviews, this isn't very surprising.

> **NOTE**
>
> To access the source code for this specific section, please refer to https://packt.live/2VVNEg
>
> You can also run this example online at https://packt.live/38Gr54r. You must execute the entire Notebook in order to get the desired result.

In this exercise, we created word vectors using different datasets and saw that the representations for the same terms and the associations that were learned are very affected by the underlying data. So, choose your data wisely.

USING PRE-TRAINED WORD VECTORS

So far, we've trained our own word embeddings using the small datasets we had access to. The folks at the Stanford NLP group have trained word embeddings on 6 billion tokens with 400,000 terms in the vocabulary. Individually, we will not have the resources to handle this scale. Fortunately, the Stanford NLP group has been benevolent enough to make these trained embeddings available to the general public so that people like us can benefit from their work. The trained embeddings are available on the GloVe page (https://nlp.stanford.edu/projects/glove/).

A quick note on GloVe: the method that's used for training is slightly different. The objective is modified to make the similar terms occur closer in space, in a little more explicit fashion. You can read about the details on the project page for GloVe (https://nlp.stanford.edu/projects/glove/), which also has a link to the original paper proposing it. The end result, however, is very similar in performance to word2vec.

We'll download the **glove.6B.zip** file from the GloVe project page. The file contains 50D, 100D, 200D, and 300D vectors. We'll work with the 100D vectors here. Please unzip the file and make sure you have the text files in your working directory. The trained vectors are available as a text file, and the format is slightly different. We'll use the **glove2word2vec** utility that's available in Gensim to convert into a format that Gensim can easily load:

```
from gensim.scripts.glove2word2vec import glove2word2vec
glove_input_file = 'glove.6B.100d.txt'
word2vec_output_file = 'glove.6B.100d.w2vformat.txt'
glove2word2vec(glove_input_file, word2vec_output_file)
```

We specified the input and the output file and ran the **glove2word2vec** utility. As the name suggests, the utility takes in word vectors in GloVe format and converts them into **word2vec** format. After this, the **word2vec** models can understand these embeddings easily. Now, let's load the **keyed** word vectors from the text file (reformatted):

```
from gensim.models.keyedvectors import KeyedVectors
glove_model = KeyedVectors.load_word2vec_format\
            ("glove.6B.100d.w2vformat.txt", binary=False)
```

With this done, we have the GloVe embeddings in the model, along with all the handy utilities we had for the embeddings model from word2vec. Let's check out the top terms similar to **"money"**:

```
glove_model.most_similar("money", topn=5)
```

The output is as follows:

```
[('funds', 0.8508071899414062),
 ('cash', 0.848483681678772),
 ('fund', 0.7594833374023438),
 ('paying', 0.7415367364883423),
 ('pay', 0.740767240524292)]
```

For closure, let's also check how this model performs on the king and queen tasks:

```
glove_model.most_similar(positive=['woman', 'king'], \
                        negative=['man'], topn=5)
```

The following is the output of the preceding code:

```
[('queen', 0.7698541283607483),
 ('monarch', 0.6843380928039551),
 ('throne', 0.6755737066268921),
 ('daughter', 0.6594556570053101),
 ('princess', 0.6520533561706543)]
```

Now that we have these embeddings in a model, we can work with them the same way we worked with the embeddings we created previously and can benefit from the larger dataset and vocabulary and the processing power used by the contributing organization.

BIAS IN EMBEDDINGS — A WORD OF CAUTION

When discussing regularities and analogies, we saw the following example:

king – man + woman = queen

It's great that the embeddings are capturing these regularities by learning from the text data. Let's try something similar to a profession. Let's see the term closest to *doctor – man + woman*:

```
model.wv.most_similar(positive=['woman', 'doctor'], \
                      negative=['man'], topn=5)
```

The output regarding the top five results will be as follows:

```
[('nurse', 0.6464251279830933),
 ('child', 0.5847542881965637),
 ('teacher', 0.569127082824707),
 ('detective', 0.5451491475105286),
 ('boyfriend', 0.5403486490249634)]
```

That's not the kind of result we want. Doctors are males, while females are nurses? Let's try another example. This time, let's try what the model thinks regarding females as corresponding to "smart" for "males":

```
model.wv.most_similar(positive=['woman', 'smart'], \
                      negative=['man'], topn=5)
```

We get the following top five results:

```
[('cute', 0.6156168580055237),
 ('dumb', 0.6035820245742798),
 ('crazy', 0.5834532976150513),
 ('pet', 0.582811713218689),
 ('fancy', 0.5697714636393738)]
```

We can see that the top terms are **'cute'**, **'dumb'**, and **'crazy'**. That's not good at all.

What's happening here? Is this seemingly great representation approach sexist? Is the word2vec algorithm sexist? There definitely is bias in the resulting word vectors, but think about where the bias is coming from. It's the underlying data that uses **'nurse'** for females in contexts where **'doctor'** is used for males. It is, therefore, the underlying text that contains the bias, not the algorithm.

This topic has recently gained significant attention, and there is ongoing research around ways to assess and get rid of biases from the learned embeddings, but a good approach is to avoid biases in the data to begin with. If you trained word embeddings on YouTube comments, don't be surprised if they contain all kinds of extreme biases. You're better off avoiding text data that you suspect to have biases.

OTHER NOTABLE APPROACHES TO WORD EMBEDDINGS

We worked with the word2vec approach primarily, and we briefly looked at the GloVe approach. While these are the most popular approaches, there are a few other approaches worth mentioning:

FastText: Created by **Facebook's AI Research** (**FAIR**) lab, it uses subword information to enrich the word embeddings. You can read more about it on the official page (https://research.fb.com/downloads/fasttext/).

WordRank: Treats the embeddings problem as a word-ranking problem. Its performance is similar to word2vec in several tasks. You can read more about this at https://arxiv.org/abs/1506.02761.

Other than these, some popular libraries now have pre-trained embeddings available (SpaCy is a good example). The choices are aplenty. We can't do a detailed treatment of these choices here, but please do explore the options.

We've discussed a lot of ideas around representation in this chapter. Now, let's implement these ideas with the help of an activity.

ACTIVITY 4.02: TEXT REPRESENTATION FOR ALICE IN WONDERLAND

In the previous activity, we tokenized and performed basic preprocessing of the text. In this activity, we will advance this process by using representation approaches for the text. You will create your own embeddings from the data and see the kind of relations we have. You will also utilize pre-trained embeddings to represent the data in the text.

> **NOTE**
>
> Note that you'll need to have completed *Activity 4.01*, *Text Preprocessing of the 'Alice in Wonderland' Text*, to proceed with this activity. In that activity, we performed stop word removal on the text.

You need to perform the following steps:

We'll continue using the same Jupyter Notebook that we used for *Activity 4.01, Text Preprocessing of the 'Alice in Wonderland' Text*. We'll work on the result of the stop word removal step we got in that activity (let's say it is stored in a variable called **alice_words_nostop**). Print the first three sentences from the result.

1. Import **word2vec** from Gensim and train your word embeddings with default parameters.

2. Find the terms most similar to **rabbit**.

3. Using a window size 2, retrain the word vectors.

4. Find the terms most similar to **rabbit**.

5. Retrain the word vectors using the Skip-gram method with a window size of **5**.

6. Find the terms most similar to **rabbit**.

7. Find the representation for the phrase **white rabbit** by averaging the vectors for **white** and **rabbit**.

8. Find the representation for **mad hatter** by averaging the vectors for **mad** and **hatter**.

9. Find the cosine similarity between these two phrases.

10. Load pre-trained GloVe embeddings of size 100D.

11. Find representations for **white rabbit** and **mad hatter**.

12. Find the cosine similarity between the two phrases. Has the cosine similarity changed?

As a result of this activity, we will have our own word vectors that have been trained on "Alice's Adventures in Wonderland" and have representation for the terms available in the text.

> **NOTE**
>
> The detailed steps for this activity, along with the solutions and additional commentary, are presented on page 407.

SUMMARY

In this chapter, we began by discussing the peculiarities of text data and how ambiguity makes NLP difficult. We discussed that there are two key ideas in working with text – preprocessing and representation. We discussed the many tasks involved in preprocessing, that is, getting your data cleaned up and ready for analysis. We saw various approaches to removing imperfections from the data.

Representation was the next big aspect – we understood the considerations in representing text and converting text into numbers. We looked at various approaches, beginning with classical approaches, which included one-hot encoding, the count-based approach, and the TF-IDF method.

Word embeddings are a whole new approach to representing text that leverage ideas from distributional semantics – terms that appear in similar contexts have similar meanings. The word2vec algorithm smartly exploits this idea by formulating a prediction problem: predict a target word given the context. It uses a neural network for the prediction and, in the process, learns vector representations for the terms.

We saw that these representations are amazing as they seem to capture meaning, and simple arithmetic operations gave some very interesting and meaningful results. You can even create representations for phrases or even sentences/documents using word vectors. This sets the stage for later when we use word embeddings in more sophisticated deep learning architectures for NLP.

In the next chapter, we'll continue our exploration of sequences by applying deep learning approaches such as recurrent neural networks and one-dimensional convolutions to them.

5

DEEP LEARNING FOR SEQUENCES

OVERVIEW

In this chapter, we will implement deep learning-based approaches to sequence modeling, after understanding the considerations of dealing with sequences. We will begin with **Recurrent Neural Networks** (**RNNs**), an intuitive approach to sequence processing that has provided state-of-the-art results. We will then discuss and implement 1D convolutions as another approach and see how it compares with RNNs. We will also combine RNNs with 1D convolutions in a hybrid model. We will employ all of these models on a classic sequence processing task – stock price prediction. By the end of this chapter, you will become adept at implementing deep learning approaches for sequences, particularly plain RNNs and 1D convolutions, and you will have laid the foundations for more advanced RNN-based models.

INTRODUCTION

Let's say you're working with text data and your objective is to build a model that checks whether a sentence is grammatically correct. Consider the following sentence: *"words? while sequence be this solved of can the ignoring"*. The question didn't make sense, right? Well, how about the following? *"Can this be solved while ignoring the sequence of words?"*

Suddenly, the text makes complete sense. What do we acknowledge, then, about working with text data? That sequence matters.

In the task of assessing whether a given sentence is grammatically correct, the sequence is important. Sequence-agnostic models would fail terribly at the task. The nature of the task requires you to analyze the sequence of the terms.

In the previous chapter, we worked with text data, discussing ideas around representation and creating our own word vectors. Text and natural language data have another important characteristic – they have a sequence to them. While text data is one example of sequence data, sequences are everywhere: from speech to stock prices, from music to global temperatures. In this chapter, we'll start working with sequential data in a way that considers the order of the elements. We will begin with RNNs, a deep learning approach that exploits the sequence of data to provide insightful results of tasks such as machine translation, sentiment analysis, recommender systems, and time series prediction, to name a few. We will then look at using convolutions for sequence data. Finally, we will see how these approaches can be combined in a single, powerful deep learning architecture. Along the way, we will also build an RNN-based model for stock price prediction.

WORKING WITH SEQUENCES

Let's look at another example to make the importance of sequence modeling clearer. The task is to predict the stock price for a company for the next 30 days. The data provided to you is the stock price for today. You can see this in the following plot, where the *y-axis* represents the stock price and the *x-axis* denotes the date. Is this data sufficient?

Figure 5.1: Stock price with just 1 day's data

Surely, one data point, that is, the price on a given day, is not sufficient to predict the price for the next 30 days. We need more information. Particularly, we need information about the past – how the stock price has been moving for the past few days/months/years. So, we ask for, and get, data for three years:

Figure 5.2: Stock price prediction using historical data

This seems much more useful, right? Looking at the past trend and some patterns in the data, we can make predictions on the future stock prices. Thus, by looking at the past trend, we get a rough idea of how the stock will move over the next few days. We can't do this without a sequence. Again, sequence matters.

In real-world use cases, say, machine translation, you need to consider the sequence in the data. Sequence-agnostic models can only get you so far in some tasks; you need an approach that truly exploits the information contained in the sequence. But before talking about the workings of those architectures, we need to answer an important question: *what are sequences, anyway?*

While the definition of a "*sequence*" from the dictionary is rather self-explanatory, we need to be able to identify sequences for ourselves and decide whether we need to consider the sequence. To understand this idea, let's go back to the first example we saw: "*words? while sequence be this solved of can the ignoring*" versus "*can this be solved while ignoring the sequence of words?*"

When you jumbled the terms of the meaningful sentence text, it stopped making sense and lost all/most of the information. This can be a simple and effective test for a sequence: If you jumbled the elements, does it still make sense? If the answer is "no," then you have a sequence at hand. While sequences are everywhere, here are some examples of sequence data: language, music, movie scripts, music videos, time-series data (stock prices, commodity prices, and more), and the survival probability of a patient.

TIME SERIES DATA — STOCK PRICE PREDICTION

We will start working on our own model for predicting stock prices. The objective of the stock price prediction task is to build a model that can predict the next day's stock price based on historical prices. As we saw in the previous section, the task requires us to consider the sequence in the data. We will predict the stock price for Apple Inc.

> **NOTE**
>
> We will use a cleaned-up version of Apple's historical stock data that's been sourced from the Nasdaq website: https://www.nasdaq.com/market-activity/stocks/aapl/historical. The dataset can be downloaded from the following link: https://packt.live/325WSKR.
>
> Make sure to place the file (**AAPL.csv**) in your working directory and start a new Jupyter Notebook for the code. It is important that you run all the code in the exercises and the topic sections in a single Jupyter Notebook.

Let's begin by understanding the data. We will load the required libraries and then load and plot the data. You can use the following commands to load the necessary libraries and use the cell magic command (**%matplotlib inline**) to plot the images inline:

```
import pandas as pd, numpy as np
import matplotlib.pyplot as plt
%matplotlib inline
```

Next, we'll load the **.csv** file, using the **read_csv()** method from Pandas, into a DataFrame (**inp0**) and have a look at a few records using the **head** method of the pandas DataFrame:

```
inp0 = pd.read_csv('AAPL.csv')
inp0.head()
```

You should get the following output:

	Date	Close	Open	High	Low	Volume
0	1/17/2020	138.31	136.54	138.330	136.16	5623336
1	1/16/2020	137.98	137.32	138.190	137.01	4320911
2	1/15/2020	136.62	136.00	138.055	135.71	4045952
3	1/14/2020	135.82	136.28	137.139	135.55	3683458
4	1/13/2020	136.60	135.48	136.640	135.07	3531572

Figure 5.3: The first five records of the AAPL dataset

We can see that the first record is for January 17, 2020 and is the most recent date in the data (the latest data at the time of writing this book). As is the convention for pandas DataFrames, the first record has an index of 0 (the index is simply the identifier for the row, and each row has an index value). **Open** refers to the value of a particular stock at the opening of the trade, **High** refers to the highest value of the stock during the day, while **Low** and **Close** represent the lowest price and closing price, respectively. We also have the volume traded on the day.

Let's also look at the last few records of the dataset using the following command:

```
inp0.tail()
```

The records look as follows:

	Date	Close	Open	High	Low	Volume
2509	1/29/2010	122.39	124.32	125.000	121.90	11571890
2510	1/28/2010	123.75	127.03	127.040	123.05	9616132
2511	1/27/2010	126.33	125.82	126.960	125.04	8719147
2512	1/26/2010	125.75	125.92	127.750	125.41	7135190
2513	1/25/2010	126.12	126.33	126.895	125.71	5738455

Figure 5.4: Bottom five records of the AAPL dataset

From the preceding tables, we can see that we have daily opening, high, low, and closing prices, and volumes, from January 25, 2010 to January 17, 2020. For our purpose, we are concerned with the closing price.

EXERCISE 5.01: VISUALIZING OUR TIME-SERIES DATA

In this exercise, we will extract the closing price from the data, perform the necessary formatting, and plot the time series to gain a better understanding of the data. Make sure that you have read through the preceding section and loaded the data, as well as imported the relevant libraries. Perform the following steps to complete this exercise:

1. Use the following command to import the necessary libraries if you haven't already:

```
import pandas as pd, numpy as np
import matplotlib.pyplot as plt
%matplotlib inline
```

2. Download the file titled **AAPL.csv** from GitHub (https://packt.live/325WSKR) and load it into a DataFrame:

```
inp0 = pd.read_csv('AAPL.csv')
```

3. Plot the **Close** column as a line plot to see the pattern using the **plot** method of the DataFrame, specifying the **Date** column as the *X-axis*:

```
inp0.plot("Date", "Close")
plt.show()
```

The plot for this will be as follows, with the *X-axis* showing the closing price and the *Y-axis* representing the dates:

Figure 5.5: Plot of the closing price

From the plot, we can see that the latest values are getting plotted first (on the left). We'll reverse the data for convenience of plotting and handling. We'll achieve this by sorting the DataFrame by the index (remember that the index was 0 for the latest record) in descending order.

4. Reverse the data by sorting the DataFrame on the index. Plot the closing price again and supply **Date** as the *X-axis*:

```
inp0 = inp0.sort_index(ascending=False)
inp0.plot("Date", "Close")
plt.show()
```

The closing price will be plotted as follows:

Figure 5.6: The trend after reversing the data

That worked as expected. We can see that the latest values are plotted to the right.

5. Extract the values for **Close** from the DataFrame as a **numpy** array, reshaped to specify one column using **array.reshape(-1,1)**:

```
ts_data = inp0.Close.values.reshape(-1,1)
```

6. Plot the values as a line plot using matplotlib. Don't worry about marking the dates; the order of the data is clear (matplotlib will use an index instead, beginning with 0 for the first point):

```
plt.figure(figsize=[14,5])
plt.plot(ts_data)
plt.show()
```

The resulting trend is as follows, with the *X-axis* representing the index and the *Y-axis* showing the closing price:

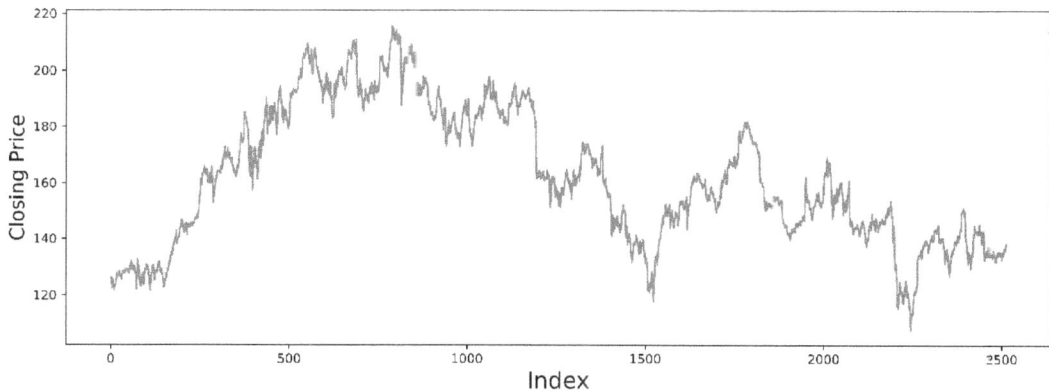

Figure 5.7: The daily stock price trend

That's what our sequence data looks like. There is no continuous clear trend; the prices rose for a period, after which the stock waxed and waned. The pattern isn't straightforward. We can see that there is some seasonality at a small duration (maybe monthly). Overall, the pattern is rather complex and there are no obvious and easy-to-identify cyclicities in the data that we can exploit. This complex sequence is what we will work with – predicting the stock price for a day using historical values.

> **NOTE**
>
> To access the source code for this specific section, please refer to https://packt.live/2ZctArVi.
>
> You can also run this example online at https://packt.live/38EDOEA. You must execute the entire Notebook in order to get the desired result.

In this exercise, we loaded the stock price data. After reversing the data for ease of handling, we extracted the closing price (the **Close** column). We plotted the data to visually examine the trend and patterns in the data, acknowledging that there aren't any obvious patterns in the data for us to exploit.

> **NOTE**
>
> Whether you treat the data as a sequence also depends on the task at hand. If the task doesn't need the information in the sequence, then maybe you don't need to treat it as such.

In this chapter, we'll be focusing on tasks that require/greatly benefit from exploiting the sequence in the data. How is that done? We'll find out in the following sections, where we'll discuss the intuition and the approach behind RNNs.

RECURRENT NEURAL NETWORKS

How does our brain process a sentence? Let's try to understand how our brain processes a sentence as we read it. You see some terms in a sentence, and you need to identify the sentiment contained in the sentence (positive, negative, neutral). Let's look at the first term – "**I**":

Figure 5.8 Sentiment analysis for the first term

"**I**" is neutral, so our classification (neutral) is appropriate. Let's look at another term:

Figure 5.9: Sentiment analysis with two terms

With the term "**can't**," we need to update our assessment of the sentiment. "**I**" and "**can't**" together typically have a negative connotation, so our current assessment is updated as "negative" and is marked with a cross. Let's look at the next couple of words:

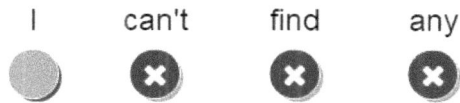

Figure 5.10: Sentiment analysis with four terms

After the two additional terms, we maintain our prediction that the sentence has a negative sentiment. With all the information so far, "**I can't find any**," is a good assessment. Let's look at the final term:

Figure 5.11: Sentiment analysis with the final term added

With that last term coming in, our prediction is completely overturned. Suddenly, we now agree that this is a positive expression. Your assessment is updated with each new term coming in, is it not? Your brain gathers/collects all the information it has at hand and makes an assessment. On arrival of the new term, the assessment so far is updated. This process is exactly what an RNN mimics.

So, what makes a network "recurrent"? The key idea is to *not only process new information but also retain the information received so far*. This is achieved in RNNs by making the output depend not only on the new input value but also on the current "state" (information captured so far). To understand this better, let's see how a standard feedforward neural network would process a simple sentence and compare it with how an RNN would process it.

Consider the task of sentiment classification (positive or negative) for an input sentence, "*life is good*." In a standard feedforward network, the inputs corresponding to all the terms in the sentence are passed to the network together. As depicted in the following diagram, the input data is the combined representation of all the terms in the sentence that have been passed to the hidden layers of the network. All the terms are considered together to classify the sentiment in the sentence as positive:

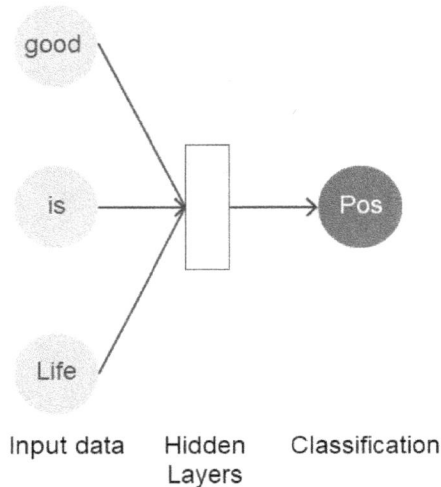

Figure 5.12: Standard feedforward network for sentiment classification

In contrast, an RNN would process the sentence word by word. As shown in the following diagram, the first input for the term "*life*" is passed to the hidden layers at time *t=0*. The hidden layers provide some output values, but this isn't the final classification of the sentence and is rather an intermediate value of the hidden layers. No classification is done yet:

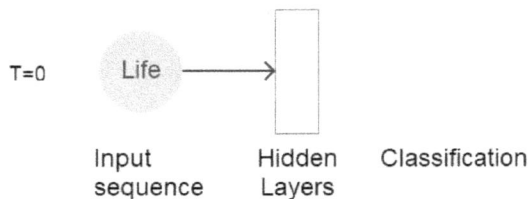

Figure 5.13: RNN processing the first term at time t=0

The next term "**is**"), along with its corresponding input, is processed at time $t=1$ and then fed to the hidden layers. As shown in the following diagram, this time, the hidden layer also considers the intermediate output from the hidden layer at time $t=0$, which is essentially the output corresponding to the term "**life**." The output from the hidden layers will now effectively consider the new input ("**is**") and the input at the previous time step ("**life**"):

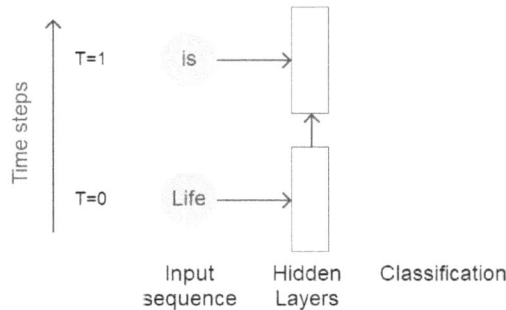

Figure 5.14: The network at t=1

After time step $t=1$, the output of the hidden layers effectively contains information from the terms "**life**" and "**is**,", effectively holding information corresponding to the inputs so far. At time $t=2$, the data corresponding to the next term, that is, "**good**," is fed into the hidden layers. The following diagram shows that the hidden layers use this new input data, along with the output from hidden layers from time $t=1$, to provide an output. This output effectively considers all the inputs so far, in the order in which they appear in the input text. It is when the entire sentence is processed that the final classification of the sentence is made ("positive", in this case):

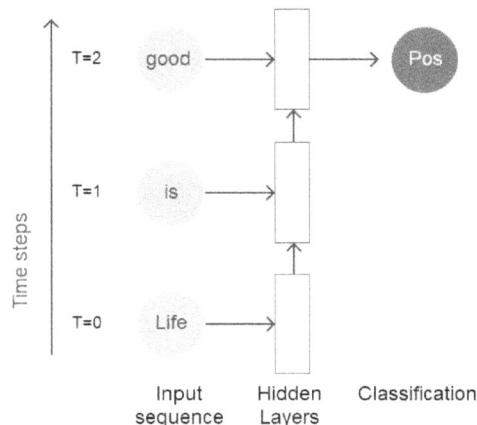

Figure 5.15: Output at t=2 when the entire sentence is processed

LOOPS – AN INTEGRAL PART OF RNNS

A common part of RNNs is using "loops," as shown in the following diagram. By loops, we mean a mechanism of retaining the "state" value containing the information so far and using it along with the new input:

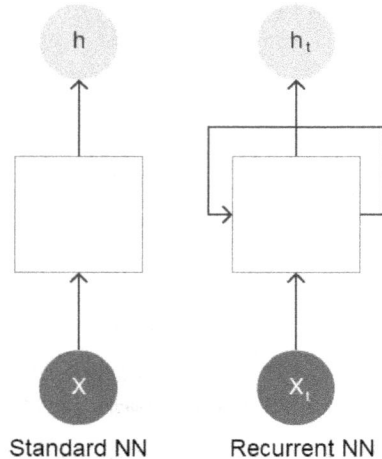

Figure 5.16: RNNs depicted with a loop

As shown in the following diagram, this is done by simply making a virtual copy of the hidden layer and using it at the next time step, that is, when processing the next input. If processing a sentence term by term, this would mean, for each term, saving the hidden layer output (time $t-1$), and when the new term comes in at time t, processing the hidden layer output (time t) along with its previous state (time $t-1$). That's all there really is to it:

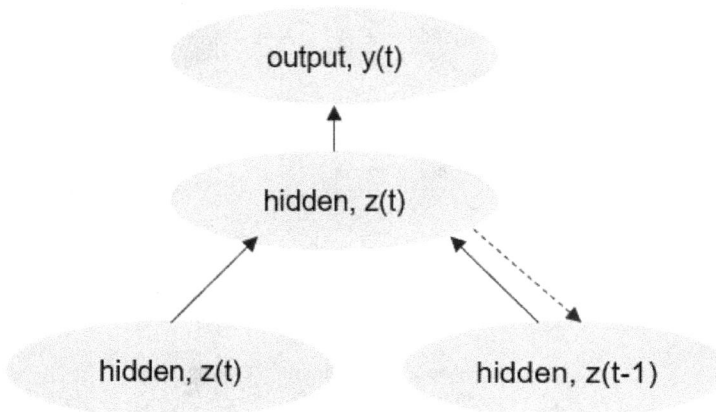

Figure 5.17: Copying the hidden layer state

To make the workings of RNNs even more clear, let's expand the view from *Figure 5.15*, where we saw how the input sentence is processed term by term. We'll understand how different an RNN is from a standard feedforward network.

The part highlighted by the dotted box should be familiar to you – it represents the standard feedforward network with hidden layers (rectangles with dotted lines). The data for an input flows from left to right across the depth of the network, using feedforward weights, W_F, to provide an output -- exactly as in a standard feedforward network. The recurrent part is the flow of data from bottom to top, across the time steps:

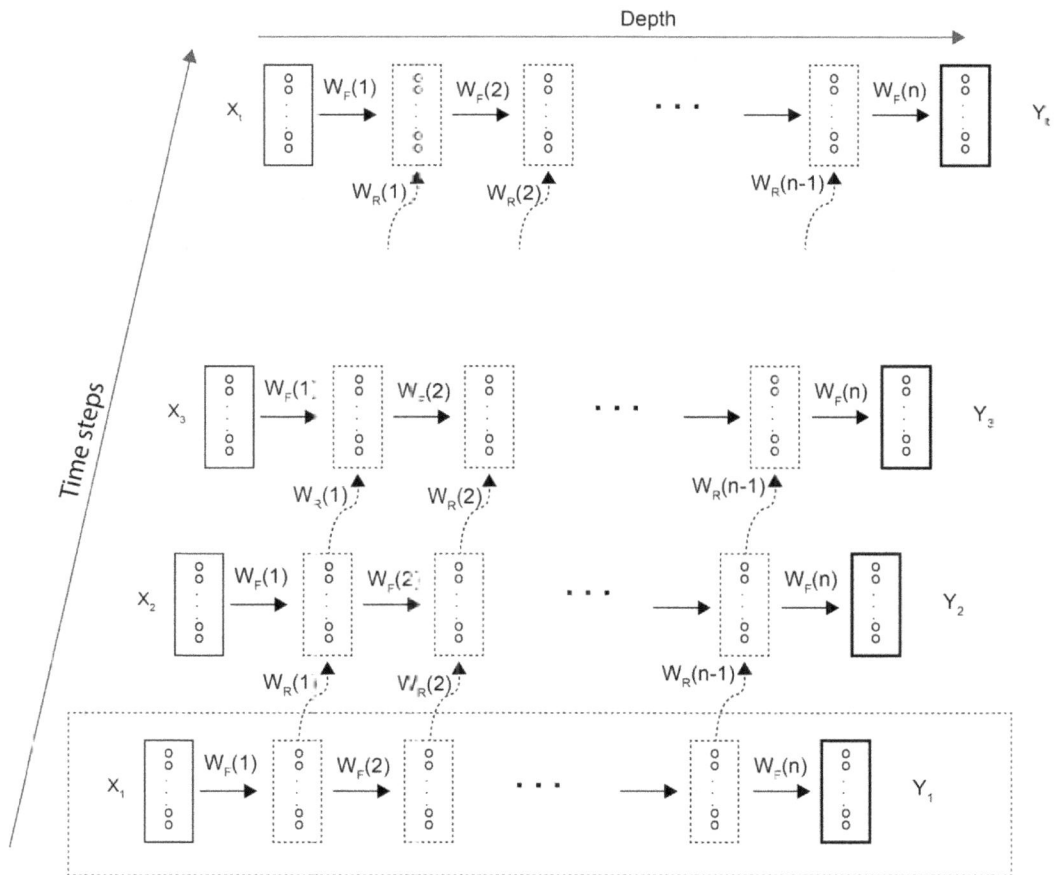

Figure. 5.18: RNN architecture

For all the hidden layers, the output propagates along the time dimension too, to the next time step. Alternately, for a hidden layer at time step t and depth l, the inputs are as follows:

- Data from the previous hidden layer at the same time step

- Data from the same hidden layer at the previous time step

Have a good look at the preceding diagram to understand the workings of an RNN. The output from the hidden layer can be derived as follows:

$$z_t^{(l)} = W_F^{(l)} a_t^{(l-1)} + W_R^{(l)} a_{t-1}^{(l)} + b^{(l)}$$
$$a_t^{(l)} = f^{(l)}\left(z_t^{(l)}\right)$$

Figure 5.19: Calculating activations in an RNN

The first part of the formula, $W_F^{(l)} a_t^{(l-1)}$, corresponds to the result of the feedforward calculation, that is, applying feedforward weights (W_F) to the output ($a_t^{(l-1)}$) from the previous layer. The second part corresponds to the recurrent calculation, that is, applying recurrent weights ($W_R^{(l)}$) to the output from the same layer from the previous time step ($a_{t-1}^{(l)}$). Additionally, as with all neural network layers, there is a bias term as well. This result, on applying the activation function, becomes the output from the layer at time t and depth l ($a_t^{(l)}$).

To make the idea more concrete, let's implement the feedforward steps of a simple RNN using TensorFlow.

EXERCISE 5.02: IMPLEMENTING THE FORWARD PASS OF A SIMPLE RNN USING TENSORFLOW

In this exercise, we will use TensorFlow to perform one pass of the operations in a simple RNN with one hidden layer and two time steps. By performing one pass, we mean calculating the activation of the hidden layer at time step $t=0$, then using this output along with the new input at $t=1$ (applying the appropriate recurrent and feedforward weights) to obtain the output at time $t=1$. Initiate a new Jupyter Notebook for this exercise and perform the following steps:

1. Import TensorFlow and NumPy. Set a random seed of **0** using **numpy** to make the results reproducible:

```
import numpy as np
import tensorflow as tf
np.random.seed(0)
tf.random.set_seed(0)
```

2. Define the **num_inputs** and **num_neurons** constants that will be holding the number of inputs (2) and the number of neurons in the hidden layer (3), respectively:

```
num_inputs = 2
num_neurons = 3
```

We will have two inputs at each time step. Let's call them **xt0** and **xt1**.

3. Define the variables for the weight matrices. We need two of them – one for the feedforward weights and another for the recurrent weights. Initialize them randomly:

```
Wf = tf.Variable(tf.random.normal\
                (shape=[num_inputs, num_neurons]))
Wr = tf.Variable(tf.random.normal\
                (shape=[num_neurons, num_neurons]))
```

Notice the dimensions for the recurrent weights – it is a square matrix, with as many rows/columns as the number of neurons in the hidden layer.

4. Add the bias variable (to make the activations fit the data better), with as many zeros as the number of neurons in the hidden layer:

```
b = tf.Variable(tf.zeros([1,num_neurons]))
```

5. Create the data – three examples for **xt0** (two inputs, three examples) as **[[0,1], [2,3], [4,5]]** and **xt1** as **[[100,101], [102,103], [104,105]]** – as **numpy** arrays of the **float32** type (consistent with **dtype** for TensorFlow's default float representation):

```
xt0_batch = np.array([[0,1],[2,3],[4,5]]).astype(np.float32)
xt1_batch = np.array([[100, 101],[102, 103],\
                     [104,105]]).astype(np.float32)
```

6. Define a function named **forward_pass** to apply a forward pass to the given data, that is, **xt0, xt1**. Use **tanh** as the activation function. The output at $t=0$ should be derived from **Wf** and **xt0** alone. The output at $t=1$ must use **yt0** with the recurrent weights, **Wf**, and use the new input, **xt1**. The function should return outputs at the two time steps:

```
def forward_pass(xt0, xt1):
    yt0 = tf.tanh(tf.matmul(xt0, Wf) + b)
    yt1 = tf.tanh(tf.matmul(yt0, Wr) + tf.matmul(xt1, Wf) + b)
    return yt0, yt1
```

Note that there is no recurrent weight here at time step 0; it comes into play only after the first time step.

7. Perform the forward pass by calling the **forward_pass** function with the created data (**xt0_batch, xt1_batch**) and put the output into variables, **yt0_output** and **yt1_output**:

```
yt0_output, yt1_output = forward_pass(xt0_batch, xt1_batch)
```

8. Print the output values, **yt0_output** and **yt1_output**, using the **print** function from TensorFlow:

```
tf.print(yt0_output)
```

The output at *t=0* gets printed out like so. Note that this result could be slightly different for you because of the random initialization that's done by TensorFlow:

```
[[-0.776318431 -0.844548464 0.438419849]
 [-0.0857750699 -0.993522227 0.516408086]
 [0.698345721 -0.999749422 0.586677969]]
```

9. Now, print the values of yt1_output:

```
tf.print(yt1_output)
```

The output at *t=1* gets is printed as follows. Again, this could be slightly different for you because of the random initial values, but all the values should be close to 1 or -1:

```
[[1 -1 0.999998629]
 [1 -1 0.999998331]
 [1 -1 0.999997377]]
```

We can see that the final output at time *t=1* is a 3x3 matrix – representing the outputs for the three neurons in the hidden layer for the three instances of data.

> **NOTE**
>
> To access the source code for this specific section, please refer
> to https://packt.live/2ZctArW.
>
> You can also run this example online at https://packt.live/38EDOEA.
> You must execute the entire Notebook in order to get the desired result.

> **NOTE**
>
> Despite having set the seeds for **numpy** as well as **tensorflow** to
> achieve reproducible results, there are a lot more causes for the variation
> in results. While the values you see may be different, the output you see
> should largely agree with ours.

In this exercise, we manually performed the forward pass for two time steps in a simple RNN. We saw that it's merely using the hidden layer output from the previous time step as an input to the next. Now, you don't really need to perform any of this manually – Keras makes making RNNs very simple. We will use Keras for our stock price prediction model.

THE FLEXIBILITY AND VERSATILITY OF RNNS

In *Exercise 5.2, Implementing the Forward Pass of a Simple RNN Using TensorFlow*, we used two inputs at each time step and we had an output at each time step. But it doesn't always have to be that way. RNNs have a lot of flexibility to offer. For starters, you can have single/multiple inputs as well as outputs. Additionally, you needn't have inputs and outputs at each time step.

You could have the following:

- Inputs at different time steps with the output only at the final step

- A single input with outputs at multiple time steps

- Both inputs and outputs (equal or unequal lengths) at multiple time steps

There is enormous flexibility in RNN architectures, and this flexibility makes them very versatile. Let's take a look at some possible architectures and what some potential applications can be:

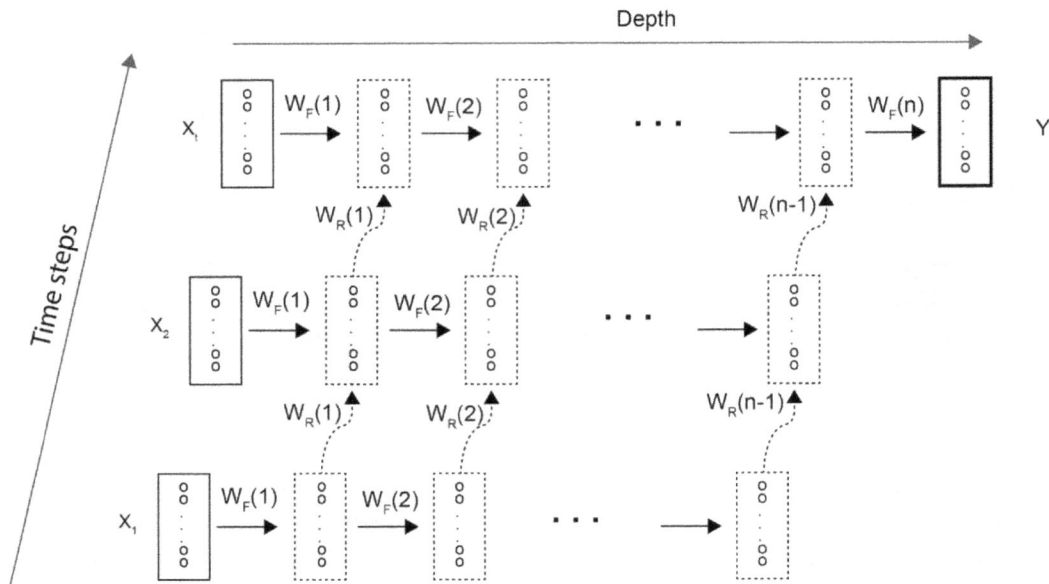

Figure 5.20: Inputs at multiple steps with the output at the final step

You can have inputs at multiple time steps, such as in a sequence (or single or more inputs) with the output only at the final time step, when the prediction is made, as shown in the preceding diagram. At each time step, the hidden layers operate on the feedforward output from the previous layer and the recurrent output from its copy from the previous time step. But there is no prediction for the intermediate time steps. Prediction is made only after processing the entire input sequence – the same process we saw in *Figure. 5.15* (the "*life is good*" example). Text classification applications extensively use this architecture – sentiment classification into positive/negative, classifying an email into spam/ham, identifying hate speech in comments, automatically moderating product reviews on a shopping platform, and many more.

Time series prediction (for example, stock prices) also utilizes this architecture, where the past few values are processed to predict a single future value:

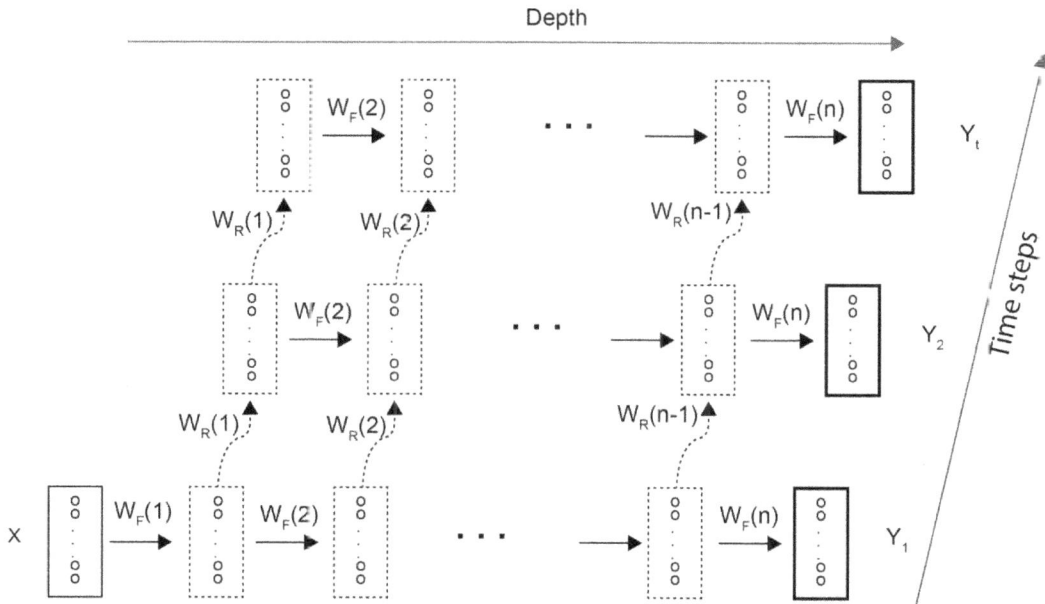

Figure 5.21: Input in a single step, output in multiple steps

The preceding diagram illustrates another architecture in which the input is received in a single step, but the output is obtained at multiple time steps. Applications around generation – generating images for a given keyword, generating music for a given keyword (composer), or generating a paragraph of text for a given keyword – are based on this architecture.

You could also have an output at each time step corresponding to the input, as depicted in the following diagram. Essentially, this model will help you make a prediction for each incoming element of the sequence. An example of such a task would be the Parts-of-Speech tagging of terms – for each term in a sentence, we identify whether the term is a noun, verb, adjective, or another part of speech.

Another example from natural language processing would be **Named Entity Recognition** (**NER**) where, for each term in the text, the objective is to detect whether it represents a named entity and then classify it as an organization, person, place, or another category if it does:

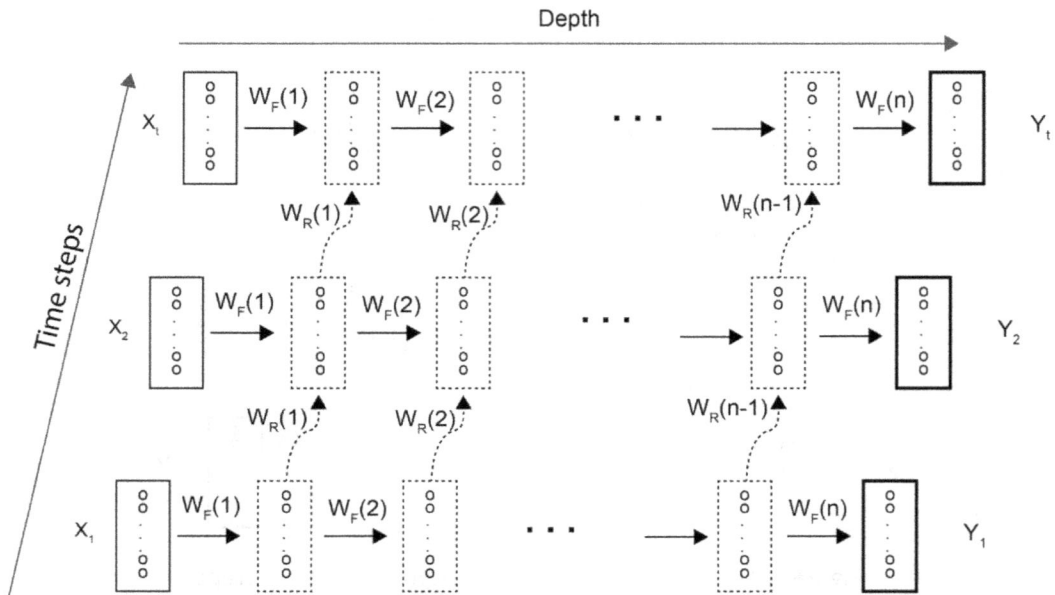

Figure 5.22: Multiple outputs at each time step

In the previous architecture, we had an output for each incoming element. In many situations, this doesn't work, and we need an architecture that has different lengths for input and output, as shown in the following diagram. Think of translation between languages. Does a sentence in English necessarily have the same number of terms in its German translation? More often than not, the answer is no. For such cases, the architecture in the following diagram provides the notion of an "encoder" and a "decoder." The information corresponding to the input sequence is stored in the final hidden layer of the encoder network, which in itself has recurrent layers.

This representation/information is processed by the decoder network (again, this is recurrent), which outputs the translated sequence:

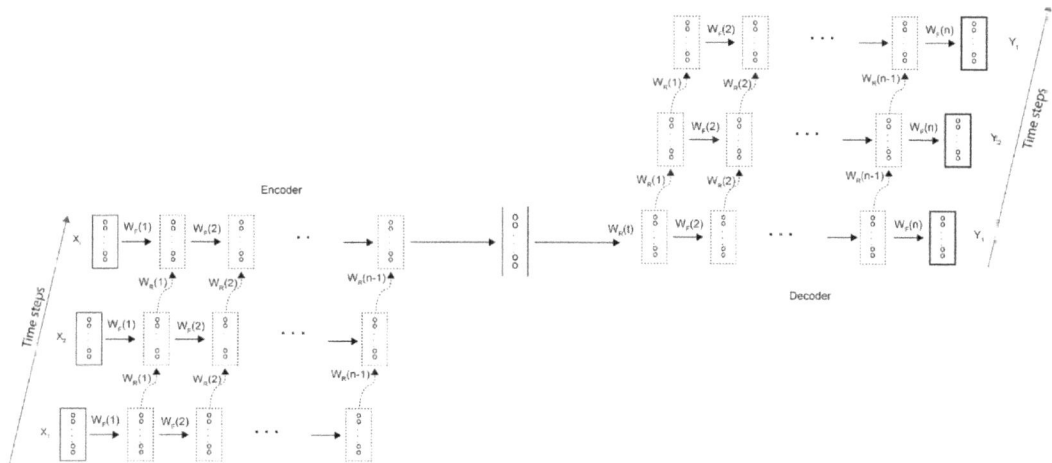

Figure 5.23: Architecture with different lengths for input and output

For all of these architectures, you could also have multiple inputs, making RNN models even more versatile. For example, when making stock price predictions, you could provide multiple inputs (previous stock prices of company, the stock exchange index, crude oil price, and whatever you think is relevant) over multiple time steps, and the RNN will be able to accommodate and utilize all of these. This is one of the reasons RNNs are very popular and have changed the way we work with sequences today. Of course, you also have all the predictive power of deep learning to add.

PREPARING THE DATA FOR STOCK PRICE PREDICTION

For our stock price prediction task, we will predict the value of a given stock on any day by using the past few days' data and feeding it to an RNN. Here, we have a single input (single feature), over multiple time steps, and a single output. We will employ the RNN architecture from *Figure 5.20*.

> **NOTE**
>
> Continue in the same Jupyter Notebook that we plotted our time-series data in throughout this chapter (unless specified otherwise).

So far, we've looked at the data and understood what we're dealing with. Next, we need to prepare the data for the model. The first step is to create a train-test split of the data. Since this is time-series data, we can't just randomly pick points to assign to our train and test sets. We need to maintain the sequence. For time-series data, we typically reserve the first portion of the data to train on and utilize the last part of the data for our test set. In our case, we will take the first 75% records as our training data and the last 25% as our test data. The following command will help us get the size of the train set needed:

```
train_recs = int(len(ts_data) * 0.75)
```

This is the number of records we'll have in the train set. We can separate the sets as follows:

```
train_data = ts_data[:train_recs]
test_data = ts_data[train_recs:]
len(train_data), len(test_data)
```

The lengths of the train and test sets will be as follows:

```
(1885, 629)
```

Next, we need to scale the stock data. For that, we can employ the min-max scaler from **sklearn**. The **MinMaxScaler** scales the data so that it's in a range between 0 and 1 (inclusive) – the highest value in the data being mapped to 1. We'll fit and transform the scaler on the train data and only transform the test data:

```
from sklearn.preprocessing import MinMaxScaler
scaler = MinMaxScaler()
train_scaled = scaler.fit_transform(train_data)
test_scaled = scaler.transform(test_data)
```

The next important step is to format the data to get the "features" for each instance. We need to define a "lookback period" – the number of days from the history that we want to use to predict the next value. The following code will help us define a function that returns the target value of **y** (stock price for a day) and **X** (values for each day in the lookback period):

```
def get_lookback(inp, look_back):
    y = pd.DataFrame(inp)
    dataX = [y.shift(i) for i in range(1, look_back+1)]
    dataX = pd.concat(dataX, axis=1)
    dataX.fillna(0, inplace = True)
    return dataX.values, y.values
```

The function takes in a dataset (a series of numbers, rather) and, for the provided lookback, adds as many values from the history. It does so by shifting the series, each time concatenating it to the result. The function returns the stock price for the day as *y* and the values in lookback period (shifted values) as our features. Now, we can define a lookback period and see the result of applying the function to our data:

```
look_back = 10
trainX, trainY = get_lookback(train_scaled, look_back=look_back)
testX, testY = get_lookback(test_scaled, look_back= look_back)
```

Try the following command to examine the shape of the outcome datasets:

```
trainX.shape, testX.shape
```

The output is as follows:

```
((1885, 10), (629, 10))
```

As expected, there are 10 features for each example, corresponding to the past 10 days. We have this history for the train data as well as the test data. With that, data preparation is complete. Before we move on to building our first RNN on this data, let's understand RNNs a little more.

> **NOTE**
>
> The **trainX** and **trainY** variables we created here will be used throughout the exercises that follow. So, make sure you are running this chapter's code in the same Jupyter Notebook.

PARAMETERS IN AN RNN

To calculate the number of parameters in an RNN layer, let's take a look at a generic hidden layer:

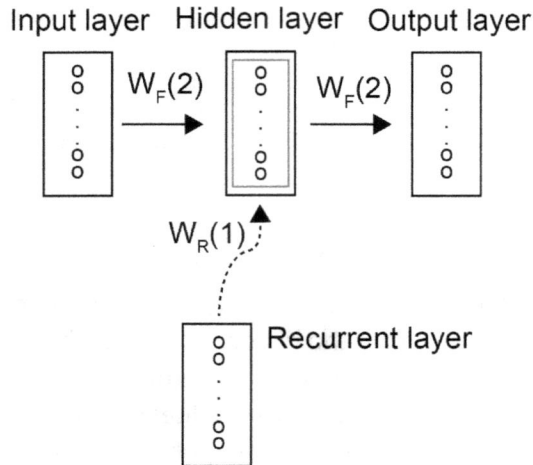

Figure 5.24: Parameters of the recurrent layer

The hidden layer takes inputs from the previous hidden layer at the same time step, and also from itself from a previous time step. If the input layer (the previous hidden layer) to the RNN layer is m-dimensional, we would need $n \times m$ weights/parameters, where n is the number of neurons in the RNN layer. For the output layer, the dimensionality if the weights would be $n \times k$, if k is the dimensionality of the output. The recurrent weight is always a square matrix of dimensionality $n \times n$ – since the dimensionality of the input is the same as the layer itself.

The number of parameters for any RNN layer would therefore be $n^2 + nk + nm$, where we have the following:

- **n**: Dimension of the hidden (current) layer
- **m**: Dimension of the input layer
- **k**: Dimension of the output layer

TRAINING RNNS

How to forward propagate information in an RNN should be clear by now. If not, please refer to *Figure 5.19* with the equations. The new information propagates along the depth of the network as well along the time steps, using the previous hidden states at each step. The additional two key aspects of training RNNs are as follows:

- **Defining loss**: We know how loss is defined for a standard neural network; that is, it has a single output. With RNNs, in the case that there is a single time step at the output (for example, text classification), the loss is calculated the same way as in standard neural networks. But we know that RNNs could have outputs over multiple time steps (for example, in Part-of-Speech tagging or machine translation). How is loss defined across multiple time steps? A very simple and popular approach is summing up the loss at all the steps. The loss for the entire sequence is calculated as the sum of the loss at all time steps.

- **Backpropagation**: Backpropagation of the errors now needs us to work across time steps, since there is a time dimension as well. We have already seen that loss is defined as the sum of loss at each time step. The usual chain rule application helps us out; we also need to sum the gradients at each time step over time. This has a very catchy name: **Backpropagation Through Time** (**BPTT**).

> **NOTE**
>
> A detailed treatment of the training process and the involved math is beyond the scope of this book. The basic concept is all we need to understand the considerations involved.

Now, let's continue building our first RNN model using Keras. We will introduce two new layers that are available in Keras in this chapter and understand their function and utility. The first layer we need is the `SimpleRNN` layer.

To import all the necessary utilities from Keras, you can use the following code:

```
from tensorflow.keras.models import Sequential
from tensorflow.keras.layers \
import SimpleRNN, Activation, Dropout, Dense, Reshape
```

The SimpleRNN layer is the simplest plain vanilla RNN layer. It takes in a sequence, and the output of the neuron is fed back as input. Additionally, if we want to follow this RNN layer with another RNN layer, we have the option of returning sequences as output. Let's have a look at some of the options.

- **?SimpleRNN**: The signature for the SimpleRNN layer is as follows:

```
Init signature:
SimpleRNN(
    units,
    activation='tanh',
    use_bias=True,
    kernel_initializer='glorot_uniform',
    recurrent_initializer='orthogonal',
    bias_initializer='zeros',
    kernel_regularizer=None,
    recurrent_regularizer=None,
    bias_regularizer=None,
    activity_regularizer=None,
    kernel_constraint=None,
    recurrent_constraint=None,
```

Figure 5.25: Signature of the SimpleRNN layer

We can see that the layer also has all the usual options of regular/standard layers in Keras that let you specify the activations, initialization, dropout, and more.

The RNN layers expect the input data to be in a certain format. Since we can have input data as multiple time steps for multiple features, the input format is expected to make that specification unambiguous. The expected input shape is (look_back, number of features). It expects a matrix with the same lookback history for each feature.

In our case, we have one feature, and the lookback period is 10. So, the expected input shape is (10, 1). Note that we currently have each input as a list of 10 values, so we need to make sure it is understood as (10,1). We'll use the reshape layer for this purpose. The reshape layer needs the input shape and the target shape. Let's start building our model by instantiating and adding a reshape layer.

> **NOTE**
>
> Even though we have set the seeds for **numpy** as well as **tensorflow** to achieve reproducible results, there are a lot more causes for variation owing to which you may get a result that's different from ours. This applies to all the models we'll use here. While the values you see may be different, the output you see should largely agree with ours. If the model performance is very different, you may want to tweak the number of epochs – the reason for this being that the weights in neural networks are initialized randomly, so the starting point for you and us could be slightly different, and we may reach a similar position when training a different number of epochs.

EXERCISE 5.03: BUILDING OUR FIRST PLAIN RNN MODEL

In this exercise, we will build our first plain RNN model. We will have a reshape layer, followed by a **SimpleRNN** layer, followed by a dense layer for the prediction. We will use the formatted data for **trainX** and **trainY** that we created earlier, along with the initialized layers from Keras. Perform the following steps to complete this exercise:

1. Let's gather the necessary utilities from Keras. Use the following code to do so:

```
from tensorflow.keras.models import Sequential
from tensorflow.keras.layers \
import SimpleRNN, Activation, Dropout, Dense, Reshape
```

2. Instantiate the **Sequential** model:

```
model = Sequential()
```

3. Add a **Reshape** layer to get the data in the format (**look_back**, **1**):

```
model.add(Reshape((look_back,1), input_shape = (look_back,)))
```

Note the arguments to the **Reshape** layer. The target shape is (**lookback, 1**), as we discussed.

4. Add a **SimpleRNN** layer with 32 neurons and specify the input shape. Note that we took an arbitrary number of neurons, so you're welcome to experiment with this number:

```
model.add(SimpleRNN(32, input_shape=(look_back, 1)))
```

5. Add a **Dense** layer of size 1:

```
model.add(Dense(1))
```

6. Add an **Activation** layer with a linear activation:

```
model.add(Activation('linear'))
```

7. Compile the model with the **adam** optimizer and **mean_squared_error** (since we're predicting a real-values quantity):

```
model.compile(loss='mean_squared_error', optimizer='adam')
```

8. Print a summary of the model:

```
model.summary()
```

The summary will be printed as follows:

```
Layer (type)                 Output Shape              Param #
=================================================================
reshape (Reshape)            (None, 10, 1)             0
_____
simple_rnn (SimpleRNN)       (None, 32)                1088
_____
dense (Dense)                (None, 1)                 33
_____
activation (Activation)      (None, 1)                 0
=================================================================
Total params: 1,121
Trainable params: 1,121
Non-trainable params: 0
_____
```

Figure 5.26: Summary of the SimpleRNN model

Pay attention to the number of parameters in the **SimpleRNN** layer. It works out to be as we expected.

> **NOTE**
>
> To access the source code for this specific section, please refer to https://packt.live/2ZctArW.
>
> You can also run this example online at https://packt.live/38EDOEA. You must execute the entire Notebook in order to get the desired result.

In this exercise, we defined our model architecture using a single-layer plain RNN architecture. This is indeed a very simple model, in comparison to the kinds of models we built earlier for image data. Next, let's see how this model performs on the task at hand.

MODEL TRAINING AND PERFORMANCE EVALUATION

We have defined and compiled the model. The next step is to learn the model parameters by fitting the model on the train data. We can do this by using a batch size of 1 and a validation split of 10%, and by training for only three epochs. We tried different values of epochs and found that the model gave the best result at three epochs. The following code will help us train the model using the **fit()** method:

```
model.fit(trainX, trainY, epochs=3, batch_size=1, \
          verbose=2, validation_split=0.1)
```

The output is as follows:

```
Train on 1696 samples, validate on 189 samples
Epoch 1/3
1696/1696 - 7s - loss: 0.0036 - val_loss: 0.0016
Epoch 2/3
1696/1696 - 6s - loss: 0.0012 - val_loss: 4.9603e-04
Epoch 3/3
1696/1696 - 6s - loss: 0.0010 - val_loss: 3.6961e-04
```

Figure 5.27: Training output

We can see that the loss is already pretty low. We trained the model here without doing any careful hyperparameter tuning. You can see that for this dataset, three epochs was sufficient, and we're trying to keep it simple here. With the model training done, we now need to assess the performance on the train and test sets.

To make our code a little more modular, we'll define two functions – one to print the RMS error on the train and test sets and the other function to plot the predictions for the test data along with the original values in the data. Let's begin by defining our first function, using the **sqrt** function from **math** to get the root of the **mean_squared_error** provided to us by the model's **evaluate** method. The function definition is as follows:

```
import math
def get_model_perf(model_obj):
    score_train = model_obj.evaluate(trainX, trainY, verbose=0)
    print('Train RMSE: %.2f RMSE' % (math.sqrt(score_train)))
    score_test = model_obj.evaluate(testX, testY, verbose=0)
    print('Test RMSE: %.2f RMSE' % (math.sqrt(score_test)))
```

To see how our model did, we need to supply our **model** object to this method. This can be done as follows:

```
get_model_perf(model)
```

The output will be as follows:

```
Train RMSE: 0.02 RMSE
Test RMSE: 0.03 RMSE
```

The values seem rather low (admittedly, we don't really have a benchmark here, but these values do seem to be good considering that our outcome values are ranging from 0 to 1). But this is a summary statistic, and we already know that the values in the data change considerably. A better idea would be to visually assess the performance, comparing the actual values to the predicted for the test period. The following code will help us define a function that plots the predictions for a given model object:

```
def plot_pred(model_obj):
    testPredict = \
    scaler.inverse_transform(model_obj.predict(testX))
    pred_test_plot = ts_data.copy()
    pred_test_plot[:train_recs+look_back,:] = np.nan
```

```
pred_test_plot[train_recs+look_back:,:] = \
testPredict[look_back:]
plt.plot(ts_data)
plt.plot(pred_test_plot, "--")
```

First, the function makes predictions on the test data. Since this data is scaled, we apply the inverse transform to get the data back to its original scale before plotting it. The function plots the actual values as a solid line and the predicted values as dotted lines. Let's use this function to visually assess how our model performs. We need to simply pass the model object to the **plot_pred** function, as demonstrated in the following code:

```
%matplotlib inline
plt.figure(figsize=[10,5])
plot_pred(model)
```

The plot that's displayed is as follows:

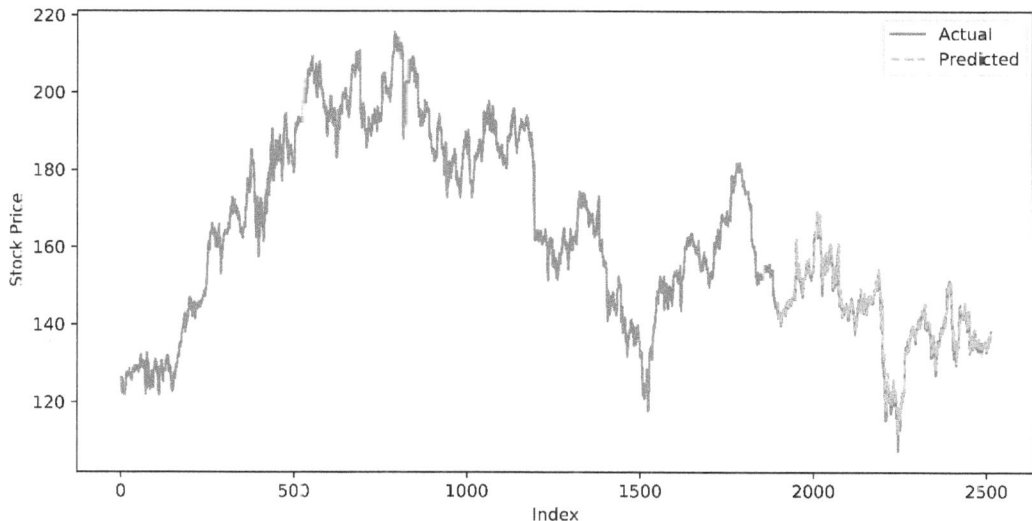

Figure 5.28: Predictions versus actuals

The preceding diagram visualizes the predictions (dotted lines) from the model juxtaposed with the actual values (solid lines). That looks pretty good, doesn't it? At this scale, it looks like overlap between the predicted and the actual is very high – the prediction curve fits the actual values almost perfectly. Prima facie, it does seem that the model has done a great job.

But before congratulating ourselves, let's recall the granularity at which we worked – we're working with 10 points to predict the next day's stock price. Of course, at this scale, even if we took simple averages, the plot would look impressive. We need to zoom in, a lot, to understand this better. Let's zoom in so that the individual points are visible. We'll use the **%matplotlib notebook** cell magic command for interactivity in the chart and zoom in on the values corresponding to indices **2400 – 2500** in the plot:

```
%matplotlib notebook
plot_pred(model)
```

> **NOTE**
>
> If the graph presented below is not displayed properly for some reason, run the cell containing **%matplotlib notebook** for a couple of times. Alternatively, you can also use **%matplotlib inline** instead of **%matplotlib notebook**.

The output is as follows, with the dotted lines showing the predictions and the solid line depicting the actual values:

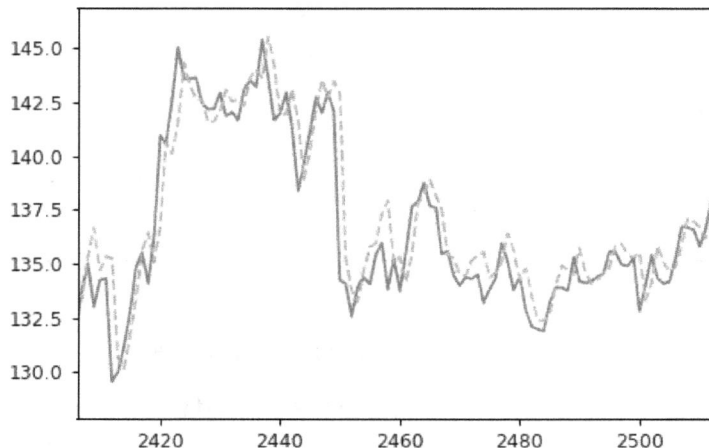

Figure 5.29: Zoomed-in view of predictions

Even after zooming in, the result is pretty good. All the variations have been captured well. A single RNN layer with just 32 neurons giving us this kind of result is great. Those who have worked with time series prediction using classical methods would be elated (as we were) to see the efficacy of RNNs for this task.

We saw what RNNs are and, through our stock price prediction model, also saw the predictive power of even a very simple model for a sequence prediction task. We mentioned earlier that using an RNN is one approach to sequence processing. There is another noteworthy approach to handling sequences that employs convolutions. We'll explore it in the next section.

1D CONVOLUTIONS FOR SEQUENCE PROCESSING

In the previous chapters, you saw how deep neural networks benefit from convolutions – you saw convnets and how they can be used for working with images, and how they help with the following:

- Reducing the number of parameters

- Learning the "local features" for the image

Interestingly, and this is something that is not very obvious, convnets can also be very helpful for sequence processing tasks. Instead of 2D, we could use 1D convolutions for sequence data. How does 1D convolution work? Let's take a look:

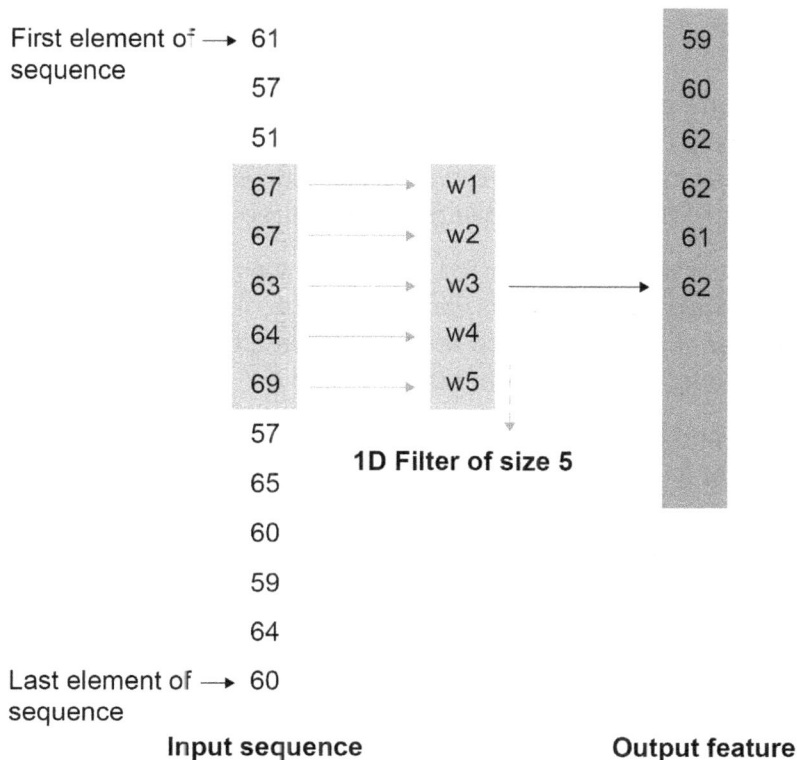

Figure 5.30: Feature generation using 1D convolutions

In *Chapter 3, Image Classification with Convolutional Networks*, we saw how a filter works for the case of images, extracting "patches" from the input image to provide us with output "features." In the case of 1D, a filter extracts subsequences from the input sequence and multiplies them by the weights to give us a value for the output features. As shown in the preceding diagram, the filter moves from the beginning to the end of the sequence (top to bottom). This way, the 1D convnet extracts local patches. As in the 2D case, the patches/features learned here can be recognized later in a different position in the sequence. Of course, as with 2D convolutions, you can choose the filter size and the stride for 1D convolutions as well. If used with a stride more than 1, the 1D convnet can also significantly reduce the number of features.

> **NOTE**
>
> When employed on text data as the first layer, the "local features" that 1D convolutions extract are features for groups of words. A filter size of 2 would help extract two-word combos (called bi-grams), 3 would extract three-word combos (tri-grams), and so on. Larger filter sizes would learn larger groups of terms.

You could also apply pooling to 1D – max or average pooling to further subsample the features. So, you could greatly reduce the effective length of sequence that you're dealing with. A long input sequence can be brought down to a much smaller, more manageable length. This should certainly help with speed.

We understand that we benefit in speed. But do 1D convnets perform well for sequences? 1D convnets have shown very good results in tasks around translation and text classification. They have also shown great results for audio generation and other tasks regarding predicting from sequences.

Will 1D convnets perform well for our task of stock price prediction? Ponder it – think about what kind of features we get and how we're handling the sequence. If you aren't sure, then don't worry – we're going to employ a 1D convnet-based model for our task and see for ourselves in the next exercise.

EXERCISE 5.04: BUILDING A 1D CONVOLUTION-BASED MODEL

In this exercise, we will build our first model using 1D convnets and evaluate its performance. We'll employ a single **Conv1D** layer, followed by **MaxPooling1D**. We'll continue using the same dataset and notebook we've been using so far. Perform the following steps to complete this exercise:

1. Import the 1D convolution-related layers from Keras:

```
from tensorflow.keras.layers import Conv1D, MaxPooling1D, Flatten
```

2. Initialize a **Sequential** model and add a **Reshape** layer to reshape each instance as a vector (**look_back, 1**):

```
model_conv = Sequential()
model_conv.add(Reshape((look_back,1), \
               input_shape = (look_back,)))
```

3. Add a Conv1D layer with five filters of size 5 and **relu** as the activation function:

```
model_conv.add(Conv1D(5, 5, activation='relu'))
```

Note that we're using fewer filters than the sequence length. In many other applications, the sequence can be much longer than in our example. The filters are generally much lower in number than the input sequence.

4. Add a Maxpooling1D layer with a pool size of 5:

```
model_conv.add(MaxPooling1D(5))
```

5. Flatten the output with a **Flatten** layer:

```
model_conv.add(Flatten())
```

6. Add a **Dense** layer with a single neuron and add a linear activation layer:

```
model_conv.add(Dense(1))
model_conv.add(Activation('linear'))
```

7. Print out the summary of the model:

```
model_conv.summary()
```

The model's summary is as follows:

```
Layer (type)                    Output Shape            Param #
=================================================================
reshape_1 (Reshape)             (None, 10, 1)              0
_____
conv1d (Conv1D)                 (None, 6, 5)              30
_____
max_pooling1d (MaxPooling1D) (None, 1, 5)                 0
_____
flatten (Flatten)               (None, 5)                 0
_____
dense_1 (Dense)                 (None, 1)                 6
_____
activation_1 (Activation)       (None, 1)                 0
=================================================================
Total params: 36
Trainable params: 36
Non-trainable params: 0
_____
```

Figure 5.31: Summary of the model

Notice the dimensions of the output from the Conv1D layer – 6 x 5. This is expected – for a filter size of 5, we get 6 features. Also, take a look at the overall number of parameters. It's just 36, which is indeed a very small number.

8. Compile the model with the loss as **mean_squared_error** and **adam** as the **optimizer**, and then fit it on the train data for 5 epochs:

```
model_conv.compile(loss='mean_squared_error', optimizer='adam')
model_conv.fit(trainX, trainY, epochs=5, \
               batch_size=1, verbose=2, validation_split=0.1)
```

You should see the following output:

```
Train on 1696 samples, validate on 189 samples
Epoch 1/5
1696/1696 - 3s - loss: 0.0351 - val_loss: 0.0013
Epoch 2/5
1696/1696 - 3s - loss: 0.0022 - val_loss: 0.0020
Epoch 3/5
1696/1696 - 3s - loss: 0.0021 - val_loss: 0.0012
Epoch 4/5
1696/1696 - 3s - loss: 0.0021 - val_loss: 0.0014
Epoch 5/5
1696/1696 - 3s - loss: 0.0020 - val_loss: 0.0011
```

Figure 5.32: Training and validation loss

From the preceding screenshot, we can see that the validation loss is pretty low for the 1D convolution model too. We need to see whether this performance is comparable to that of the plain RNN. Let's evaluate the performance of the model and see whether it aligns with our expectations.

9. Use the **get_model_perf** function to get the RMSE for the train and test sets:

```
get_model_perf(model_conv)
```

The output is as follows:

```
Train RMSE: 0.04 RMSE
Test RMSE: 0.05 RMSE
```

This is marginally higher than that of the plain RNN model. Let's visualize the predictions next.

10. Using the **plot_pred** function, plot the predictions and the actual values:

```
%matplotlib inline
plt.figure(figsize=[10,5])
plot_pred(model_conv)
```

The model output would be as follows, with the dotted lines showing the predictions and solid lines depicting the actual values:

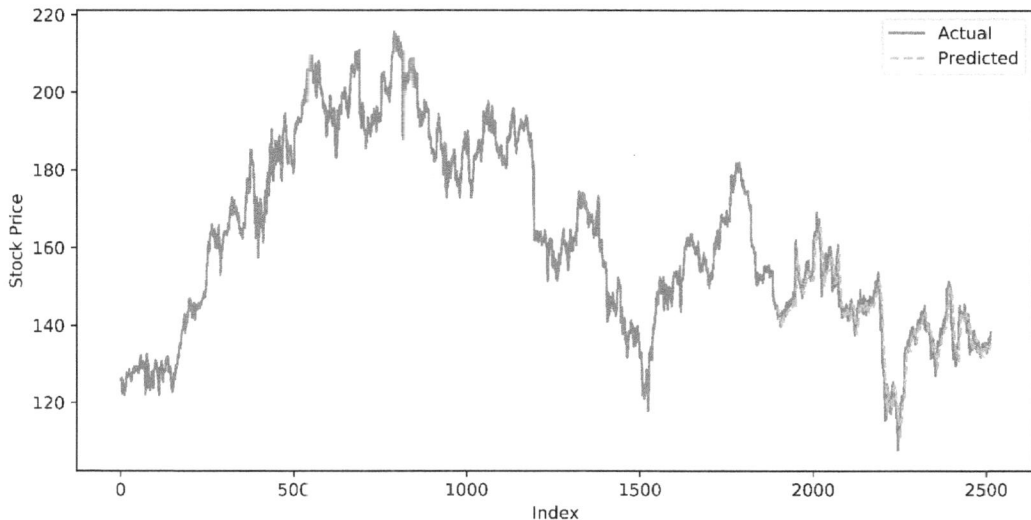

Figure 5.33: Plotting the predictions and actual values

This is very similar to the plot from the predictions from the RNN model (*Figure 5.29*). But we now acknowledge that a better assessment would need interactive visualization and zooming in to a scale where the individual points are visible. Let's zoom in using the interactive plotting features of matplotlib using the notebook backend by using the **%matplotlib** cell magic command.

11. Plot again with interactivity and zoom into the last 100 data points:

```
%matplotlib notebook
plot_pred(model_conv)
```

The output will be as follows:

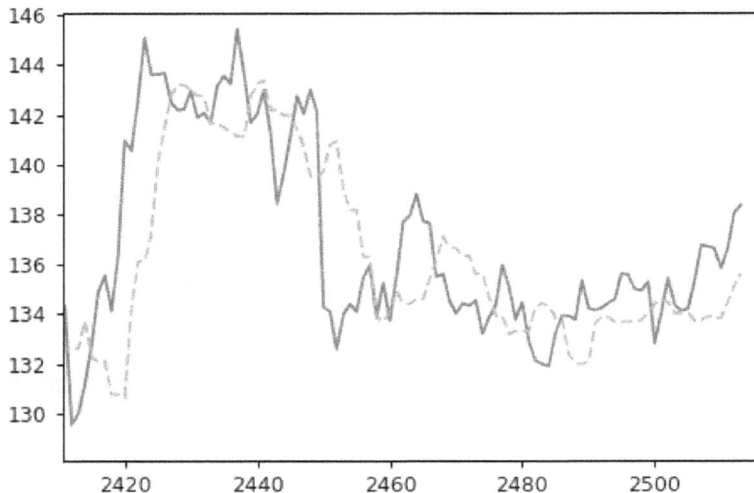

Figure 5.34: Zoomed-in view of predictions

> **NOTE**
>
> If the preceding graph is not displayed properly for some reason, run the cell containing **%matplotlib notebook** for a couple of times. Alternatively, you can also use **%matplotlib inline** instead of **%matplotlib notebook**.

The preceding diagram shows a closer view of the predictions (dotted lines) and the actual values (solid lines). Things aren't looking too good at this scale. The output is very smooth, and almost looks like some kind of averaging is going on. What happened? Is this in line with your expectations? Can you explain this output?

> **NOTE**
>
> To access the source code for this specific section, please refer to https://packt.live/2ZctArW.
>
> You can also run this example online at https://packt.live/38EDOEA. You must execute the entire Notebook in order to get the desired result.

In this exercise, we built and trained our 1D convolution-based model for stock price prediction. We saw that the number of parameters is very low, and that the training time was much lower.

PERFORMANCE OF 1D CONVNETS

To explain the result in the previous exercise, we need to understand what is happening after we extract the subsequences using the Conv1D layer. The sequence in the data is being captured, that is, in the individual filters. But is the sequence being retained after that, and are we really exploiting the sequence in the data? No, we are not. Once the patches have been extracted, they are being treated independently. It is for this reason that the performance is not great.

So, why did we state that 1D convnets do great on sequence tasks previously? How do you make them perform well for our task? 1D convnets do very well on tasks regarding text, especially classification, where the short, local sequence has very high importance and following the order in the entire sequence (say, 200 terms) doesn't provide a huge benefit. For time series tasks, we need order in the entire sequence. There are ways to induce order consideration for tasks such as time series tasks, but they aren't great.

USING 1D CONVNETS WITH RNNS

We saw the benefits of 1D convnets – speed, feature reduction, lower number of parameters, learning local features, and much more. We also saw that RNNs provide very powerful and flexible architectures for handling sequences but have a lot of parameters and are expensive to train. One possible approach can be to combine both – the benefit of the representation and feature reduction from 1D convnets in the initial layers, and the benefit of the sequence processing power of RNNs in the following layers. Let's try it out for our task.

EXERCISE 5.05: BUILDING A HYBRID (1D CONVOLUTION AND RNN) MODEL

In this exercise, we will build a model that will employ both 1D convolutions and RNNs and assess the change in performance. Making a hybrid model is straightforward – we'll begin with the convolution layer, the output of which is features in a sequence. The sequence can be fed straight into the RNN layer. Therefore, combining the 1D convolutions with RNNs is as simple as following the Conv1D layer with an RNN layer. We'll continue this exercise in the same Jupyter Notebook. Perform the following steps to complete this exercise:

1. Initialize a sequential model, add a **Reshape** layer (as in the preceding exercise), and add a **Conv1D** layer with five filters and a filter size 3:

```
model_comb = Sequential()
model_comb.add(Reshape((look_back,1), \
                      input_shape = (look_back,)))
model_comb.add(Conv1D(5, 3, activation='relu'))
```

2. Next, add a **SimpleRNN** layer with 32 neurons, followed by a **Dense** layer and an **Activation** layer:

```
model_comb.add(SimpleRNN(32))
model_comb.add(Dense(1))
model_comb.add(Activation('linear'))
```

3. Print out the model summary:

```
model_comb.summary()
```

The output will be as follows:

```
Layer (type)                 Output Shape              Param #
=================================================================
reshape_3 (Reshape)          (None, 10, 1)             0

conv1d_2 (Conv1D)            (None, 8, 5)              20

simple_rnn_2 (SimpleRNN)     (None, 32)                1216

dense_3 (Dense)              (None, 1)                 33

activation_3 (Activation)    (None, 1)                 0
=================================================================
Total params: 1,269
Trainable params: 1,269
Non-trainable params: 0
_____
```

Figure 5.35: Summary of the hybrid (1D convolution and RNN) model

The output from the Conv1D ayer is 8 × 5 – 8 features from 5 filters. The overall number of parameters is slightly higher than the plain RNN model. This is because the sequence size we're dealing with is very low. If we were dealing with larger sequences, we would have seen a reduction in the parameters. Let's compile and fit the model.

4. Compile and fit the model on the training data for three epochs:

```
model_comb.compile(loss='mean_squared_error', optimizer='adam')
model_comb.fit(trainX, trainY, epochs=3, \
                batch_size=1, verbose=2, validation_split=0.1)
```

The model training output is as follows:

```
Train on 1696 samples, validate on 189 samples
Epoch 1/3
1696/1696 - 7s - loss: 0.0026 - val_loss: 0.0023
Epoch 2/3
1696/1696 - 6s - loss: 0.0011 - val_loss: 6.0237e-04
Epoch 3/3
1696/1696 - 6s - loss: 0.0010 - val_loss: 5.4069e-04
```

Figure 5.36: Training and validation loss

Let's assess the performance first by looking at RMSE. We don't expect this to be very useful for our example, but let's print it out as good practice.

5. Print the RMSE for the train and test set using the **get_model_perf** function:

```
get_model_perf(model_comb)
```

You'll get the following output:

```
Train RMSE: 0.02 RMSE
Test RMSE: 0.03 RMSE
```

The values seem lower, but only a very close look will help us assess the performance of the model.

6. Plot the prediction versus actual in interactive mode and zoom in on the last 100 points:

```
%matplotlib notebook
plot_pred(model_comb)
```

The output of the preceding command will be as follows:

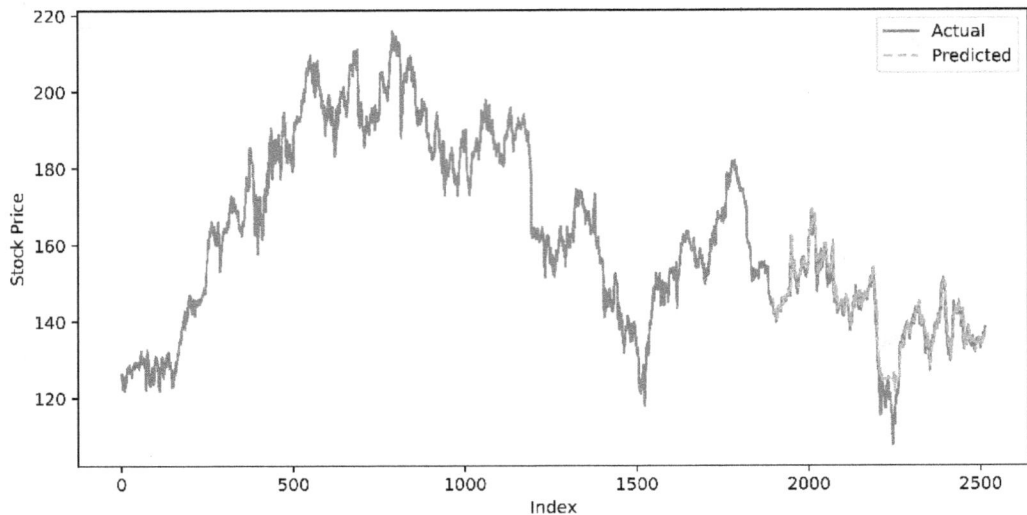

Figure 5.37: Plot of the combined model

Following is a zoomed-in view of the predictions:

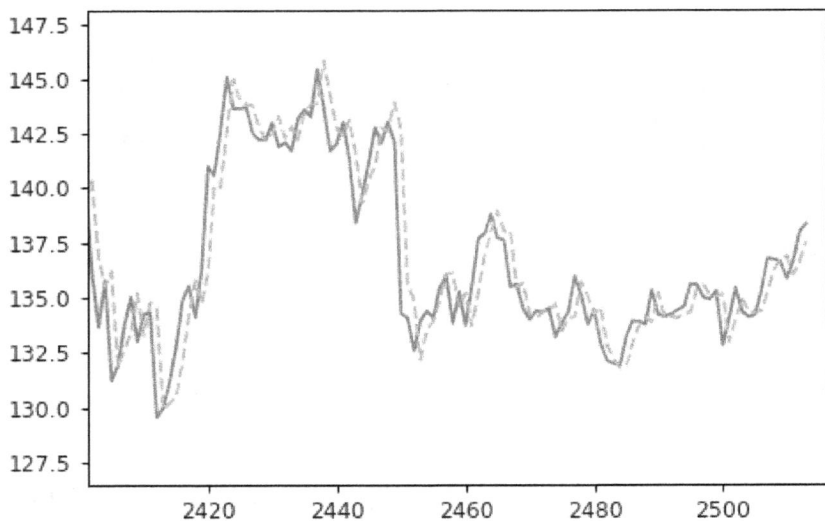

Figure 5.38: Zoomed-in view of predictions

> **NOTE**
>
> If the graphs presented below are not displayed properly for some reason, run the cell containing `%matplotlib notebook` for a couple of times. Alternatively, you can also use `%matplotlib inline` instead of `%matplotlib notebook`.

This result is extremely good. The prediction (dotted lines) is extremely close to the actual (solid lines) for the test data – capturing not only the level but also the minute variations very well. There is also some effective regularization going on when the 1D convnet is extracting patches from the sequence. These features are being fed in sequence to the RNN, which is using its raw power to provide the output we see. There is indeed merit in combining 1D convnets with RNNs.

> **NOTE**
>
> To access the source code for this specific section, please refer to https://packt.live/2ZctArW.
>
> You can also run this example online at https://packt.live/38EDOEA. You must execute the entire Notebook in order to get the desired result.

In this exercise, we saw how we can combine 1D convnets and RNNs to form a hybrid model that can provide high performance. We acknowledge that there is merit in trying this combination for sequence processing tasks.

ACTIVITY 5.01: USING A PLAIN RNN MODEL TO PREDICT IBM STOCK PRICES

We have seen RNNs in action and can now appreciate the kind of power they bring in sequence prediction tasks. We also saw that RNNs in conjunction with 1D convnets provide great results. Now, let's employ these ideas in another stock price prediction task, this time predicting the stock price for IBM. The dataset can be downloaded from https://packt.live/3fgmqlL. You will visualize the data and understand the patterns. From your understanding of the data, choose a lookback period and build an RNN-based model for prediction. The model will have a 1D convnet as well as a plain RNN layer. You will also employ dropout to prevent overfitting.

Perform the following steps to complete this exercise:

1. Load the `.csv` file, reverse the index, and plot the time series (the **Close** column) for visual inspection.

2. Extract the values for **Close** from the DataFrame as a **numpy** array and plot them using **matplotlib**.

3. Assign the final 25% data as test data and the first 75% as train data.

4. Using **MinMaxScaler** from **sklearn**, scale the train and test data.

5. Using the **get_lookback** function we defined in this chapter, get lookback data for the train and test data using a lookback period of 15.

6. From Keras, import all the necessary layers for employing plain RNNs (**SimpleRNN**, **Activation**, **Dropout**, **Dense**, and **Reshape**) and 1D convolutions (Conv1D). Also, import **mean_squared_error**.

7. Build a model with a 1D convolution layer (5 filters of size 3) and an RNN layer with 32 neurons. Add 25% dropout after the RNN layer. Print the model's summary.

8. Compile the model with the **mean_squared_error** loss and the **adam** optimizer. Fit this on the train data in five epochs with a validation split of 10% and a batch size of 1.

9. Using the **get_model_perf** method, print the RMSE of the model.

10. Plot the predictions – the entire view, as well as a zoomed-in one for a close assessment of the performance.

The zoomed-in view of the predictions (dotted lines) versus the actuals (solid lines) should be as follows:

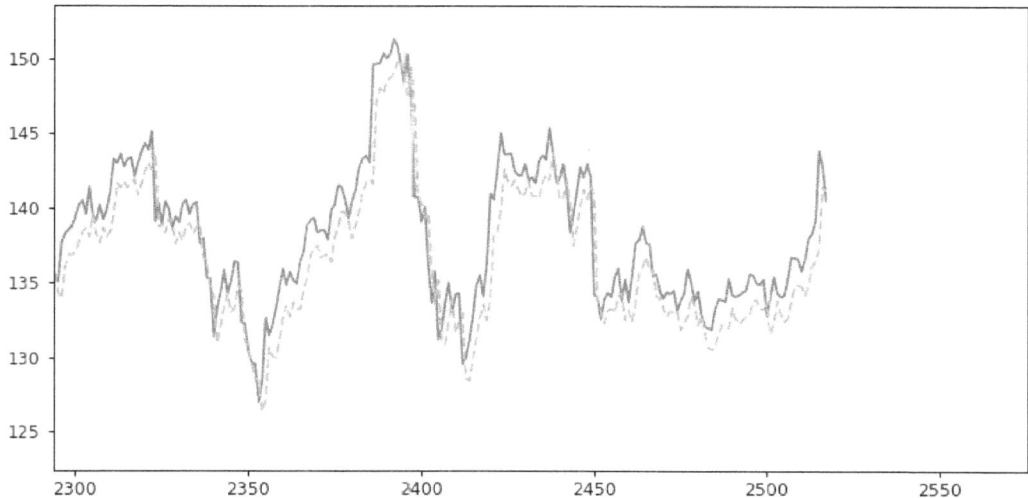

Figure 5.39: Zoomed-in view of predictions

NOTE

The detailed steps for this activity, along with the solutions and additional commentary, are presented on page 410.

SUMMARY

In this chapter, we looked at the considerations of working with sequences. There are several tasks that require us to exploit information contained in a sequence, where sequence-agnostic models would fare poorly. We saw that using RNNs is a very powerful approach to sequence modeling – the architecture explicitly processes the sequence and considers the information accumulated so far, along with the new input, to generate the output. Even very simple RNN architectures performed very well on our stock price prediction task. We got the kind of results that would take a lot of effort to get using classical approaches.

We also saw that 1D convolutions can be employed in sequence prediction tasks. 1D convolutions, like their 2D counterparts for images, learn local features in a sequence. We built a 1D convolution model that didn't fare too well on our task. The final model that we built combined 1D convolutions and RNNs and provided excellent results regarding the stock price prediction task.

In the next chapter, we will discuss models that are variations of RNNs that are even more powerful. We will also discuss architectures that extract the latent power of the idea of the RNN. We will apply these "RNNs on steroids" to an important task in natural language processing – sentiment classification.

6

LSTMS, GRUS, AND ADVANCED RNNS

OVERVIEW

In this chapter, we will study and implement advanced models and variations of the plain **Recurrent Neural Network** (**RNN**) that overcome some of RNNs' practical drawbacks and are among the best performing deep learning models at the moment. We will start by understanding the drawbacks of plain RNNs and see how the novel idea of **Long Short-Term Memory** overcomes them. We will then see and implement a **Gated Recurrent Unit** based model. We will also work with bidirectional and stacked RNNs and explore attention-based models. By the end of this chapter, you will have built and assessed the performance of these models on a sentiment classification task, observing for yourself the trade-offs in choosing the different models.

INTRODUCTION

Let's say you're working with product reviews for a mobile phone and your task is to classify the sentiment in the reviews as being positive or negative. You encounter a review that says: *"The phone does not have a great camera, or an amazingly vivid display, or an excellent battery life, or great connectivity, or other great features that make it the best."* Now, when you read this, you can easily identify that the sentiment in the review is negative, despite the presence of many positive phrases such as *"excellent battery life"* and *"makes it the best"*. You understand that the presence of the term *"not"* right toward the beginning of the text negates everything else that comes after.

Will the models we've created so far be able to identify the sentiment in such a case? Probably not, because if your models don't realize that the term *"not"* toward the beginning of the sentences is important and needs to be connected strongly to the output several terms later, they won't be able to identify the sentiment correctly. This, unfortunately, is a major drawback of plain RNNs.

In the previous chapter, we looked at a couple of deep learning approaches for dealing with sequences, that is, one-dimensional convolutions and RNNs. We saw that RNNs are extremely powerful models that provide us with a great amount of flexibility to handle different sequence tasks. The plain RNNs that we saw have been subject to plenty of research. Now, we will look at some approaches that have been built on top of RNNs to create new, powerful models that overcome the drawbacks of RNNs. We will look at LSTM, GRUs, stacked and bidirectional LSTMs, and attention-based models. We will apply these models to a sentiment classification task, thereby bringing together the concepts discussed in *Chapter 4, Deep Learning for Text – Embeddings*, and *Chapter 5, Deep Learning for Sequences*, as well.

LONG-RANGE DEPENDENCE/INFLUENCE

The sample mobile phone review we saw in the previous section was an example of a long-range dependence/influence – where a term/value in a sequence has an influence on the assessment of a lot of the subsequent terms/values. Consider the following example, where you need to fill in the blank with a missing country name: *"After a top German university granted her admission for her Masters in Dentistry, Hina was extremely excited to start this new phase of her career with international exposure and couldn't wait till the end of the month to book her flight to ___."*

The correct answer, of course, is **Germany**, arriving at which would require you to understand the importance of the term "German", which appears at the beginning of the sentence, on the outcome at the end of the sentence. This is another example of long-range dependence. The following figure shows the long-range dependence of the answer, "*Germany*", on the term "*German*" appearing early in the sentence:

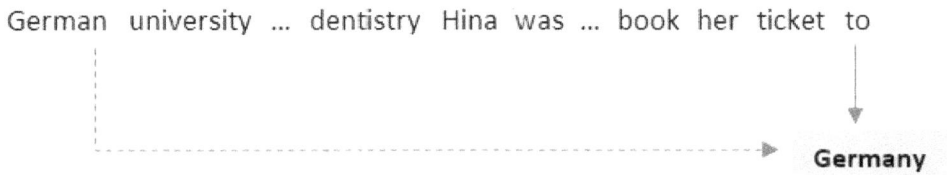

German university ... dentistry Hina was ... book her ticket to

Germany

Figure 6.1: Long-range dependence

To get the best outcome, we need to be able to handle long-range dependencies. In the context of deep learning models and RNNs, this would mean that learning (or the backpropagation of errors) needs to happen smoothly and effectively over many time steps. This is easier said than done, primarily because of the vanishing gradient problem.

THE VANISHING GRADIENT PROBLEM

One of the biggest challenges while training standard feedforward deep neural networks is the vanishing gradient problem (as discussed in *Chapter 2, Neural Networks*). As the model gets more and more layers, backpropagating the errors all the way back to the initial layers becomes increasingly difficult. Layers close to the output will be "learning"/updated at a good pace, but by the time the error propagates to the initial layers, its value will have diminished greatly and have little or no effect on the parameters for the initial layers.

With RNNs, this problem is further compounded, as the parameters need to be updated not only along the depth but also for the time steps. If we have one hundred time steps in the inputs (which isn't uncommon, especially when working with text), the network needs to propagate the error (calculated at the 100th time step) all the way back to the first time step. For plain RNNs, this task can be a bit too much to handle. This is where RNN variants can come in useful.

> **NOTE**
>
> Another practical issue with training deep networks is the exploding gradient problem, where the gradient values get very high – too high to be represented by the system. This issue has a rather simple workaround called "**Gradient Clipping**", which means capping the values of the gradient.

SEQUENCE MODELS FOR TEXT CLASSIFICATION

In *Chapter 5, Deep Learning for Sequences*, we learned that RNNs perform extremely well on sequence-modeling tasks and provide high performance on text-related tasks. In this chapter, we will use plain RNNs and variants of RNNs on a sentiment classification task: processing the input sequence and predicting whether the sentiment is positive or negative.

We'll use the IMDb reviews dataset for this task. The dataset contains 50,000 movie reviews, along with their sentiment – 25,000 highly polar movie reviews for training and 25,000 for testing. A few reasons for using this dataset are as follows:

- It is very conveniently available to load Keras (tokenized version) with a single command.

- The dataset is commonly used for testing new approaches/models. This should help you compare your results with other approaches easily.

- Longer sequences in the data (IMDb reviews can get very long) help us assess the differences between the variants of RNNs better.

Let's get started by building our first model using plain RNNs and then benchmark the future model performances against that of the plain RNN. Let's start with data preprocessing and formatting the model.

LOADING DATA

To begin, you need to start a new Jupyter Notebook and import the **imdb** module from the Keras datasets. Note that unless mentioned otherwise, the code and exercises for the rest of this chapter should continue in the same Jupyter Notebook:

```
from tensorflow.keras.datasets import imdb
```

With the module imported, importing the dataset (tokenized and separated into train and test sets) is as easy as running **imdb.load_data**. The only parameter we need to provide is the vocabulary size we wish to use. Recall that the vocabulary size is the total number of unique terms we wish to consider for the modeling process. When we specify a vocabulary size, V, we work with the top V terms in the data. Here, we will specify a vocabulary size of 8,000 for our models (an arbitrary choice; you can modify this as desired) and load the data using the **load_data** method, as shown here:

```
vocab_size = 8000
(X_train, y_train), (X_test, y_test) = imdb.load_data\
                                      (num_words=vocab_size)
```

Let's inspect the **X_train** variable to see what we are working with. Let's print the type of it and the type of constituting elements, and also have a look at one of the elements:

```
print(type(X_train))
print(type(X_train[5]))
print(X_train[5])
```

We will see the following output:

```
<class 'numpy.ndarray'>
<class 'list'>
[1, 778, 128, 74, 12, 630, 163, 15, 4, 1766, 7982, 1051,
 2, 32, 85, 156, 45, 40,
 148, 139, 121, 664, 665, 10, 10, 1361, 173, 4, 749, 2, 16,
 3804, 8, 4, 226, 65,
 12, 43, 127, 24, 2, 10, 10]
```

The **X_train** variable is a **numpy** array – each element of the array is a list representing the text for a single review. The terms in the text are present as numerical tokens instead of raw tokens. This is a very convenient format.

The next step is to define an upper limit on the length of the sequences that we'll work with and limit all sequences to the defined maximum length. We'll use **200** – an arbitrary choice, in this case – to quickly get started with the model-building process*****. For our purpose, we'll pick **200** steps so that the networks don't get too heavy, and because **200** time steps are sufficient to demonstrate the different RNN approaches. Let's define the **maxlen** variable:

```
maxlen = 200
```

The next step is to get all our sequences to the same length using the **pad_sequences** utility from Keras.

> **NOTE**
>
> *Ideally, we would analyze the lengths of the sequences and identify one that covers most of the reviews. We'll perform these steps in the activity at the end of the chapter, in which we'll use ideas from not only the current chapter, but also from *Chapter 4, Deep Learning for Text – Embeddings*, and *Chapter 5, Deep Learning for Sequences*, bringing this all together in a single activity.

STAGING AND PREPROCESSING OUR DATA

The **pad_sequences** utility from the **sequences** module in Keras helps us in getting all the sequences to a specified length. If the input sequence is shorter than the specified length, the utility pads the sequence with a reserved token (indicating a blank/missing). If the input sequence is longer than the specified length, the utility truncates the sequence to limit it. In the following example, we will apply the **pad_sequences** utility to our test and train datasets:

```
from tensorflow.keras import preprocessing
X_train = preprocessing.sequence.pad_sequences\
        (X_train, maxlen=maxlen)
X_test = preprocessing.sequence.pad_sequences\
        (X_test, maxlen=maxlen)
```

To understand the result of the steps, let's see the output for a particular instance in the training data:

```
print(X_train[5])
```

The processed instance is as follows:

```
[    0    0    0    0    0    0    0    0    0    0    0    0    0    0
     0    0    0    0    0    0    0    0    0    0    0    0    0    0
     0    0    0    0    0    0    0    0    0    0    0    0    0    0
     0    0    0    0    0    0    0    0    0    0    0    0    0    0
     0    0    0    0    0    0    0    0    0    0    0    0    0    0
     0    0    0    0    0    0    0    0    0    0    0    0    0    0
     0    0    0    0    0    0    0    0    0    0    0    0    0    0
     0    0    0    0    0    0    0    0    0    0    0    0    0    0
     0    0    0    0    0    0    0    0    0    0    0    0    0    0
     0    0    0    0    0    0    0    0    0    0    0    0    0    0
     0    0    0    0    0    0    0    0    0    0    0    0    0    0
     0    0    0    1  778  128   74   12  630  163   15    4 1766 7982
  1051    2   32   85  156   45   40  148  139  121  664  665   10   10
  1361  173    4  749    2   16 3804    8    4  226   65   12   43  127
    24    2   10   10]
```

Figure 6.2: Result of pad_sequences

We can see that there are plenty of **0**s at the beginning of the result. As you may have inferred, this is the padding that's done by the **pad_sequence** utility because the input sequence was shorter than **200**. Padding at the beginning of the sequence is the default behavior of the utility. For a sequence that is less than the specified limit, the truncation, by default, is done from the left – that is, the last **200** terms would be retained. All instances in the output now have **200** terms. The dataset is now ready for modeling.

> **NOTE**
>
> The default behavior of the utility is to pad the beginning of the sequence and truncate from the left. These can be important hyperparameters. If you believe that the first few terms are the most important for the prediction, you may want to truncate the last terms by specifying the "**truncating**" parameter as "**post**". Similarly, to have padding toward the end of the sequence, you can set "**padding**" to "**post**".

THE EMBEDDING LAYER

In *Chapter 4, Deep Learning for Text – Embeddings*, we discussed that we can't feed text directly into a neural network, and therefore need good representations. We discussed that embeddings (low-dimensional, dense vectors) are a great way of representing text. To pass the embeddings into the neural network's layers, we need to employ the embedding layer.

The functionality of the embedding layer is two-fold:

- For any input term, perform a lookup and return its word embedding/vector

- During training, learn these word embeddings

The part about looking up is straightforward – the word embeddings are stored as a matrix of the $V \times D$ dimensionality, where V is the vocabulary size (the number of unique terms considered) and D is the length/dimensionality of each vector. The following figure illustrates the embedding layer. The input term, "**life**", is passed to the embedding layer, which performs a lookup and returns the corresponding vector of length D. This vector, which is the representation for the term **life**, is fed to the hidden layer.

What do we mean by learning these embeddings while training the predictive model? Aren't word embeddings learned by using an algorithm such as **word2vec**, which tries to predict the center word based on context terms (remember the CBOW architecture we discussed in *Chapter 4, Deep Learning for Text – Embeddings*)? Well yes and no:

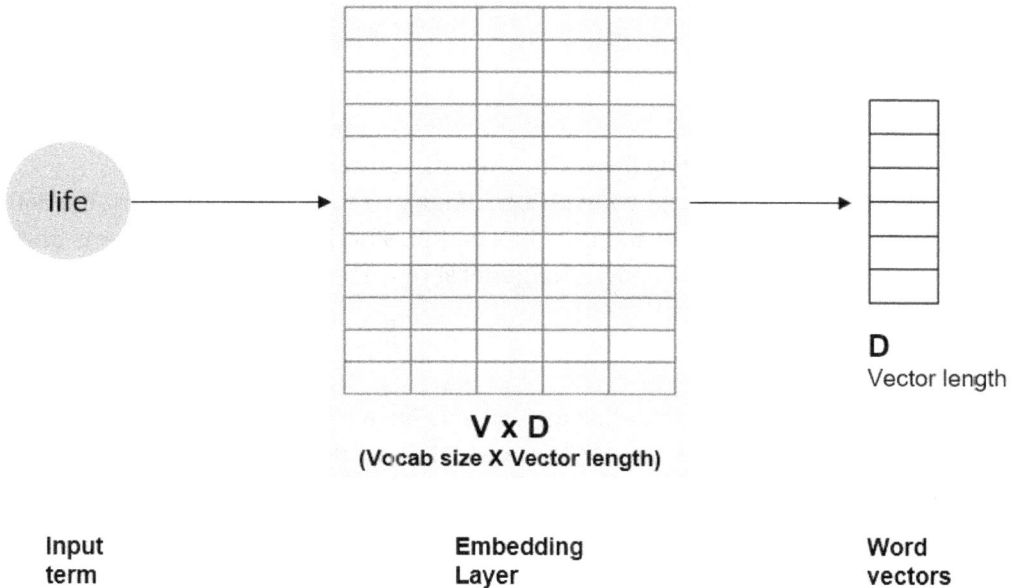

V x D
(Vocab size X Vector length)

Input   **Embedding**   **Word**
term   **Layer**   **vectors**

Figure 6.3: Embedding layer

The **word2vec** approach had the objective of learning a representation that captures the meaning of the term. Therefore, predicting the target word based on context was a perfect formulation for the objective. In our case, the objective is different – we wish to learn representations that help us best predict the sentiment in the text. It makes sense, then, to learn the representation that works explicitly toward our objective.

The embedding layer is always the first layer in the model. You can follow it up with any architecture of your choice (RNNs, in our case). We randomly initialize the vectors, essentially the weights in the embedding layer. While the model trains, the weights are updated in a way that predicts the outcome in a better way. The weights learned and therefore the word vectors, are then tuned to the task. This is a very useful step – why use generic representations when you can tune them to your task?

The embedding layer in Keras has two main parameters:

- **`input_dim`** : The number of unique terms in the vocabulary, that is, the vocabulary size

- **`output_dim`** : The dimension of the embedding/the length of the word vector

The **`input_dim`** parameter needs to be set to the vocabulary size being employed. The **`output_dim`** parameter specifies the length of the embedding vector for each term.

Note that the embedding layer in Keras also allows you to use your own specified weight matrix in the embedding layer. This means you can use pre-trained embeddings (such as **GloVe**, or even embeddings you trained in a different model) in the embedding layer. The **GloVe** model has been trained on billions of tokens and it could be useful to leverage this powerful general representation.

> **NOTE**
>
> If you use pre-trained embeddings, you also have the option to make them trainable in your model – essentially, use **GloVe** embeddings as a starting point and fine-tune them for your task. This is a great example of transfer learning for text.

BUILDING THE PLAIN RNN MODEL

In the next exercise, we will build our first model for the sentiment classification task using plain RNNs. The model architecture we'll use is depicted in the following figure, which demonstrates how the model would process an input sentence "**Life is good**", with the term "**Life**" coming in at time step **T=0** and "**good**" appearing at time step **T=2**. The model will process the inputs one by one, using the embedding layer to look up the word embeddings that will be passed to the hidden layers. The classification will be done when the final term, "**good**", is processed at time step **T=2**. We'll use Keras to build and train our models:

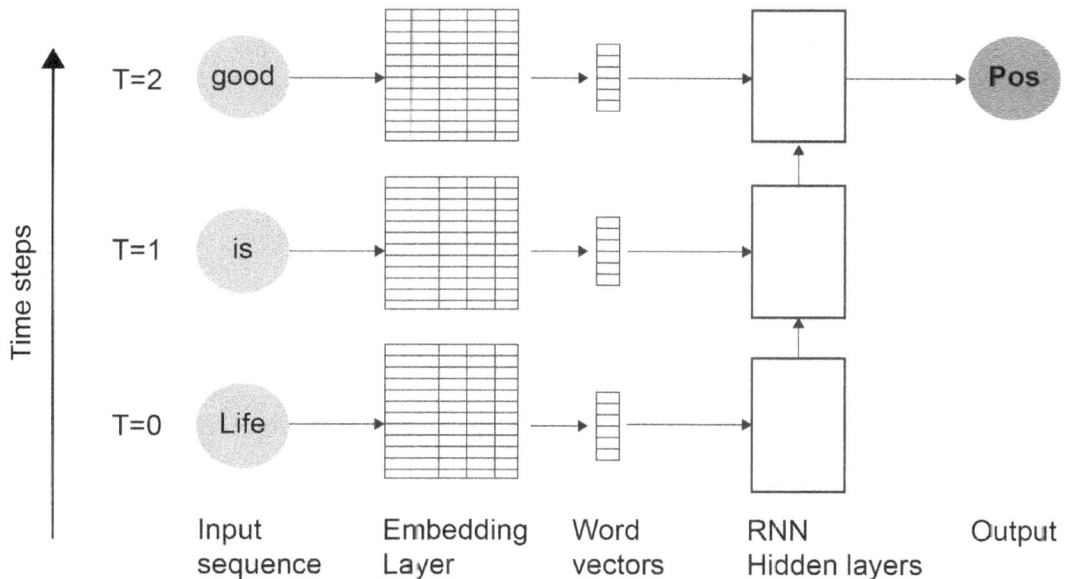

Figure 6.4: Architecture using an embedding layer and RNN

EXERCISE 6.01: BUILDING AND TRAINING AN RNN MODEL FOR SENTIMENT CLASSIFICATION

In this exercise, we will build and train an RNN model for sentiment classification. Initially, we will define the architecture for the recurrent and prediction layers, and we will assess the model's performance on the test data. We will add the embedding layer and some dropout and complete the model definition by adding the RNN layer, dropout, and a dense layer to finish. Then, we'll check the accuracy of the predictions on the test data to assess how well the model generalizes. Follow these steps to complete this exercise:

1. Let's begin by setting the seed for **numpy** and **tensorflow** random number generation, to get, to the best extent possible, reproducible results. We'll import **numpy** and **tensorflow** and set the seed using the following commands:

```
import numpy as np
import tensorflow as tf
np.random.seed(42)
tf.random.set_seed(42)
```

> **NOTE**
>
> Even though we have set the seeds for **numpy** and **tensorflow** to achieve reproducible results, there are a lot more causes for variation, owing to which you may get a result that's different from ours. This applies to all the models we'll use from now on. While the values you see may be different, the output you see should largely agree with ours. If the model's performance is very different, you may want to tweak the number of epochs – the reason for this being that the weights in neural networks are initialized randomly, so the starting points for you and us could be slightly different, and we may reach a similar position when training a different number of epochs.

2. Now, let's continue by importing all the necessary packages and layers and initializing a sequential model named **model_rnn** using the following commands:

```
from tensorflow.keras.models import Sequential
from tensorflow.keras.layers \
import SimpleRNN, Flatten, Dense, Embedding, \
SpatialDropout1D, Dropout
model_rnn = Sequential()
```

3. Now, we need to specify the embedding layer. The **input_dim** parameter needs to be set to the **vocab_size** variable. For the **output_dim** parameter, we'll choose **32**. Recall from *Chapter 4, Deep Learning for Text – Embeddings*, that this is a hyperparameter and you may want to experiment with this to get better results. Let's specify the embedding layer and use dropout (to minimize overfitting) using the following commands:

```
model_rnn.add(Embedding(vocab_size, output_dim=32))
model_rnn.add(SpatialDropout1D(0.4))
```

Note that the dropout employed here is **SpatialDropout1D** – this version performs the same function as regular dropout layer, but instead of dropping individual elements, t drops entire one-dimensional feature maps (vectors, in our case).

4. Add a **SimpleRNN** layer with **32** neurons to the model (chosen arbitrarily; another hyperparameter to tune):

```
model_rnn.add(SimpleRNN(32))
```

5. Next, add a dropout layer with **40%** dropout (again, an arbitrary choice):

```
model_rnn.add(Dropout(0.4))
```

6. Add a dense layer with a **sigmoid** activation function to complete the model architecture. This is the output layer that makes the prediction:

```
model_rnn.add(Dense(1, activation='sigmoid'))
```

7. Compile the model and view the model summary:

```
model_rnn.compile(loss='binary_crossentropy', \
                  optimizer='rmsprop', metrics=['accuracy'])
model_rnn.summary()
```

The model summary is as follows:

```
Model: "sequential"

_____
Layer (type)                     Output Shape             Param #
=================================================================
embedding (Embedding)            (None, None, 32)         256000

spatial_dropout1d (SpatialDr     (None, None, 32)         0

simple_rnn (SimpleRNN)           (None, 32)               2080

dropout (Dropout)                (None, 32)               0

dense (Dense)                    (None, 1)                33
=================================================================
Total params: 258,113
Trainable params: 258,113
Non-trainable params: 0
_____
```

Figure 6.5: Summary of the plain RNN model

We can see that there are **258,113** parameters, most of which are present in the embedding layer. The reason for this is that the word embeddings are being learned during the training – so we're learning the embedding matrix, which is of dimensionality **vocab_size(8000)** × **output_dim(32)**.

Let's proceed and train the model (with the hyperparameters that we've observed to provide the best result with this data and architecture).

8. Fit the model on the train data with a batch size of **128** for **10** epochs (both of these are hyperparameters that you can tune). Use a validation split of **0.2** – monitoring this will give us a sense of the model performance on unseen data:

```
history_rnn = model_rnn.fit(X_train, y_train, \
                            batch_size=128, \
                            validation_split=0.2, \
                            epochs = 10)
```

The training output for the last five epochs will be as follows. Depending on your system configuration, this step could take more or less time than it did here for us:

```
Epoch 6/10
20000/20000 [==============================] - 9s 453us/sample - loss: 0.2532 - acc: 0.9022 - val_loss: 0.
3756 - val_acc: 0.8462
Epoch 7/10
20000/20000 [==============================] - 9s 438us/sample - loss: 0.2240 - acc: 0.9171 - val_loss: 0.
5079 - val_acc: 0.7708
Epoch 8/10
20000/20000 [==============================] - 9s 456us/sample - loss: 0.1995 - acc: 0.9262 - val_loss: 0.
5041 - val_acc: 0.7828
Epoch 9/10
20000/20000 [==============================] - 9s 452us/sample - loss: 0.1807 - acc: 0.9327 - val_loss: 0.
5454 - val_acc: 0.7642
Epoch 10/10
20000/20000 [==============================] - 9s 439us/sample - loss: 0.1627 - acc: 0.9420 - val_loss: 0.
4306 - val_acc: 0.8606
```

Figure 6.6: Training the plain RNN model – the final five epochs

From the training output, we can see that the validation accuracy goes up to about 86%. Let's make predictions on the test set and check the performance of the model.

9. Make predictions on the test data using the **predict_classes** method of the model and use the **accuracy_score** method from **sklearn**:

```
y_test_pred = model_rnn.predict_classes(X_test)
from sklearn.metrics import accuracy_score
print(accuracy_score(y_test, y_test_pred))
```

The accuracy of the test is as follows:

```
0.85128
```

We can see that the model does a decent job. We used a simple architecture with **32** neurons and used a vocabulary size of just **8000**. Tweaking these and other hyperparameters may get you better results and you are encouraged to do so.

> **NOTE**
>
> To access the source code for this specific section, please refer to https://packt.live/31ZPO2g.
>
> You can also run this example online at https://packt.live/2Oa2trm. You must execute the entire Notebook in order to get the desired result.

In this exercise, we have seen how to build an RNN-based model for text. We saw how an embedding layer can be used to derive word vectors for the task at hand. These word vectors are the representations for each incoming term, which are passed to the RNN layer. We have seen that even a simple architecture can give us good results. Now, let's discuss how this model can be used to make predictions on new, unseen reviews.

MAKING PREDICTIONS ON UNSEEN DATA

Now that you've trained your model on some data and assessed its performance on the test data, the next thing is to learn how to use this model to predict the sentiment for new data. That is the purpose of the model, after all – being able to predict the sentiment for data previously unseen by the model. Essentially, for any new review in the form of raw text, we should be able to classify its sentiment.

The key step for this would be to create a process/pipeline that converts the raw text into a format the predictive model understands. This would mean that the new text would need to undergo exactly the same preprocessing steps that were performed on the text data that was used to train the model. The function for preprocessing needs to return formatted text for any input raw text. The complexity of this function depends on the steps performed on the train data. If tokenization was the only preprocessing step performed, then the function only needs to perform tokenization.

Our model (**model_rnn**) was trained on IMDb reviews that were tokenized, had their case lowered, had punctuation removed, had a defined vocabulary size, and were converted into a sequence of indices. Our function/pipeline for preparing data for the RNN model needs to perform the same steps. Let's work toward creating our own function. To begin, let's create a new variable called "**inp_review**" containing the text "*An excellent movie*" using the following code. This is the variable containing the raw review text:

```
inp_review = "An excellent movie!"
```

The sentiment in the text is positive. If the model is working well enough, it should predict the sentiment as positive.

First, we must tokenize this text into its constituent terms, normalize its case, and remove punctuation. To do so, we need to import the **text_to_word_sequence** utility from Keras using the following code:

```
from tensorflow.keras.preprocessing.text \
import text_to_word_sequence
```

To check if it works as we expect, we can apply this to the **inp_review** variable, as shown in the following code:

```
text_to_word_sequence inp_review)
```

The tokenized sentence will be as follows:

```
['an', 'excellent', 'movie']
```

We can see that it works just as expected – the case has been normalized, the sentences have been tokenized, and punctuation has been removed from the input text. The next step would be to use a defined vocabulary for the data. This would require using the same vocabulary that was used by TensorFlow when we loaded the data. The vocabulary and the term-to-index mapping can be loaded using the **get_word_index** method from the **imdb** module (that we employed to load the code). The following code can be used to load the vocabulary into a dictionary named **word_map**:

```
word_map = imdb.get_word_index()
```

This dictionary contains the mapping for about 88.6 K terms that were available in the raw reviews data. We loaded the data with a vocabulary size of **8000**, thereby using the first **8000** indices from the mapping. Let's create our mapping with limited vocabulary so that we can use the same terms/indices that the training data used. We'll limit the mapping to **8000** terms by sorting the **word_map** variable on the index and picking the first **8000** terms, as follows:

```
vocab_map = dict(sorted(word_map.items(), \
                  key=lambda x: x[1])[:vocab_size])
```

The vocab map will be a dictionary containing the term for index mapping for the **8000** terms in the vocabulary. Using this mapping, we'll convert the tokenized sentence into a sequence of term indices by performing a lookup for each term and returning the corresponding index. Using the following code, we'll define a function that accepts raw text, applies the **text_to_word_sequence** utility to it, performs a lookup from **vocab_map**, and returns the corresponding sequence of integers:

```
def preprocess(review):
    inp_tokens = text_to_word_sequence(review)
    seq = []
    for token in inp_tokens:
        seq.append(vocab_map.get(token))
    return seq
```

We can apply this function to the **inp_review** variable, like so:

```
preprocess(inp_review)
```

The output is as follows:

```
[32, 318, 17]
```

This is the sequence of term indices corresponding to the raw text. Note that the data is now in the same format as the IMDb data we loaded. This sequence of indices can be fed to the RNN model (using the **predict_classes** method) to classify the sentiment, as shown in the following code. If the model is working well enough, it should predict the sentiment as positive:

```
model_rnn.predict_classes([preprocess(inp_review)])
```

The output prediction is **1** (positive), just as we expected:

```
array([[1]])
```

Let's apply the function to another raw text review and supply it to the model for prediction. Let's update the **inp_review** variable so that it contains the text "**Don't watch this movie – poor acting, poor script, bad direction.**" The sentiment in the review is negative. We expect the model to classify it as such:

```
inp_review = "Don't watch this movie"\
            " - poor acting, poor script, bad direction."
```

Let's apply our preprocessing function to the **inp_review** variable and make a prediction using the following code:

```
model_rnn.predict_classes([preprocess(inp_review)])
```

The prediction is **0**, as shown here:

```
array([[0]])
```

The predicted sentiment is negative, just as we would expect the model to behave.

We applied this pipeline in the form of a function on a single review, but you can very easily apply this to a whole collection of reviews to make predictions using the model. You are now ready to classify the sentiment of any new review using the RNN model we trained.

> **NOTE**
>
> The pipeline we built here is specifically for this dataset and model. This is not a generic processing function that you can utilize for predictions from any model. The vocabulary used, the cleanup that was done, the patterns the model learned – these were all specific to this task and dataset. For any other model, you need to create your pipeline accordingly.

The higher-level approach can be employed to make processing pipelines for other models too. Depending on the data, the preprocessing steps, and setting up where the model will be deployed, the pipeline can vary. All these factors also affect the steps you may want to include in the model building process. Therefore, we encourage to you to start thinking about these aspects right away when you begin the whole modeling process.

We saw how to make predictions on unseen data using the trained RNN model, thereby giving us an understanding of the end-to-end process. In the next section, we'll begin working with variants of RNNs. The implementation-related ideas we've discussed so far are applicable to all the subsequent models.

LSTMS, GRUS, AND OTHER VARIANTS

The idea behind plain RNNs is very powerful and the architecture has shown tremendous promise. Due to this, researchers have experimented with the architecture of RNNs to find ways to overcome the one major drawback (the vanishing gradient problem) and exploit the power of RNNs. This led to the development of LSTMs and GRUs, which have now practically replaced RNNs. Indeed, these days, when we refer to RNNs, we usually refer to LSTMs, GRUs, or their variants.

This is because these variants are designed specifically to handle the vanishing gradient problem and learn long-range dependencies. Both approaches have outperformed plain RNNs significantly in most tasks around sequence modeling, and the difference is especially higher for long sequences. The paper titled *Learning Phrase Representations using RNN Encoder-Decoder for Statistical Machine Translation* (available at https://arxiv.org/abs/1406.1078) performs an empirical analysis of the performance of plain RNNs, LSTMs, and GRUs. How have these approaches overcome the drawbacks of plain RNNS? We'll understand this in the next section, where we'll discuss LSTMs in detail.

LSTMS

Let's think about this for a moment. Knowing the architecture of the plain RNN, how can we tweak it, or what can be done differently to capture long-range influences? We can't add more layers; that would be counterproductive for sure, as every added layer would compound the problem. One idea (available at https://pubmed.ncbi.nlm.nih.gov/9377276), proposed in 1997 by Sepp Hochreiter and Jurgen Schmidhuber, is to use an explicit value (state) that does not pass through activations. If we had a cell (corresponding to a neuron for plain RNNs) value flowing freely and not through activations, this value could potentially help us model long-range dependence. This is the first key difference in an LSTM – an explicit cell state.

The cell state can be thought of as a way to identify and store information over multiple time steps. Essentially, we are identifying some value as the long-term memory of the network that helps us predict the output better and taking care to retain this value as long as required.

But how do we regulate the flow of this cell state? How do we decide when to update the value and by how much? For this, Hochreiter and Schmidhuber proposed the use of *gating mechanisms* as a way to regulate how and when to update the value of the cell state. This is the other key difference in an LSTM. The freely flowing cell state, together with the regulatory mechanisms, allow the LSTM to perform extremely well on longer sequences and provide it with all its predictive power.

> ### NOTE
> A detailed treatment of the inner workings of the LSTM and the associated math is beyond the scope of this book. For those interested in reading further, https://packt.live/3gL42lb is a good reference that provides a good visual understanding of LSTMs.

Let's understand the intuition behind the working of the LSTM. The following figure shows the internals of the LSTM cell. Apart from the usual outputs, that is, the hidden state, the LSTM cell also outputs a cell "state". The hidden state holds the short-term memory, while the cell state holds the long-term memory:

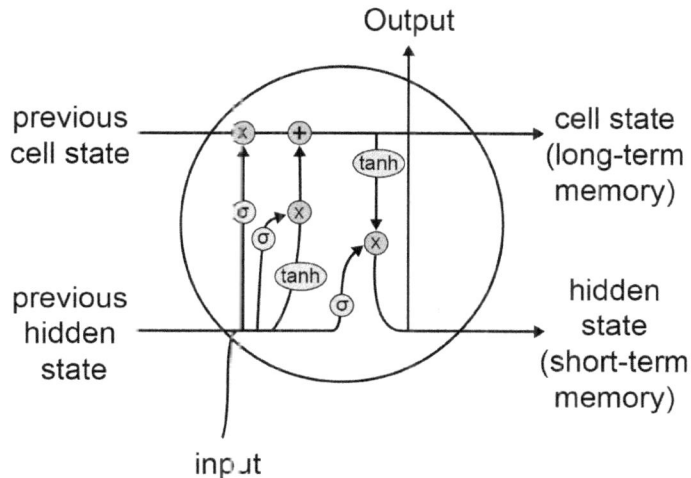

Figure 6.7: The LSTM cell

This view of the internals can be intimidating, which is why we'll look at a more abstracted view, as can be seen in *Figure 6.8*. The first thing to notice is that the only operations that take place on the cell state are two linear operations – a multiplication and an addition. The cell state does not pass through any activation function. This is why we said that the cell state flows freely. This free-flow setup is also called a "Constant Error Carousel" – a moniker you *don't* need to remember.

The output of the **FORGET** block is multiplied by the cell state. Because the output of this block is between **0** and **1** (modeled by the sigmoid activation), a multiplication of this with the cell state will regulate how much of the previous cell state is to be forgotten. If the **FORGET** block outputs **0**, the previous cell state is completely forgotten; while for output **1**, the cell state is completely retained. Note that the inputs to the **FORGET** gate are the output from the hidden layer from the previous time step (h_{t-1}) and the new input at the present time step, x_t (for a layer deep in the network, this could be the output from the previous hidden layer):

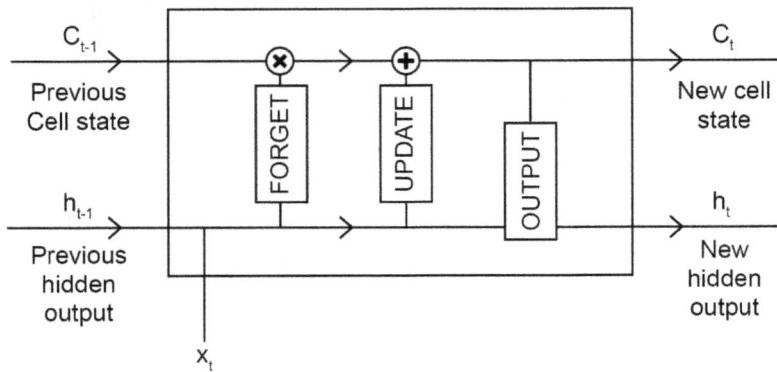

Figure 6.8: Abstracted view of the LSTM cell

In the preceding figure, we can see that after the cell state is multiplied by the **FORGET** block's result, the next decision is how much to update the cell state by. This comes from the **UPDATE** block's output, which is added (note the plus sign) to the processed cell state. This way, the processed cell state is updated. That's all the operations that are performed on the previous cell state, (C_{t-1}), to give us the new cell state, (C_t), as an output. This is how the long-term memory of the cell is regulated. The cell also needs to update the hidden state. This operation takes place in the **OUTPUT** block and is pretty much the same as the update in a plain RNN. The only difference is that the explicit cell state is multiplied by the output from the sigmoid to form the final hidden state, h_t.

Now that we understand the individual blocks/gates, let's see them marked on the following detailed figure. This should clarify how these gating mechanisms come together to regulate the flow of information in an LSTM:

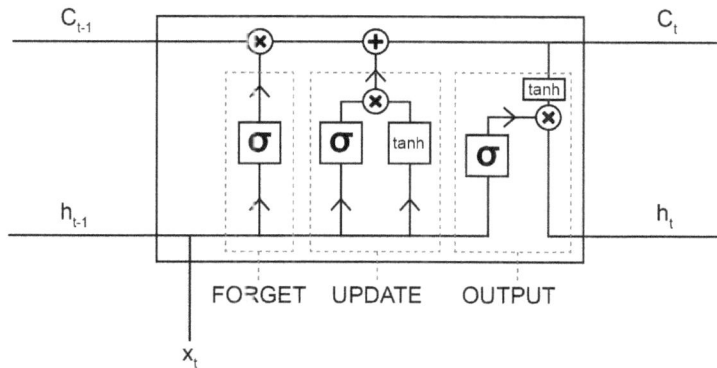

Figure 6.9: The LSTM cell explained

To make this example more concrete, let's take a look at the following figure and understand how the cell state is updated. We can assume the previous cell state, (C_{t-1}), was **5**. How much of this value should be propagated is decided by the output of the **FORGET** gate. The output value of the **FORGET** gate is multiplied by the previous cell state, C_{t-1}. In this case, the output of the forget block is **0.5**, resulting in **2.5** as the processed cell state being passed. This value (**2.5**) then encounters the addition from the **UPDATE** gate. Since the **UPDATE** gate output value of **-0.8**, the result of the addition is **1.7**. This is the final, updated cell state, C_t, that is passed to the next time step:

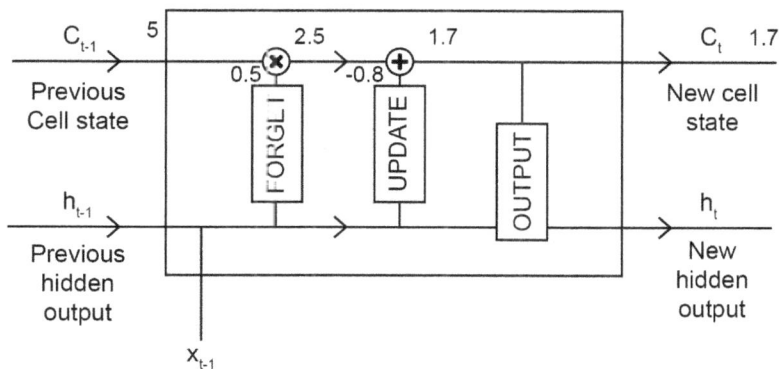

Figure 6.10: LSTM cell state update example

PARAMETERS IN AN LSTM

LSTMs are built on plain RNNs. If you simplified the LSTM and removed all the gates, retaining only the tanh function for the hidden state update, you would have a plain RNN. The number of activations that the information – the new input data at time **t** and the previous hidden state at time **t−1** ($\mathbf{x}_t$ and $\mathbf{h}_{t-1}$) – passes through in an LSTM is four times the number that it passes through in a plain RNN. The activations are applied once in the forget gate, twice in the update gate, and once in the output gate. The number of weights/parameters in an LSTM is, therefore, four times the number of parameters in a plain RNN.

In *Chapter 5, Deep Learning For Sequences,* in the section titled *Parameters in an RNN,* we calculated the number of parameters in a plain RNN and saw that we already have a quite a few parameters to work with ($\mathbf{n}^2$ + $\mathbf{nk}$ + $\mathbf{nm}$, where **n** is the number of neurons in the hidden layer, **m** is the number of inputs, and **k** is the dimension of the output layer). With LSTMs, we saw that the number is four times this. Needless to say, we have a lot of parameters in an LSTM, and that isn't necessarily a good thing, especially when working with smaller datasets.

EXERCISE 6.02: LSTM-BASED SENTIMENT CLASSIFICATION MODEL

In this exercise, we will build a simple LSTM-based model to predict sentiment on our data. We will continue with the same setup we used previously (that is, the number of cells, embedding dimensions, dropout, and so on). Thus, you must continue this exercise in the same Jupyter Notebook. Follow these steps to complete this exercise:

1. Import the LSTM layer from Keras **layers**:

```
from tensorflow.keras.layers import LSTM
```

2. Instantiate the sequential model, add the embedding layer with the appropriate dimensions, and add a 40% spatial dropout:

```
model_lstm = Sequential()
model_lstm.add(Embedding(vocab_size, output_dim=32))
model_lstm.add(SpatialDropout1D(0.4))
```

3. Add an LSTM layer with **32** cells:

```
model_lstm.add(LSTM(32))
```

4. Add the dropout (**40%** dropout) and dense layers, compile the model, and print the model summary:

```
model_lstm.add(Dropout(0.4))
model_lstm.add(Dense(1, activation='sigmoid'))
model_lstm.compile(loss='binary_crossentropy', \
                   optimizer='rmsprop', metrics=['accuracy'])
model_lstm.summary()
```

The model summary is as follows:

```
Layer (type)                   Output Shape              Param #
=================================================================
embedding_2 (Embedding)        (None, None, 32)          256000

spatial_dropout1d_2 (Spatial   (None, None, 32)          0

lstm (LSTM)                    (None, 32)                8320

dropout_2 (Dropout)            (None, 32)                0

dense_2 (Dense)                (None, 1)                 33
=================================================================
Total params: 264,353
Trainable params: 264,353
Non-trainable params: 0
```

Figure 6.11: Summary of the LSTM model

We can see from the model summary that the number of parameters in the LSTM layer is **8320**. A quick check can confirm that this is exactly four times the number of parameters in the plain RNN layer we saw in *Exercise 6.01, Building and Training an RNN Model for Sentiment Classification*, which is in line with our expectations. Next, let's fit the model on the training data.

5. Fit on the training data for **5** epochs (this gives us the best result for the model) with a batch size of **128**:

```
history_lstm = model_lstm.fit(X_train, y_train, \
                              batch_size=128, \
                              validation_split=0.2, \
                              epochs=5)
```

The output from the training process is as follows:

```
Train on 20000 samples, validate on 5000 samples
Epoch 1/5
20000/20000 [==============================] - 37s 2ms/sample - loss: 0.5569 - acc: 0.7070 - val_loss: 0.3
665 - val_acc: 0.8546
Epoch 2/5
20000/20000 [==============================] - 37s 2ms/sample - loss: 0.3336 - acc: 0.8676 - val_loss: 0.3
020 - val_acc: 0.8760
Epoch 3/5
20000/20000 [==============================] - 37s 2ms/sample - loss: 0.2695 - acc: 0.8954 - val_loss: 0.2
859 - val_acc: 0.8794
Epoch 4/5
20000/20000 [==============================] - 37s 2ms/sample - loss: 0.2417 - acc: 0.9099 - val_loss: 0.3
143 - val_acc: 0.8692
Epoch 5/5
20000/20000 [==============================] - 37s 2ms/sample - loss: 0.2173 - acc: 0.9207 - val_loss: 0.2
886 - val_acc: 0.8788
```

Figure 6.12: LSTM training output

Notice that training the LSTM took much longer than it does with plain RNNs. Again, considering the architecture of the LSTM and the sheer number of parameters, this was expected. Also, note that the validation accuracy is significantly higher than that of the plain RNN. Let's check the performance on the test data in terms of the accuracy score.

6. Make predictions on the test set and print the accuracy score:

```
y_test_pred = model_lstm.predict_classes(X_test)
print(accuracy_score(y_test, y_test_pred))
```

The accuracy is printed out as follows:

```
0.87032
```

The accuracy we got (87%) is a significant improvement from the accuracy we got using plain RNNs (85.1%). It looks like the extra parameters and the extra predictive power from the cell state came in handy for our task.

> **NOTE**
>
> To access the source code for this specific section, please refer to https://packt.live/31ZPO2g.
>
> You can also run this example online at https://packt.live/2Oa2trm. You must execute the entire Notebook in order to get the desired result.

In this exercise, we saw how we can employ LSTMs for sentiment classification of text. The training time was significantly higher, and the number of parameters is higher too. But in the end, even this simple architecture (without any hyperparameter tuning) gave better results than the plain RNN. You are encouraged to tune the hyperparameters further to get the most out of the powerful LSTM architecture.

LSTM VERSUS PLAIN RNNS

We saw that LSTMs are built on top of plain RNNs, with the primary goal of addressing the vanishing gradient problem to enable modeling long-range dependencies. Looking at the following figure tells us that a plain RNN passes only the hidden state (the short-term memory), whereas an LSTM passes the hidden state as well as the explicit cell state (the long-term memory), giving it more power. So, when the term "**good**" is being processed in the LSTM, the recurrent layer also passes the cell states holding the long-term memory:

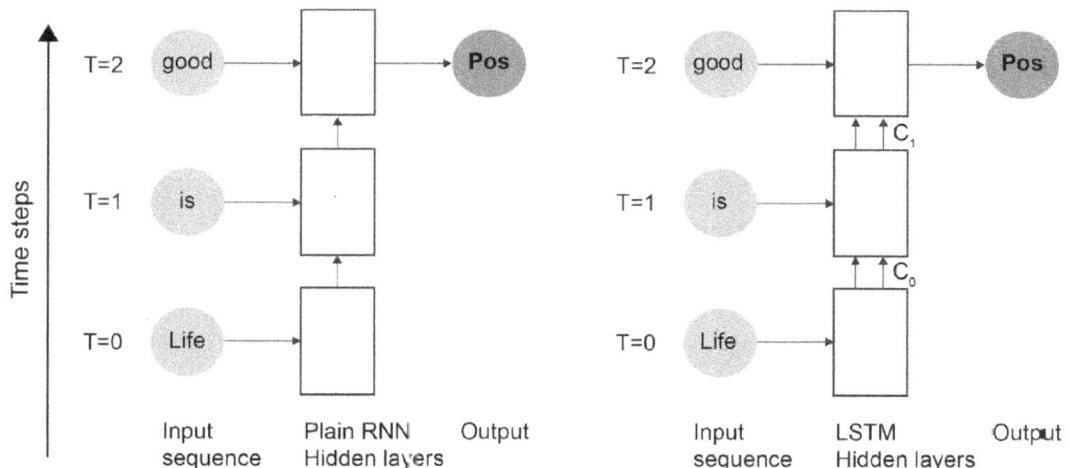

Figure 6.13: Plain RNNs (left) and LSTMs (right)

In practice, does this mean that you always need an LSTM? The answer to this question, as with most questions in data science and especially deep learning, is, "it depends". To understand these considerations, we need to understand the benefits and drawbacks of LSTMs compared to plain RNNs.

Benefits of LSTMs:

- More powerful, as it uses more parameters and an explicit cell state

- Models long-range dependencies better

Drawbacks of LSTMs:

- Many more parameters

- Takes more time to train

- More prone to overfitting

If you have long sequences to work with, LSTM would be a good choice. If you have a small dataset and the sequences you are dealing with are short (<10), then you're probably okay to use a plain RNN, owing to there being a lower number of parameters (although you could also try LSTMs, making sure to use regularization to avoid overfitting). A larger dataset with long sequences would probably extract the most out of powerful models such as LSTMs. Note that training LSTMs is computationally expensive and time-consuming, so if you have an extremely large dataset, training LSTMs may not be the most practical approach. Of course, all these statements should serve merely as guidance – the best approach would be what works best for your data and your task.

GATED RECURRENCE UNITS

In the previous section, we saw that LSTMs have a lot of parameters and seem much more complex than the regular RNN. You may be wondering, are all these apparent complications really necessary? Can the LSTM be simplified a little without it losing significant predictive power? Researchers wondered the same for a while, and in 2014, Kyunghyun Cho and their team proposed the GRU as an alternative to LSTMs in their paper (https://arxiv.org/abs/1406.1078) on machine translation.

GRUs are simplified forms of LSTMs and aim at reducing the number of parameters while retaining the power of the LSTM. In tasks around speech modeling and language modeling, GRUs provide the same performance as LSTMs, but with fewer parameters and faster training times.

One major simplification done in a GRU is the omission of the explicit cell state. This sounds counterintuitive considering that the freely flowing cell state was what gave the LSTM its power, right? What really gave LSTMs all that power was the freely flowing nature of the cell state and not the cell state itself? Indeed, if the cell state were also subject to activations, LSTMs probably wouldn't have had the success they did:

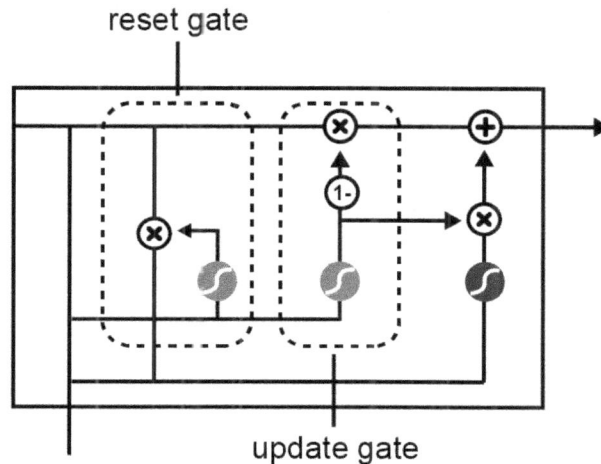

Figure 6.14: Gated Recurrent Unit

So, the freely flowing values is the key differentiating idea. GRUs retain this idea, by allowing the hidden state to flow freely. Let's look at the preceding figure to understand what this means. GRUs allow the hidden state to pass through freely. Another way to look at this is that GRUs effectively bring the idea of the cell state (as in LSTMs) to the hidden state.

We still need to regulate the flow of the hidden state, though, so we still have gates. GRUs combine the forget gate and update gate into a single update gate. To understand the motivation behind this, consider this – if we forget a cell state, and don't update it, what are we really doing? Maybe there is merit in having a single update operation. This is the second major difference in the architecture.

As a result of these two changes, GRUs have the data pass through three activations instead of four, as in LSTMs, reducing the number of parameters. While GRUs still have three times the number of parameters of a plain RNN, these have 75% of the parameters of LSTMs, and that is a welcome change. We still have information flowing freely through the network and this should allow us to model long-range dependencies.

Let's see how a GRU-based model performs on our task of sentiment classification.

EXERCISE 6.03: GRU-BASED SENTIMENT CLASSIFICATION MODEL

In this exercise, we will build a simple GRU-based model to predict sentiments in our data. We will continue with the same setup that we used previously (that is, the number of cells, embedding dimensions, dropout, and so on). Using GRUs instead of LSTMs in the model is as simple as replacing "**LSTM**" with "**GRU**" when adding the layer. Follow these steps to complete this exercise:

1. Import the **GRU** layer from Keras **layers**:

```
from tensorflow.keras.layers import GRU
```

2. Instantiate the sequential model, add the embedding layer with the appropriate dimensions, and add 40% spatial dropout:

```
model_gru = Sequential()
model_gru.add(Embedding(vocab_size, output_dim=32))
model_gru.add(SpatialDropout1D(0.4))
```

3. Add a GRU layer with 32 cells. Set the **reset_after** parameter to **False** (this is a minor TensorFlow 2 implementation detail in order to maintain consistency with the implementation of plain RNNs and LSTMs):

```
model_gru.add(GRU(32, reset_after=False))
```

4. Add the dropout (40%) and dense layers, compile the model, and print the model summary:

```
model_gru.add(Dropout(0.4))
model_gru.add(Dense(1, activation='sigmoid'))
model_gru.compile(loss='binary_crossentropy', \
                  optimizer='rmsprop', metrics=['accuracy'])
model_gru.summary()
```

The model summary is as fol ows:

```
Layer (type)                    Output Shape              Param #
=================================================================
embedding_4 (Embedding)         (None, None, 32)          256000

spatial_dropout1d_4 (Spatial    (None, None, 32)          0

gru_1 (GRU)                     (None, 32)                6240

dropout_4 (Dropout)             (None, 32)                0

dense_4 (Dense)                 (None, 1)                 33
=================================================================
Total params: 262,273
Trainable params: 262,273
Non-trainable params: 0
```

Figure 6.′5: Summary of the GRU model

From the summary of the GRU model, we can see that the number of parameters in the GRU layer is **6240**. You can check that this is exactly three times the number of parameters in the plain RNN layer we saw in *Exercise 6.01, Building and Training an RNN Model for Sentiment Classification*, and **0.75** times the parameters of the LSTM layer we saw in *Exercise 6.02, LSTM-Based Sentiment Classification Model* – again, this is in line with our expectations. Next, let's fit the model on the training data.

5. Fit on the training data for four epochs (which gives us the best result):

```
history_gru = model_gru.fit(X_train, y_train, \
                            batch_size=128, \
                            validation_split=0.2, \
                            epochs = 4)
```

The output from the training process is as follows:

```
Train on 20000 samples, validate on 5000 samples
Epoch 1/4
20000/20000 [==============================] - 33s 2ms/sample - loss: 0.5554 - acc: 0.7007 - val_loss: 0.4
152 - val_acc: 0.8116
Epoch 2/4
20000/20000 [==============================] - 30s 1ms/sample - loss: 0.3372 - acc: 0.8609 - val_loss: 0.3
966 - val_acc: 0.8448
Epoch 3/4
20000/20000 [==============================] - 30s 2ms/sample - loss: 0.2766 - acc: 0.8904 - val_loss: 0.3
375 - val_acc: 0.8656
Epoch 4/4
20000/20000 [==============================] - 30s 1ms/sample - loss: 0.2478 - acc: 0.9042 - val_loss: 0.2
981 - val_acc: 0.8792
```

Figure 6.16: GRU training output

Notice that training the GRUs also took much longer than plain RNNs but was faster than LSTMs. The validation accuracy is better than the plain RNN and seems close to that of the LSTM. Let's see how the model fares on the test data.

6. Make predictions on the test set and print the accuracy score:

```
y_test_pred = model_gru.predict_classes(X_test)
accuracy_score(y_test, y_test_pred)
```

The accuracy is printed out as follows:

```
0.87156
```

We can see that the accuracy of the GRU model (87.15%) is very similar to that of the LSTM (87%) and is higher than the plain RNN. This is an important point – GRUs are simplifications of LSTMs that aim to provide similar accuracy with fewer parameters. Our exercise here shows this is true.

> **NOTE**
>
> To access the source code for this specific section, please refer to https://packt.live/31ZPO2g.
>
> You can also run this example online at https://packt.live/2Oa2trm. You must execute the entire Notebook in order to get the desired result.

In this exercise, we saw how we can employ GRUs for the sentiment classification of text. The training time was slightly lower than the LSTM model and the number of parameters is lower. Even this simple architecture (without any hyperparameter tuning) gave better results than the plain RNN model and gave results similar to the LSTM model.

LSTM VERSUS GRU

So, which one should you choose? The LSTM has more parameters and an explicit cell state designed to store long-term memory. The GRU has fewer parameters, which means faster training, and also has a free-flowing cell state to allow it to model long-range dependencies.

An empirical evaluation (available at https://arxiv.org/abs/1412.3555) by Junyoung Chung, Yoshua Bengio, and their team in 2014 on music-modeling and speech-modeling tasks showed that both LSTMs and GRUs are markedly superior to plain RNNs. They also found that GRUs are on par with LSTMs in terms of performance. They remarked that tuning hyperparameters such as layer size is probably more important than choosing between LSTM and GRU.

In 2018, Gail Weiss, Yoav Goldberg, and their team demonstrated and concluded that LSTMs outperform GRUs in tasks that require unbounded counting, that is, those that need to handle sequences of an arbitrary length. The Google Brain team, in 2018, also showed that the performance of LSTMs is superior to GRUs when it comes to machine translation. This leads us to think that the extra power that LSTMs bring may be very useful in certain applications.

BIDIRECTIONAL RNNS

The RNN models we've just looked at – LSTMs, GRUs – are powerful indeed and provide extremely good results when it comes to sequence-processing tasks. Now, let's discuss how to make them even more powerful, and the methods that yield the amazing successes in deep learning that you have been hearing about.

Let's begin with the idea of bidirectional RNNs. The idea applies to all variants of RNNs, including, but not limited to, LSTMs and GRUs. Bidirectional RNNs process the sequence in both directions, allowing the network to have both backward and forward information about the sequence, providing it with a much richer context:

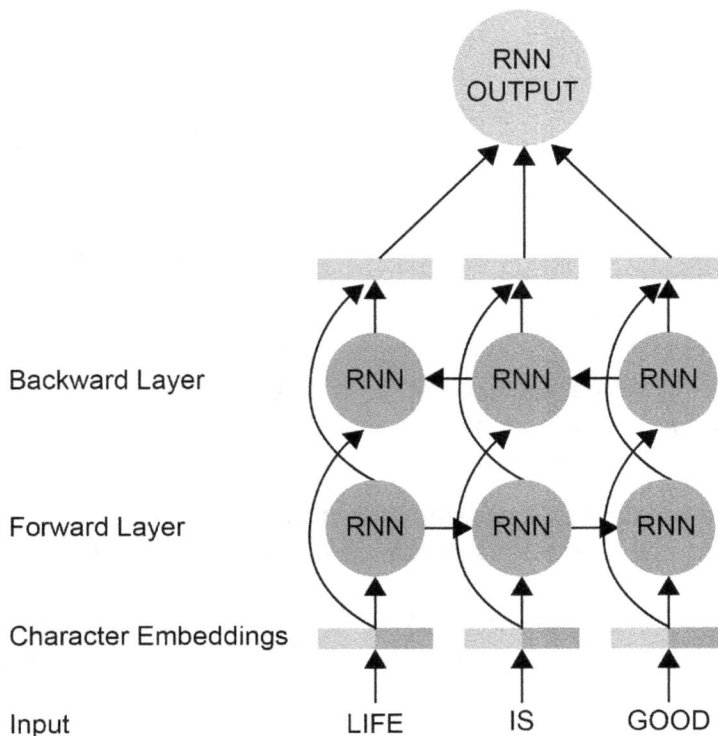

Figure 6.17: Bidirectional LSTM

The bidirectional model essentially employs two RNNs in parallel – one as the **"forward layer"** and the other as the **"backward layer"**. As shown in the preceding figure, the forward layer processes the sequence in the order of its elements. For the sentence, "*Life is good*", the forward layer will process the term "*Life*" first, followed by "*is*", followed by "*good*" – no different from the usual RNN layer. The backward layer reverses this order – it processes "good" first, followed by "is", followed by "Life". At each step, the states of the forward and the backward layers are concatenated to form the output.

What kind of tasks benefit the most from this architecture? Looking at both sides of the context helps resolve any ambiguity about the term at hand. When we read a statement such as "*The stars*", we're not sure as to what "*stars*" we're reading about – is it the stars in the sky or movie stars? But when we also see the terms coming later in the sequence and read "*The stars at the movie premiere*", we're confident that this sentence is about movie stars. The tasks that can benefit the most from such a setup are machine translation, parts-of-speech tagging, named entity recognition, and word prediction tasks, to list a few. Bidirectional RNNs show performance gains for general text classification tasks as well. Let's apply a bidirectional LSTM-based model to our sentiment classification task.

EXERCISE 6.04: BIDIRECTIONAL LSTM-BASED SENTIMENT CLASSIFICATION MODEL

In this exercise, we will use bidirectional LSTMs to predict sentiment on our data. We'll be using the bidirectional wrapper from Keras to create bidirectional layers on LSTMs (you could create a bidirectional GRU model by simply replacing **LSTM** with **GRU** in the wrapper). Follow these steps to complete this exercise:

1. Import the **Bidirectional** layer from Keras **layers**. This layer is essentially a wrapper you can use around other RNNs:

    ```
    from tensorflow.keras.layers import Bidirectional
    ```

2. Instantiate the sequential model, add the embedding layer with the appropriate dimensions, and add a 40% spatial dropout:

    ```
    model_bilstm = Sequential()
    model_bilstm.add(Embedding(vocab_size, output_dim=32))
    model_bilstm.add(SpatialDropout1D(0.4))
    ```

3. Add a **Bidirectional** wrapper to an LSTM layer with **32** cells:

    ```
    model_bilstm.add(Bidirectional(LSTM(32)))
    ```

4. Add the dropout (40%) and dense layers, compile the model, and print the model summary:

    ```
    model_bilstm.add(Dropout(0.4))
    model_bilstm.add(Dense(1, activation='sigmoid'))
    model_bilstm.compile(loss='binary_crossentropy', \
                         optimizer='rmsprop', metrics=['accuracy'])
    model_bilstm.summary()
    ```

The summary is as follows:

```
Layer (type)                    Output Shape              Param #
=================================================================
embedding_6 (Embedding)         (None, None, 32)          256000

spatial_dropout1d_6 (Spatial    (None, None, 32)          0

bidirectional_1 (Bidirection    (None, 64)                16640

dropout_6 (Dropout)             (None, 64)                0

dense_6 (Dense)                 (None, 1)                 65
=================================================================
Total params: 272,705
Trainable params: 272,705
Non-trainable params: 0
```

Figure 6.18: Summary of the bidirectional LSTM model

Note the parameters of the model shown in the preceding screenshot. Not surprisingly, the bidirectional LSTM layer has **16640** parameters – twice the number of parameters that the LSTM layer (**8320** parameters) had in *Exercise 6.02, LSTM-Based Sentiment Classification Model*. This is eight times the parameters of the plain RNN. Next, let's fit the model on the training data.

5. Fit the training data for four epochs with a batch size of **128**:

```
history_bilstm = model_bilstm.fit(X_train, y_train, \
                                  batch_size=128, \
                                  validation_split=0.2, \
                                  epochs = 4)
```

The output from training is as follows:

```
Train on 20000 samples, validate on 5000 samples
Epoch 1/4
20000/20000 [==============================] - 69s 3ms/sample - loss: 0.5722 - acc: 0.7048 - val_los
s: 0.3745 - val_acc: 0.8472
Epoch 2/4
20000/20000 [==============================] - 67s 3ms/sample - loss: 0.3563 - acc: 0.8561 - val_los
s: 0.3135 - val_acc: 0.8716
Epoch 3/4
20000/20000 [==============================] - 67s 3ms/sample - loss: 0.2833 - acc: 0.8929 - val_los
s: 0.3043 - val_acc: 0.8740
Epoch 4/4
20000/20000 [==============================] - 67s 3ms/sample - loss: 0.2446 - acc: 0.9084 - val_los
s: 0.2857 - val_acc: 0.8802
```

Figure 6.19: Bidirectional LSTM training output

Notice that, as we expect, training bidirectional LSTMs takes much longer than regular LSTMs, and several times longer than plain RNNs. The validation accuracy seems to be closer to the LSTM's accuracy.

6. Make predictions on the test set and print the accuracy score:

```
y_test_pred = model_bilstm.predict_classes(X_test)
accuracy_score(y_test, y_test_pred)
```

The accuracy is as follows:

```
0.877
```

The accuracy we received here (87.7%) is a slight improvement over the LSTM model's accuracy, which was 87%. Again, you can tune the hyperparameters even further to extract the most out of this powerful architecture. Note that we had twice the number of parameters compared to the LSTM model, and eight times the parameters of the plain RNN. Working with a large dataset may make the performance differences bigger.

> **NOTE**
>
> To access the source code for this specific section, please refer to https://packt.live/31ZPO2g.
>
> You can also run this example online at https://packt.live/2Oa2trm. You must execute the entire Notebook in order to get the desired result.

STACKED RNNS

Now, let's look at another approach we can follow to extract more power from RNNs. In all the models we've looked at in this chapter, we've used a single layer for the RNN layer (plain RNN, LSTM, or GRU). Going deeper, that is, adding more layers, has typically helped us for feedforward networks so that we can learn more complex patterns/features in the deeper layers. There is merit in trying this idea for recurrent networks. Indeed, stacked RNNs do seem to give us more predictive power.

The following figure illustrates a simple two-layer stacked LSTM model. Stacking RNNs simply means feeding the output of one RNN layer to another RNN layer. The RNN layers can output sequences (that is, output at each time step) and these can be fed, like any input sequence, into the subsequent RNN layer. In terms of implementation through code, stacking RNNs is as simple as returning sequences from one layer, and providing this as input to the next RNN layer, that is, the immediate next layer:

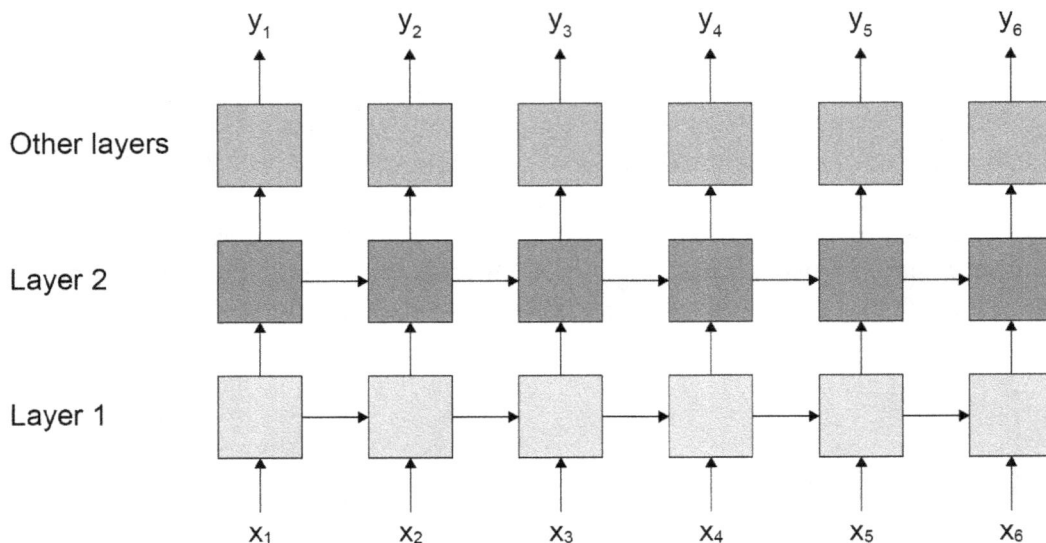

Figure 6.20: Two-layer stacked RNN

Let's see the stacked RNN (LSTM) in action by using it on our sentiment classification task.

EXERCISE 6.05: STACKED LSTM-BASED SENTIMENT CLASSIFICATION MODEL

In this exercise, we will "go deeper" into the RNN architecture by stacking two LSTM layers to predict sentiment in our data. We will continue with the same setup that we used in the previous exercises (the number of cells, embedding dimensions, dropout, and so on) for the other layers. Follow these steps to complete this exercise:

1. Instantiate the sequential model, add the embedding layer with the appropriate dimensions, and add 40% spatial dropout:

```
model_stack = Sequential()
model_stack.add(Embedding(vocab_size, output_dim=32))
model_stack.add(SpatialDropout1D(0.4))
```

2. Add an LSTM layer with **32** cells. Make sure to specify **return_sequences** as **True** in the LSTM layer. This will return the output of the LSTM at each time step, which can then be passed to the next LSTM layer:

```
model_stack.add(LSTM(32, return_sequences=True))
```

3. Add another LSTM layer with **32** cells. This time, you don't need to return sequences. You can either specify the **return_sequences** option as **False** or skip it altogether (the default value is **False**):

```
model_stack.add(LSTM(32, return_sequences=False))
```

4. Add the dropout (50% dropout; this is higher since we're building a more complex model) and dense layers, compile the model, and print the model summary:

```
model_stack.add(Dropout(0.5))
model_stack.add(Dense(1, activation='sigmoid'))
model_stack.compile(loss='binary_crossentropy', \
                    optimizer='rmsprop', \
                    metrics=['accuracy'])
model_stack.summary()
```

The summary is as follows:

```
Layer (type)                 Output Shape              Param #
=================================================================
embedding_9 (Embedding)      (None, None, 32)          256000

spatial_dropout1d_9 (Spatial (None, None, 32)          0

lstm_7 (LSTM)                (None, None, 32)          8320

lstm_8 (LSTM)                (None, 32)                8320

dropout_9 (Dropout)          (None, 32)                0

dense_9 (Dense)              (None, 1)                 33
=================================================================
Total params: 272,673
Trainable params: 272,673
Non-trainable params: 0
```

Figure 6.21: Summary of the stacked LSTM model

Note that the stacked LSTM model has the same number of parameters as the bidirectional model. Let's fit the model on the training data.

5. Fit the model on the training data for four epochs:

```
history_stack = model_stack.fit(X_train, y_train, \
                                batch_size=128, \
                                validation_split=0.2, \
                                epochs = 4)
```

The output from training is as follows:

```
Train on 20000 samples, validate on 5000 samples
Epoch 1/4
20000/20000 [==============================] - 76s 4ms/sample - loss: 0.5237 - acc: 0.7286 - val_loss: 0.3
558 - val_acc: 0.8552
Epoch 2/4
20000/20000 [==============================] - 72s 4ms/sample - loss: 0.3284 - acc: 0.8687 - val_loss: 0.3
360 - val_acc: 0.8702
Epoch 3/4
20000/20000 [==============================] - 72s 4ms/sample - loss: 0.2742 - acc: 0.8944 - val_loss: 0.3
436 - val_acc: 0.8666
Epoch 4/4
20000/20000 [==============================] - 72s 4ms/sample - loss: 0.2354 - acc: 0.9096 - val_loss: 0.2
827 - val_acc: 0.8826
```

Figure 6.22: Stacked LSTM training output

Training stacked LSTMs took less time than training bidirectional LSTMs. The validation accuracy seems to be close to that of the bidirectional LSTM model.

6. Make predictions on the test set and print the accuracy score:

```
y_test_pred = model_stack.predict_classes(X_test)
accuracy_score(y_test, y_test_pred)
```

The accuracy is printed out as follows:

```
0.87572
```

The accuracy of **87.6%** is an improvement over the LSTM model (**87%**) and is practically the same as that of the bidirectional model (**87.7%**). This is a somewhat significant improvement over the performance of the regular LSTM model, considering that we're working with a rather small dataset. The larger your dataset is, the more you can benefit from these sophisticated architectures. Try tuning the hyperparameters in order to get the most out of this powerful architecture.

> **NOTE**
>
> To access the source code for this specific section, please refer to https://packt.live/31ZPO2g.
>
> You can also run this example online at https://packt.live/2Oa2trm. You must execute the entire Notebook in order to get the desired result.

SUMMARIZING ALL THE MODELS

In this chapter, we've looked at different variants of RNNs – from plain RNNs to LSTMs to GRUs. We also looked at the bidirectional approach and the stacking approach to using RNNs. Now is a good time to take a holistic look at things and make a comparison between the models. Let's look at the following table, which compares the five models in terms of parameters, training time, and performance (that is, the level of accuracy on our dataset):

Model	RNN layer parameters	Training time	Test accuracy
Plain RNN	2,080	Low	85.1%
LSTM	8,320	High	87.0%
GRU	6,240	Medium-High	87.1%
Bi-directional LSTM	16,640	Very High	87.7%
Stacked LSTM	16,640	Very High	87.6%

Figure 6.23: Comparing the five models

> **NOTE**
>
> As mentioned earlier in the chapter, while working through the practical elements, you may have obtained values different from the ones shown above; however, the test accuracies you obtain should largely agree with ours. If the model's performance is very different, you may want to tweak the number of epochs.

Plain RNNs are the lowest on parameters and have the lowest training times but have the lowest accuracy of all the models. This is in line with our expectations – we are dealing with sequences that are 200 characters in length, and we know not to expect much from plain RNNs, and that gated RNNs (LSTMs, GRUs) are more suitable. Indeed, LSTMs and GRUs do perform significantly better than plain RNNs. But the accuracy comes at the cost of significantly higher training times, and several times the parameters, making these models more prone to overfitting.

The approaches of stacking and using bidirectional processing seem to provide an incremental benefit in terms of predictive power, but this is at the cost of significantly higher training times and several times the parameters. The stacked and bidirectional approaches gave us the highest accuracy, even on this small dataset.

While the performance results are specific to our dataset, the gradation in performance we see here is fairly common. The stacked and bidirectional models are present in many of the solutions today that provide state-of-the-art results in various tasks. With a larger dataset and when working with much longer sequences, we would expect the differences in model performances to be larger.

ATTENTION MODELS

Attention models were first introduced in late 2015 by Dzmitry Bahdanau, KyungHyun Cho, and Yoshua Bengio in their influential and seminal paper (https://arxiv.org/abs/1409.0473) that demonstrated the state-of-the-art results of English-to-French translation. Since then, this idea has been used for many sequence-processing tasks with great success, and attention models are becoming increasingly popular. While a detailed explanation and mathematical treatment is beyond the scope of this book, let's understand the intuition behind the idea that is considered by many big names in the field of deep learning as a significant development in our approach to sequence modeling.

The intuition behind attention can be best understood using an example from the task it was developed for – translation. When a novice human translates a long sentence between languages, they don't translate the entire sentence in one go. They break the original sentence down into smaller, manageable chunks, thereby generating a translation for each chunk sequentially. For each chunk, there would be a part that is the most important for the translation task, that is, where you need to pay the most attention:

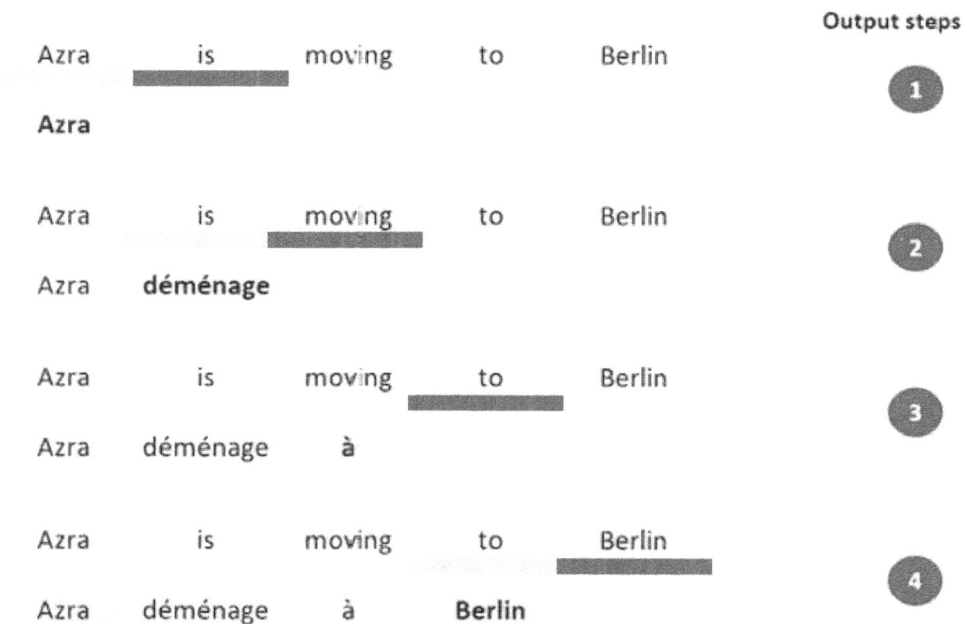

Figure 6.24: Idea of attention simplified

The preceding figure shows a simple example where we're translating the sentence, "**Azra is moving to Berlin**", into French. The French translation is, "**Azra déménage à Berlin**". To get the first term in the French translation, "**Azra**", we need to pay attention primarily to the first term in the original sentence (underscored by a light gray line) and maybe a bit to the second (underscored by a dark gray line) – these terms get higher importance (weight). The remaining parts of the sentence aren't relevant. Similarly, to generate the term "**déménage**" in the output, we need to pay attention to the terms "**is**" and "**moving**". The importance of each term toward the output term is expressed as weights. This is known as "**alignment**".

These alignments can be seen in the following figure, which was sourced from the original paper (https://arxiv.org/abs/1409.0473). It beautifully demonstrates what the model identified as most important for each term in the output. A lighter color in a cell in the grid means a higher weight for the corresponding input term in the column. We can see that for the output term "**marin**", the model correctly identifies "**marine**" as the most important input term to pay attention to. Similarly, it has identified "**environment**" as the most important term for "**environnement**", "**known**" for "**connu**", and so on. Pretty neat, isn't it?

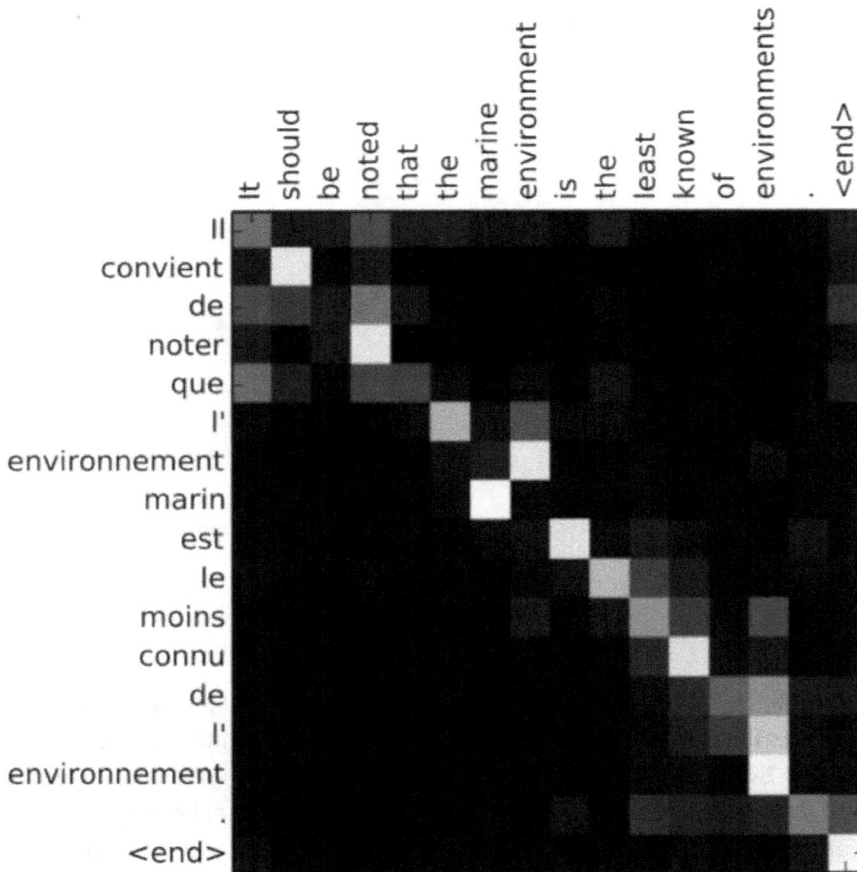

Figure 6.25: The alignment learned by the model

While attention models were originally designed for translation tasks, the models have been employed on a variety of other tasks with good success. That being said, note that the attention models have a very high number of parameters. The models are typically employed on bidirectional LSTM layers and add additional weights for the importance values. A high number of parameters makes the model more prone to overfitting, which means they will need much larger datasets to utilize their power.

MORE VARIANTS OF RNNS

We've seen quite a few variations of RNNs in this chapter – covering all the prominent ones and the major upcoming (in terms of popularity) variations. Sequence modeling and its associated architectures are a hot area of research, and we see plenty of developments coming in every year. Many variants aim to make lighter models with fewer parameters that aren't as hardware hungry as current RNNs. **Clockwork RNNs (CWRNNs)** are a recent development and show great success. There are also **Hierarchal Attention Networks** built on the idea of attention, but ultimately also propose that you shouldn't use RNNs as building blocks. There's a lot going on in this exciting area, so keep your eyes and ears open for the next big idea.

ACTIVITY 6.01: SENTIMENT ANALYSIS OF AMAZON PRODUCT REVIEWS

So far, we've looked at the variants of RNNs and used them to predict sentiment on movie reviews from the IMDb dataset. In this activity, we will build a sentiment classification model on Amazon product reviews. The data contains reviews for several categories of products. The original dataset, available at https://snap.stanford.edu/data/web-Amazon.html, is huge; therefore, we have sampled 50,000 reviews for this activity.

> **NOTE**
>
> The sampled dataset, which has been split into train and test sets, can be found at https://packt.live/3iNTUjN.

This activity will bring together the concepts and methods we discussed in this chapter and those discussed in *Chapter 4, Deep Learning for Text – Embeddings*, and *Chapter 5, Deep Learning for Sequences*. You will begin by performing a detailed text cleanup and conduct preprocessing to get it ready for the deep learning model. You will also use embeddings to represent text. For the prediction part, you will employ stacked LSTMs (two layers) and two dense layers.

For convenience (and awareness), you will also utilize the **Tokenizer** API from TensorFlow (Keras) to convert the cleaned-up text into the corresponding sequences. The **Tokenizer** combines the function of the tokenizer from **NLTK** with the **vectorizer (CountVectorizer/ TfIdfVectorizer)** by tokenizing the text first and then learning a vocabulary from a dataset. Let's see it in action by creating some toy data using the following command:

```
sents = ["life is good", "good life", "good"]
```

The **Tokenizer** can be imported, instantiated, and fit on the toy data using the following commands:

```
tok = Tokenizer()
tok.fit_on_texts(sents)
```

Once the vocabulary has been trained on the toy data (index learned for each term), we can convert the input text into a corresponding sequence of indices for the terms. Let's convert the toy data into the corresponding sequences of indices using the **texts_to_sequences** method of the tokenizer:

```
tok.texts_to_sequences(sents)
```

We'll get the following output:

```
[[2, 3, 1], [1, 2], [1]]
```

Now, the data format is the same as that of the IMDb dataset we've used throughout this chapter, and it can be processed in a similar fashion.

With this, you are now ready to get started. The following are the high-level steps you will need to follow to complete this activity:

1. Read in the data files for the train and test sets (**Amazon_reviews_train. csv** and **Amazon_reviews_test.csv**). Examine the shapes of the datasets and print out the top five records from the train data.

2. For convenience when it comes to processing, separate the raw text and the labels for the train and test set. Print the first two reviews from the train text. You should have the following four variables: **train_raw** comprising the raw text for the train data, **train_labels** with labels for the train data, **test_raw** containing raw text for the test data, and **test_labels** with labels for the test data.

3. Normalize the case and tokenize the test and train texts using NLTK's **word_tokenize** (after importing it, of course – hint: use list comprehension for cleaner code). Print the first review from the train data to check if the tokenization worked. Download **punkt** from NLTK if you haven't used the tokenizer before.

4. Import **stopwords** (built in to NLTK) and punctuation from the string module. Define a function (**drop_stop**) to remove these tokens from any input tokenized sentence. Download **stopwords** from NLTK if you haven't used it before.

5. Using the defined function (**drop_stop**), remove the redundant stop words from the train and the test texts. Print the first review of the processed train texts to check if the function worked.

6. Using **Porter Stemmer** from NLTK, stem the tokens for the train and test data.

7. Create the strings for each of the train and text reviews. This will help us work with the utilities in Keras to create and pad the sequences. Create the **train_texts** and **test_texts** variables. Print the first review from the processed train data to confirm it.

8. From the Keras preprocessing utilities for text (**keras.preprocessing. text**), import the **Tokenizer** module. Define a vocabulary size of **10000** and instantiate the tokenizer with this vocabulary.

9. Fit the tokenizer on the train texts. This works just like **CountVectorizer** did in *Chapter 4, Deep Learning for Text – Embeddings*, and trains the vocabulary. After fitting, use the **texts_to_sequences** method of the tokenizer on the train and test sets to create the sequences for them. Print the sequence for the first review in the train data.

10. We need to find the optimal length of the sequences to process in the model. Get the length of the reviews from the train set into a list and plot the histogram of the lengths.

11. The data is now in the same format as the IMDb data we used in the chapter. Using a sequence length of **100** (define the **maxlen = 100** variable), use the **pad_sequences** method from the **sequence** module in Keras' preprocessing utilities (**keras.preprocessing.sequence**) to limit the sequences to **100** for both the train and test data. Check the shape of the result for the train data.

12. To build the model, import all the necessary layers from Keras (**embedding**, **spatialdropout**, **LSTM**, **dropout**, and **dense**) and import the **Sequential** model. Initialize the **Sequential** model.

13. Add an embedding layer with **32** as the vector size (**output_dim**). Add a spatial dropout of **40%**.

14. Build a stacked LSTM model with **2** layers with **64** cells each. Add a dropout layer with **40%** dropout.

15. Add a dense layer with **32** neurons with **relu** activation, then a **50%** dropout layer, followed by another dense layer of **32** neurons with **relu** activation, and follow this up with another dropout layer with **50%** dropout.

16. Add a final dense layer with a single neuron with **sigmoid** activation and compile the model. Print the model summary.

17. Fit the model on the training data with a **20%** validation split and a batch size of **128**. Train for **5** epochs.

18. Make a prediction on the test set using the **predict_classes** method of the model. Using the **accuracy_score** method from **scikit-learn**, calculate the accuracy on the test set. Also, print out the confusion matrix.

With the preceding parameters, you should get about **86%** accuracy. With some hyperparameter tuning, you should be able to get a significantly higher accuracy.

> **NOTE**
>
> The detailed steps for this activity, along with the solutions and additional commentary, are presented on page 416.

SUMMARY

In this chapter, we started by understanding the reasons for plain RNNs not being practical for very large sequences – the main culprit being the vanishing gradient problem, which makes modeling long-range dependencies impractical. We saw the LSTM as an update that performs extremely well for long sequences, but it is rather complicated and has a large number of parameters. GRU is an excellent alternative that is a simplification over LSTM and works well on smaller datasets.

Then, we started looking at ways to extract more power from these RNNs by using bidirectional RNNs and stacked layers of RNNs. We also discussed attention mechanisms, a significant new approach that provides state-of-the-art results in translation but can also be employed on other sequence-processing tasks. All of these are extremely powerful models that have changed the way several tasks are performed and form the basis for models that produce state-of-the-art results. With active research in the area, we expect things to only get better as more novel variants and architectures are released.

Now that we've discussed a variety of powerful modeling approaches, in the next chapter, we will be ready to discuss a very interesting topic in the deep learning domain that enables AI to be creative – **Generative Adversarial Networks**.

7

GENERATIVE ADVERSARIAL NETWORKS

INTRODUCTION

In this chapter, you will embark on another interesting topic within the deep learning domain: **Generative Adversarial Networks** (**GANs**). You will get introduced to GANs and their basic components, along with some of their use cases. This chapter will give you hands-on experience of creating a GAN to generate a data distribution produced by a sine function. You will also be introduced to deep convolutional GANs and will perform an exercise to generate an MNIST data distribution. By the end of this chapter, you will have tested your understanding of GANs by generating the MNIST fashion dataset.

INTRODUCTION

The power of creativity was always the exclusive domain of the human mind. This was one of the facts touted as one of the major differences between the human mind and the artificial intelligence domain. However, in the recent past, deep learning has been making baby steps in the path to being creative. Imagine you were at the Sistine Chapel in the Vatican and were looking up with bewilderment at the frescos immortalized by Michelangelo, wishing your deep learning models were able to recreate something like that. Well, maybe 10 years back, people would have scoffed at your thought. Not anymore, though – deep learning models have made great strides in regenerating immortal works. Applications like these are made possible by a class of networks called **Generative Adversarial Networks** (**GANs**).

Many applications have been made possible with GANs. Take a look at the following image:

Figure 7.1: Image translation using GANs

> **NOTE:**
>
> The preceding image is sourced from the research paper titled *Image-to-Image Translation with Conditional Adversarial Networks*: Phillip Isola, Jun-Yan Zhu, Tinghui Zhou, Alexei A. Efros, available at https://arxiv.org/pdf/1611.07004.pdf.

The preceding image demonstrates how the input image, which has a very different color scheme, has been transformed by the GAN into an image that looks very similar to the real one. This application of GANs is called image translation.

In addition to these examples, many other use cases are finding traction. Some of the notable ones are as follows:

- Synthetic data generation for data augmentation

- Generating cartoon characters

- Text to image translation

- Three-dimensional object generation

The list goes on. As the days go by, applications of GANs increasingly become mainstream.

So, what exactly are GANs? What are the inner dynamics of GANs? How do you generate images or other data distributions from totally unconnected distributions? In this chapter, we'll find out the answers to those questions.

In the previous chapter, we learned about **recurrent neural networks** (**RNNs**), a class of deep learning networks used for sequence data. In this chapter, we will embark on a fascinating safari to the world of GANs. First, we will start with an introduction to GANs. Then, we'll focus on generating a data distribution that is similar to a known mathematical expression. We'll then move on to **deep convolutional GANs** (**DCGANs**). To see how well our generative models work, we will generate a data distribution similar to the MNIST handwritten digits. We'll start this journey by learning about GANs.

> **NOTE**
>
> Depending on your system configuration, some of the exercises and activities in this chapter may take quite a long time to execute.

KEY COMPONENTS OF GENERATIVE ADVERSARIAL NETWORKS

GANs are used to create a data distribution from random noise data and make it look similar to a real data distribution. GANs are a family of deep neural networks that comprise two networks that are competing against each other. One of these networks is called the **generator network**, while the other is called the **discriminator network**. The functions of these two networks are to compete against each other to generate a probability distribution that closely mimics an existing probability distribution. To state an example of generating a new probability distribution, let's say we have a collection of images of cats and dogs (real images). Using a GAN, we can generate a different set of images (fake images) of cats and dogs from a very random distribution of numbers. The success of a GAN is in generating the best set of cat and dog images to the point that it is difficult for people to differentiate between the fake ones and the real ones.

Another example where GANs can become useful is in data privacy. The data of companies, especially in domains such as finance and healthcare, is extremely sensitive. However, there might be instances where data has to be shared with third parties for research purposes. In such scenarios, to maintain the confidentiality of data, companies can use GANs to generate datasets that are similar in nature to their existing datasets. There is a multitude of such business use cases where GANs can come in really handy.

Let's understand GANs better by mapping out some of their components, as shown in the following diagram:

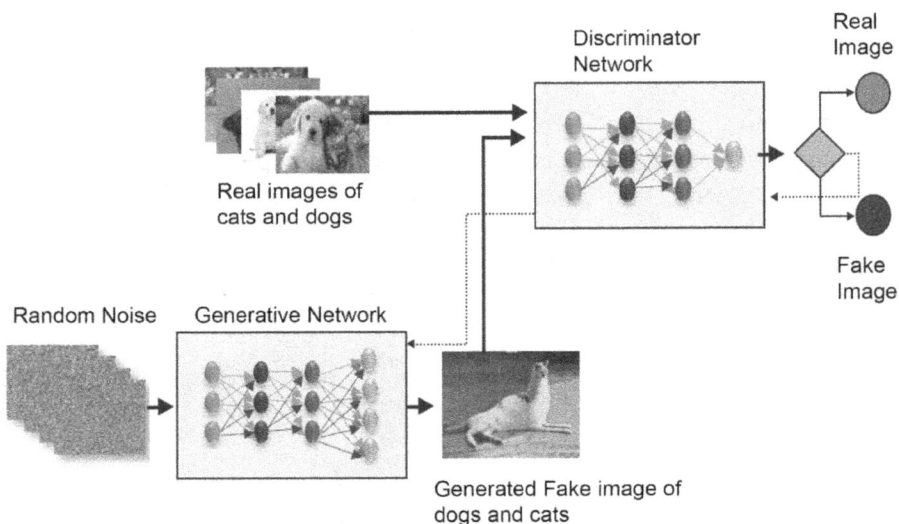

Figure 7.2: Example of GAN structure

The preceding figure provides a concise overview of the components of a GAN and how they come in handy in generating fake images from real ones. Let's understand the process in the context of the preceding diagram:

1. The set of images at the top-left corner of the preceding figure represents a probability distribution of real data (for example, MNIST, images of cats and dogs, pictures of human faces, and more).

2. The generative network shown in the bottom-left part of the diagram generates fake images (probability distributions) from a random noise distribution.

3. The trained discriminative network classifies whether the image that is fed in is fake or real.

4. A feedback loop (the diamond-shaped box) working through the backpropagation algorithm gives feedback to the generator network, thereby refining the parameters of the generator model.

5. The parameters continue to be refined until the discriminator network can't discriminate between the fake images and the real ones.

Now that we have an overview of each of the components, let's dive deeper and understand them better through a problem statement.

PROBLEM STATEMENT – GENERATING A DISTRIBUTION SIMILAR TO A GIVEN MATHEMATICAL FUNCTION

In this problem, we will use GANs to generate a distribution that is similar to a data distribution from a mathematical function. The function we will be using to generate the real data is a simple **sine wave**. We will train a GAN to generate a fake distribution of data that will be similar to the data we generated from the known mathematical function. We will progressively build each component that's required while we traverse the solution for this problem statement.

The process we will follow is explained in the following figure. We will follow a pedagogical approach as per the steps detailed in this figure:

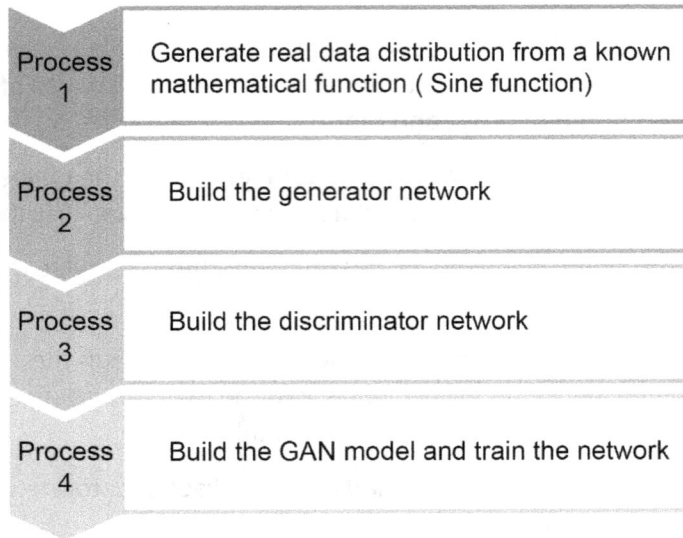

Process 1	Generate real data distribution from a known mathematical function (Sine function)
Process 2	Build the generator network
Process 3	Build the discriminator network
Process 4	Build the GAN model and train the network

Figure 7.3: Four-step process to building a GAN from a known function

Now, let's explore each of these processes.

PROCESS 1 – GENERATING REAL DATA FROM THE KNOWN FUNCTION

To begin our journey, we need a real distribution of data. This distribution of data will comprise two features – the first one is the sequence and the second one is the sine of the sequence. The first feature is a sequence of data points spaced at equal intervals. To generate this sequence, we need to randomly generate a data point from a normal distribution and then find other numbers spaced in sequence at equal intervals. The second feature will be the **sine()** of the first feature. Both these features will form our real data distribution. Before we get into an exercise that generates the real dataset, let's look at some of the functions in the **numpy** library we will use in this process.

Random Number Generation

First, we will generate a random number from a normal distribution using the following function:

```
numpy.random.normal(loc,scale,size)
```

This function takes three arguments:

- **loc**: This is the mean of the data distribution.

- **scale**: This is the standard deviation of the data distribution.

- **size**: This defines the number of data points we want.

Arranging the Data into a Sequence

To arrange data in a sequence, we use the following function:

```
numpy.arange(start,end,spacing)
```

The arguments are the following:

- **start**: This is the point that the sequence should start from.

- **end**: The point where the sequence ends.

- **spacing**: The frequency between each successive number in the sequence. For example, if we start off with 1 and generate a series with a spacing of **0.1**, the series will look as follows:

```
1, 1.1,1.2 ........
```

Generating the Sine Wave

To generate the sine of a number, we use the following command:

```
numpy.sine()
```

Let's use these concepts in the following exercise and learn how to generate a real data distribution.

EXERCISE 7.01: GENERATING A DATA DISTRIBUTION FROM A KNOWN FUNCTION

In this exercise, we will generate a data distribution from a simple sine function. By completing this exercise, you will learn how to generate a random number from a normal distribution and create a sequence of equally spaced data with the random number as its center. This sequence will be the first feature. The second feature will be created by calculating the **sine()** for the first feature. Follow these steps to complete this exercise:

1. Open a new Jupyter Notebook and name it **Exercise 7.01**. Run the following command to import the necessary library packages:

```
# Importing the necessary library packages
import numpy as np
```

2. Generate a random number from a normal distribution that has a mean of 3 and a standard deviation of 1:

```
"""
Generating a random number from a normal distribution
with mean 3 and sd = 1
"""
np.random.seed(123)
loc = np.random.normal(3,1,1)
loc
```

> **NOTE**
>
> The triple-quotes (""") shown in the code snippet above are used to denote the start and end points of a multi-line code comment. Comments are added into code to help explain specific bits of logic.

For reproducing the results, we use **random.seed(123)**.

You should get the following output:

```
array([1.9143694])
```

3. Using the previously generated random number as a midpoint, we will generate equal sequences of numbers to the right and left of the midpoint. We will generate a batch of 128 numbers. So, we take 64 numbers each to the right and left of the midpoint with a spacing of 0.1. The following code generates a sequence to the right of the midpoint:

```
# Generate numbers to right of the mid point
xr = np.arange(loc,loc+(0.1*64),0.1)
```

4. Generate 64 numbers to the left of the midpoint:

```
# Generate numbers to left of the random point
xl = np.arange(loc-(0.1*64),loc,0.1)
```

5. Concatenate both these sequences to generate the first feature:

```
# Concatenating both these numbers
X1 = np.concatenate((xl,xr))
print(X1)
```

You should get an output similar to the one shown here:

```
[-4.4856306 -4.3856306 -4.2856306 -4.1856306 -4.0856306 -3.9856306
 -3.8856306 -3.7856306 -3.6856306 -3.5856306 -3.4856306 -3.3856306
 -3.2856306 -3.1856306 -3.0856306 -2.9856306 -2.8856306 -2.7856306
 -2.6856306 -2.5856306 -2.4856306 -2.3856306 -2.2856306 -2.1856306
 -2.0856306 -1.9856306 -1.8856306 -1.7856306 -1.6856306 -1.5856306
 -1.4856306 -1.3856306 -1.2856306 -1.1856306 -1.0856306 -0.9856306
 -0.8856306 -0.7856306 -0.6856306 -0.5856306 -0.4856306 -0.3856306
 -0.2856306 -0.1856306 -0.0856306  0.0143694  0.1143694  0.2143694
  0.3143694  0.4143694  0.5143694  0.6143694  0.7143694  0.8143694
  0.9143694  1.0143694  1.1143694  1.2143694  1.3143694  1.4143694
  1.5143694  1.6143694  1.7143694  1.8143694  1.9143694  2.0143694
  2.1143694  2.2143694  2.3143694  2.4143694  2.5143694  2.6143694
  2.7143694  2.8143694  2.9143694  3.0143694  3.1143694  3.2143694
  3.3143694  3.4143694  3.5143694  3.6143694  3.7143694  3.8143694
  3.9143694  4.0143694  4.1143694  4.2143694  4.3143694  4.4143694
  4.5143694  4.6143694  4.7143694  4.8143694  4.9143694  5.0143694
  5.1143694  5.2143694  5.3143694  5.4143694  5.5143694  5.6143694
  5.7143694  5.8143694  5.9143694  6.0143694  6.1143694  6.2143694
  6.3143694  6.4143694  6.5143694  6.6143694  6.7143694  6.8143694
  6.9143694  7.0143694  7.1143694  7.2143694  7.3143694  7.4143694
  7.5143694  7.6143694  7.7143694  7.8143694  7.9143694  8.0143694
  8.1143694  8.2143694]
```

Figure 7.4: Sequence of numbers with equal spacing

The preceding is the distribution of **128** numbers equally spaced from one another. This sequence will be our first feature for the data distribution.

6. Generate the second feature, which is the **sine()** of the first feature:

```
# Generate second feature
X2 = np.sin(X1)
```

7. Plot the distribution:

```
# Plot the distribution
import matplotlib.pyplot as plot
plot.plot(X1, X2)
plot.xlabel('Data Distribution')
plot.ylabel('Sine of data distribution')
plot.show()
```

You should get the following output:

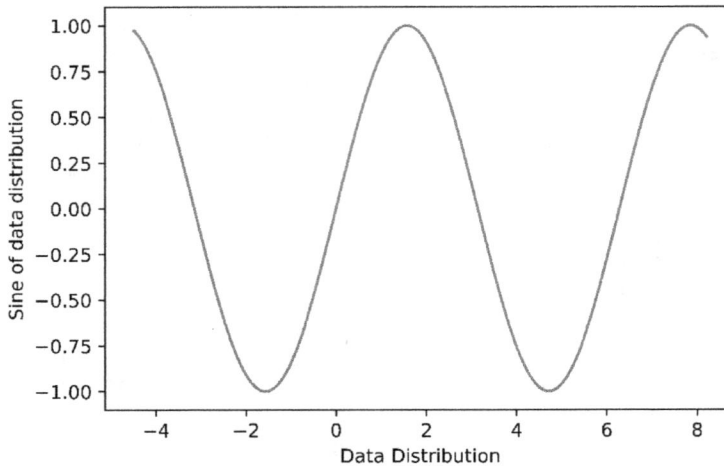

Figure 7.5: Plot for the sine function

The preceding plot shows the distribution that you would be trying to mimic using GANs.

8. Reshape each feature before concatenating them:

```
# Reshaping the individual data sets
X1 = X1.reshape(128,1)
X2 = X2.reshape(128,1)
```

9. Concatenate both features to form a single DataFrame:

```
# Concatenate both features to form the real data set
realData = np.concatenate((X1,X2),axis=1)
realData.shape
```

You should get the following output:

```
(128, 2)
```

> **NOTE**
>
> To access the source code for this specific section, please refer to https://packt.live/3gHhv42.
>
> You can also run this example online at https://packt.live/2O62M6r. You must execute the entire Notebook in order to get the desired result.

In this exercise, we created a data distribution from a mathematical function. We will be using this data distribution later to train the GAN to generate a distribution similar to this. In a production environment, you will be provided with a real dataset, similar to the MNIST or **Imagenet** datasets. In this case, our real dataset is a known mathematical function. Later in this chapter, we will use some random noise data and train the GAN to make that random noise data similar to this real data distribution.

Now that we have seen the real data distribution, the next section will be all about creating a basic generative network.

PROCESS 2 – CREATING A BASIC GENERATIVE NETWORK

In the previous process, we worked on an example that will generate a distribution from a known function. As we mentioned earlier, the purpose of the generative network is to sample data from any arbitrary distribution and then transform that data into generative samples that look similar to the known distribution.

The way the generative network achieves this is through the dynamics of the generator, the discriminator, and the training process. The success of the generative network relies on its ability to create data distributions that the discriminator can't differentiate between – in other words, it can't determine whether the distribution is fake or not. This ability of the generative network to create distributions that can fool the discriminator is acquired by the training process. We will talk more about the discriminator and the training process later in this chapter. For now, let's see how a generator network can be constructed to generate fake data distributions from some random distribution.

BUILDING THE GENERATIVE NETWORK

Generative networks are neural networks that are trained to transform an arbitrary distribution so that it looks similar to the known distribution. We can use any type of neural network for this, such as **multi-layer perceptrons (MLPs)**, **convolutional neural networks (CNNs)**, and more, to build the generator network. The input data to these networks are the samples that we take from any arbitrary distribution. In this example, we will be using an MLP to build a generative network. Before we start building the network, let's revisit some of the building blocks of a neural network that you will have learned about in the previous chapters. We will be building the network using the Keras library.

SEQUENTIAL()

As you might already know, a neural network consists of different layers of nodes that have connections between them. The **Sequential()** API is the mechanism through which you can create those layers in Keras. The **Sequential()** API is instantiated using the following code:

```
from tensorflow.keras import Sequential
Genmodel= Sequential()
```

In the first part of the code, the **Sequential()** class is imported from the **tensorflow.Keras** module. It is then instantiated as a variable model in the second line of code.

KERNEL INITIALIZERS

In *Chapter 2, Neural Networks*, you learned that the training process involves updating the weights and biases of a neural network so that the function that maps the inputs to the outputs is learned effectively. As a first step in the training process, we initialize some values for the weights and biases. These get updated more during the backpropagation stage. The initialization of the weights and biases is done through a parameter called the **kernel initializer**. Different types of kernel initializers are used in a network in Keras. We will be using a kernel initializer called **he_uniform** in the exercise that follows. A kernel initializer will be added as a parameter within the network.

DENSE LAYERS

The basic dynamics within each layer in a neural network is the matrix multiplication (dot product) between the weights for the layer and the input to the layer, and the further addition of a bias. This is represented by the **dot(X,W) + B** equation, where **X** is the input to the layer, **W** is the weight or the kernel, and **B** is the bias. This operation of the neural network is done using the dense layer in Keras. This is implemented in code as follows:

```
from tensorflow.keras.layers import Dense
Genmodel.add(Dense(hidden_layer,activation,\
                kernel_initializer,input_dim))
Genmodel.add(Dense(hidden_layer,activation,kernel_initializer))
```

> **NOTE**
>
> The above code block is solely meant to explain how the code is implemented. It may not result in a desirable output when run in its current form. For now, try to understand the syntax completely; we will be putting this code into practice in *Exercise 7.02*, *Building a Generative Network.*

As you can see, we add a dense layer to the instantiation of the **Sequential()** class (**Genmodel**) we created earlier. Some of the key parameters that need to be given to define a dense layer are as follows:

- **Hidden Layers (hidden_layer)**: As you might know, hidden layers are the intermediate layers in a neural network. The number of nodes of a hidden layer is defined as the first parameter.

- **Activation functions (activation)**: The other parameter is the type of activation function that will be used. Activation functions will be discussed in detail in the next section.

- **Kernel Initializer (kernel_initializer)**: The kind of kernel initializer that is used for the layer is defined within the dense layer.

- **Input dimensions (input_dim)**: For the first layer of the network, we have to define the dimensions of the input (**input_dim**). For the subsequent layers, this is deduced automatically based on the output dimensions of each layer.

ACTIVATION FUNCTIONS

As you might know, activation functions introduce non-linearity to the outputs of a neuron. In a neural network, activation functions are introduced just after the dense layer. The output of the dense layer is the input of the activation function. Different activation functions will be used within the following exercise. They are as follows:

- **ReLU**: This stands for **Rectified Linear Unit**. This activation function only outputs positive values. All negative values will be output as zero. This is one of the most widely used activation functions.

- **ELU**: This stands for **Exponential Linear Unit**. This is very similar to ReLU except for the fact that it outputs negative values as well.

- **Linear**: This is a straight-line activation function. In this function, the activations are proportional to the inputs.

- **SELU**: This stands for **Scaled Exponential Linear Unit**. This activation function is a relatively lesser-used one. It enables an idea called internal normalization, which ensures that the mean and variance from the previous layers are maintained.

- **Sigmoid**: This is a very standard activation function. A sigmoid function squashes any input into a value between 0 and 1. Therefore, the output from a sigmoid function can also be treated as a probability distribution as the values are between 0 and 1.

Now that we have seen some of the basic building blocks of the network, let's go ahead and build our generative network in the next exercise.

Before we start the exercise, let's see where the next exercise lies in the overall scheme of things. In *Exercise 7.01*, *Generating a Data Distribution from a Known Function*, we created a data distribution from a known mathematical function, which is a `sine()` function. We created the entire distribution by arranging the first feature with equal intervals and then creating the second feature by taking the `sine()` function of the first feature. So, we literally controlled the entire process of creating this dataset. That's why this is called a real data distribution because the data is created from a known function. The ultimate aim of a GAN is to transform a random noise distribution and make it look like a real data distribution; that is, make a random distribution look like the structured `sine()` distribution. This will be achieved in later exercises. However, as a first step, we will create a generative network that will create a random noise distribution. This is what we will do in the next exercise.

EXERCISE 7.02: BUILDING A GENERATIVE NETWORK

In this exercise, we will build a generative network. The purpose of the generative network is to generate fake data distribution from a random noise data. We'll do this by generating random data points as input to the generator network. Then, we'll build a six-layer network, layer by layer. Finally, we'll predict the output from the network and plot the output distribution. This data distribution will be our fake distribution. Follow these steps to complete this exercise:

1. Open a new Jupyter Notebook and name it **Exercise 7.02**. Import the following library packages:

```
# Importing the library packages
import tensorflow as tf
import numpy as np
```

```
from numpy.random import randn
from tensorflow.keras.models import Sequential
from tensorflow.keras.layers import Dense
from matplotlib import pyplot
```

2. In this step, we define the number of input features and output features for the network:

```
# Define the input features and output features
infeats = 10
outfeats = 2
```

We will have 10 features as input and the output will be two features. The input features of **10** are arbitrarily selected. The output features of **2** are selected because our real dataset contains two features.

3. Now, we will generate a batch of random numbers. Our batch size will be **128**:

```
# Generate a batch of random numbers
batch = 128
genInput = randn(infeats * batch)
```

We can select any batch size. A batch size of **128** is selected so that we take cognizance of the computation resources we have. Since the input size is 10, we should generate 128 × 10 random numbers. Also, in the preceding code, **randn()** is the function to generate random numbers. Inside the function, we specify how many data points we want, which is (128 × 10) in our case.

4. Let's reshape the random data into the input format we want using the following code:

```
# Reshape the data
genInput = genInput.reshape(batch,infeats)
print(genInput.shape)
```

You should get the following output:

```
(128, 10)
```

5. In this step, we will define the generator. This network will have six layers:

```
# Defining the Generator model
Genmodel = Sequential()
Genmodel.add(Dense(32,activation = 'linear',\
                    kernel_initializer='he_uniform',\
```

```
                       input_dim=infeats))
Genmodel.add(Dense(32,activation = 'relu',\
                   kernel_initializer='he_uniform'))
Genmodel.add(Dense(64,activation = 'elu',\
                   kernel_initializer='he_uniform'))
Genmodel.add(Dense(32,activation = 'elu',\
                   kernel_initializer='he_uniform'))
Genmodel.add(Dense(32,activation = 'selu',\
                   kernel_initializer='he_uniform'))
Genmodel.add(Dense(outfeats,activation = 'selu'))
```

From the network, we can see that, in the first layer, we define the dimension of the input, which is 10, and in the last layer, we define the output dimension, which is 2. This is based on the input data dimensions that we generated in *Step 4* (10) and the output features that we want, which is similar to the number of features of the real data distribution.

6. We can see the summary of this network by using the **model.summary()** function call:

```
# Defining the summary of the network
Genmodel.summary()
```

You should get the following output:

```
Model: "sequential_1"
```

Layer (type)	Output Shape	Param #
dense_6 (Dense)	(None, 32)	352
dense_7 (Dense)	(None, 32)	1056
dense_8 (Dense)	(None, 64)	2112
dense_9 (Dense)	(None, 32)	2080
dense_10 (Dense)	(None, 32)	1056
dense_11 (Dense)	(None, 2)	66

```
Total params: 6,722
Trainable params: 6,722
Non-trainable params: 0
```

Figure 7.6: Summary of the generator network

From the summary, you can see the shapes of the output from each layer. For example, the output from the dense layer has a shape of (*size of batch*, **32**) since the first hidden layer has **32** neurons. **None** in the shape layer denotes the number of examples, which in this case means the input batch size. The figure of 352 for the first layer is the size of the parameters, which includes both the weights and bias. The weight matrix will have a size of (10 × 32) as the number of inputs to the first layer is 10 and the next layer (hidden layer 1) has 32 neurons. The number of bias will be (32 × 1), which will be equivalent to the number of hidden layer neurons in the first layer. So, in total, there are 320 + 32 = 352 parameters. The second layer would be (32 × 32) + (32 × 1) = 1,056 and so on for all subsequent layers.

7. Now that we have defined the generator network, let's generate the output from the network. We can do that using the **predict()** function:

```
# Generating fake samples from network
fakeSamps = Genmodel.predict(genInput)
fakeSamps.shape
```

You should get the following output:

```
(128, 2)
```

We can see that the output from the generator function generates a sample with two features and several examples equal to the batch size we have given.

8. Plot the distribution:

```
# Plotting the fake distribution
from matplotlib import pyplot
pyplot.scatter(fakeSamps[:,0],fakeSamps[:,1])
pyplot.xlabel('Feature 1 of the distribution')
pyplot.ylabel('Feature 2 of the distribution')
pyplot.show()
```

You should get an output similar to the following. Please note that modeling will be stochastic in nature and therefore you might not get the same output:

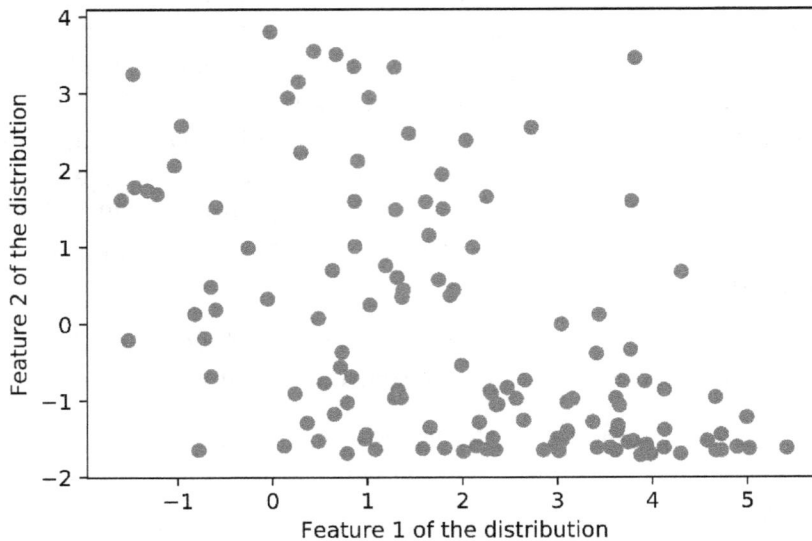

Figure 7.7: Plot of the fake data distribution

As we can see, very random data has been generated. As you will see in upcoming exercises, this random data will be transformed so that it looks like the real data distribution.

> **NOTE**
>
> To access the source code for this specific section, please refer to https://packt.live/2W0FxyZ.
>
> You can also run this example online at https://packt.live/2WhZpOn. You must execute the entire Notebook in order to get the desired result.

In this exercise, we defined the generator network, which had six layers and then generated the first fake samples from the generator network. You may be wondering how we arrived at those six layers. What about the choice of the activation functions? Well, the network architecture was arrived at after a lot of experimentation for this problem statement. There are no real shortcuts in terms of finding the right architecture. We have to arrive at the most optimal architecture after experimenting with different parameters such as the number of layers, type of activations, and more.

SETTING THE STAGE FOR THE DISCRIMINATOR NETWORK

In the previous exercise, we defined the generator network. Now, it is time to set the stage before we define the discriminator network. Looking at the output we got from the generator network, we can see that the data points are randomly distributed. Let's take a step back and assess where we are really headed. In our introduction to generative networks, we stated that we want the output from the generative network to be similar to the real distribution we are trying to mimic. In other words, we want the output from the generative network to look similar to the output from the real distribution, as shown in the following plot:

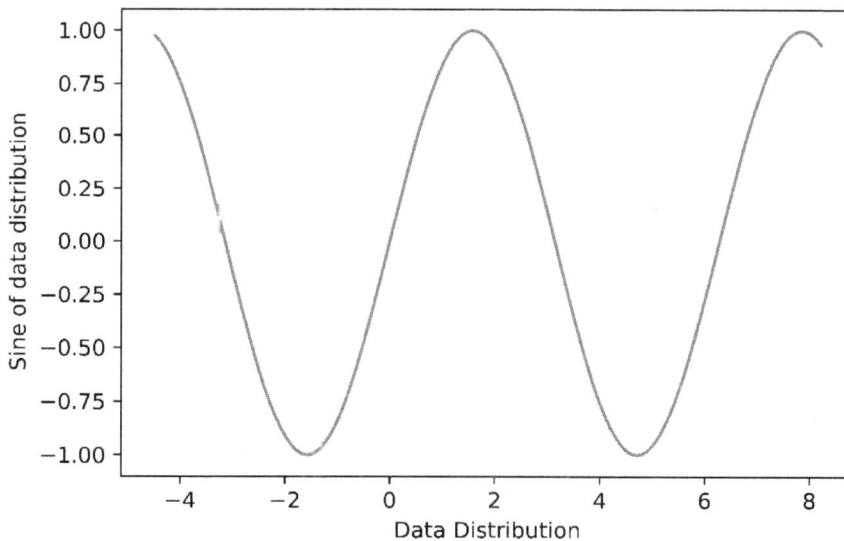

Figure 7.8: Real data distribution

We can see that the current distribution that has been generated by the generator network is nowhere near the distribution we want to mimic. Why do you think this is happening? Well, the reason is quite obvious; we have not done any training yet. You will also have noticed that we don't have an optimizer function as part of the network. The optimizer function in Keras is defined using the **compile()** function, as shown in the following code, where we define the type of loss function and what kind of optimizers we want to adopt:

```
model.compile(loss='binary_crossentropy',\
              optimizer='adam',metrics=['accuracy'])
```

We have excluded the **compile()** function on purpose. Later, when we are introduced to the GAN model, we will use the **compile()** function to optimize the generator network. So, hang on until then. For now, we will go ahead with the next step of the process, which is defining the discriminator network.

PROCESS 3 – DISCRIMINATOR NETWORK

In the previous process, we were introduced to the generative network, a neural network that generated fake samples. The discriminator network is also another neural network, albeit with different functionality from the generator network. The purpose of the discriminator function is to identify whether a given example is a real one or a fake one. Using an analogy, if the generator network is a conman who makes fake currency, then the discriminator network is the super cop who identifies that the currency is fake. Once caught by the super cop, the conman will try to perfect their craft to make better counterfeits so that they can fool the super cop. However, the super cop will also undergo lots of training to know the nuances of different currencies and work toward perfecting the craft of catching whatever the conman generates. We can see here that both these protagonists are in adversarial positions all the time. This is the reason why the network is called a *generative adversarial network.*

Taking a cue from the preceding analogy, training a discriminator is similar to the super cop undergoing more training to identify fake currency. The discriminator network is like any binary classifier you would have learned about in machine learning. As part of the training process, the discriminator will be provided with two classes of examples, one generated from the real distribution and the other from the generator distribution. Each of these sets of examples will have their respective labels too. The real distribution will have a label of "1", while the fake distribution will have a label of "0". The discriminator, after being trained, will have to correctly classify whether an example is real or fake, which is a typical binary classification problem.

IMPLEMENTING THE DISCRIMINATOR NETWORK

The core structure of the discriminator network would be similar to the generator network we implemented in the previous section. The complete process behind building the discriminator network is as follows:

1. Generate batches of real distribution.

2. Using the generator network, it generates batches of fake distribution.

3. Train the discriminator network with examples of both these distributions. The real distribution will have a label of 1, while the fake distribution will have a label of 0.

4. Evaluate the performance of the discriminator.

In *Steps 1* and *2*, we have to generate batches of both the real and fake distributions. This will necessitate making use of the real distribution we built in *Exercise 7.01, Generating a Data Distribution from a Known Function,* and the generator network we developed in *Exercise 7.02, Building a Generative Network*. Since we have to use these two distributions, it would be convenient to package them into two types of functions to efficiently train the discriminator network. Let's look at the two types of functions we will build.

FUNCTION TO GENERATE REAL SAMPLES

The content of this function, which is used to generate real samples, is the same as the code we developed in *Exercise 7.01, Generating a Data Distribution from a Known Function*. The only notable addition is the label for the input data. As we stated earlier, the real samples will have a label of 1. So, as labels, we will generate an array of 1s with the same size as the batch size. There is a utility function in **numpy** that can be used to generate a series of 1s called **np.ones((batch,1)**. This will generate an array of 1s whose size is equal to the batch size. Let's revisit the different steps in this function:

1. Generate equally spaced numbers to the right and left of a random number.

2. Concatenate both sets to get a series that is equal in length to the batch size we require. This is our first feature.

3. Generate the second feature by taking the **sine()** function of the first feature we generated in *Step 2*.

4. Reshape both features so their size is equal to **(batch,1)** and then concatenate them along the columns. This will result in an array of shape **(batch,2)**.

5. Generate the labels using the **np.ones((batch,1))** function. The label array will have a dimension of **(batch,1)**.

The arguments that we will provide to the function are the random number and the batch size. One subtle change to note in *Step 1* is that since we want a series equal in length to the batch size, we will take equally spaced numbers to the left and right equal to half of the batch size (batch size /2). In this way, when we combine both series to the left and right, we get a series equal to the batch size we want.

FUNCTIONS TO GENERATE FAKE SAMPLES

The function(s) to generate fake samples will be the same as what we developed in *Exercise 7.02, Building a Generative Network*. However, we will have to divide this into three separate functions. The reason for dividing the code we implemented in *Exercise 7.02, Building a Generative Network* into three separate functions is for convenience and efficiency during the training process. Let's take a look at these three functions:

- **Function 1**: The first of these functions is used to generate the inputs for generating fake samples. This is the part of *Exercise 7.02, Building a Generative Network* where we gave a batch size and the number of input features and generated random normal numbers using the **randn()** function. The output will be an array of size (**batch,input features**). The arguments to this function are **batch size** and **input feature size**.

- **Function 2**: The second function is the complete six-layer generator network we developed in *Exercise 7.02, Building a Generative Network*. The inputs to this function are **input feature size** and **output feature size**.

- **Function 3**: The third function is a function that calls both the first and second functions to generate fake samples. In addition to generating fake samples, the function also generates the labels, which in the case of the generator network should be 0. Just like in the discriminator network where we generated a series of 1s, we have a utility function in **numpy** to generate 0s called **np.zeros((batch,1))**.

Let's look at the complete process for these three functions:

1. Generate fake inputs using *function 1*.

2. Use the generator model function (*function 2*) to predict a fake output.

3. Generate labels, which is a series of 0s, using the **np.zeros()** function. This is part of *function 3*.

The arguments to the third function are **generator model**, **batch size**, and **input feature size**.

BUILDING THE DISCRIMINATOR NETWORK

The discriminator network will be built along the same lines as the generator network; that is, it will be created using the **Sequential()** class, the dense layer, and the activation and initialization functions. The only notable exception is that we will also have the optimization layer in the form of the **compile()** function. In the optimization layer, we will define the loss function, which in this case will be **binary_crossentropy** as the discriminator network is a binary classification network. For the optimizer, we will be using the **adam optimizer** as this is found to be very efficient and is a very popular choice.

TRAINING THE DISCRIMINATOR NETWORK

Now that we have gone through all the components for implementing the discriminator network, let's look at the steps involved in training the discriminator network:

1. Generate a random number and then generate a batch of real samples and its labels using the function to generate real samples.

2. Generate a batch of fake samples and its labels using the third function described to generate fake samples. The third function will use both the other functions to generate the fake samples.

3. Train the discriminator model using the **train_on_batch()** function with the batch of real samples and fake samples.

4. Steps *1* to *3* are repeated for the number of epochs we want the training to run for. This is done through a **for** loop over the number of epochs.

5. At every intermediate step, we calculate the accuracy of the model on the fake samples and real samples using the **evaluate()** function. The accuracy of the model is printed.

Now that we have seen the steps involved in implementing the discriminator network, we'll implement this in the next exercise.

EXERCISE 7.03: IMPLEMENTING THE DISCRIMINATOR NETWORK

In this exercise, we will build the discriminator network and train the network on both the real samples and fake samples. Follow these steps to complete this exercise:

1. Open a new Jupyter Notebook and name it **Exercise 7.03**. Import the following library packages:

```
# Import the required library functions
import tensorflow as tf
import numpy as np
from numpy.random import randn
from tensorflow.keras.models import Sequential
from tensorflow.keras.layers import Dense
from matplotlib import pyplot
```

2. Let's define a function that will generate the features of our real data distribution. The return values of this function will be the real dataset and its label:

`Exercise7.03.ipynb`

```
# Function to generate real samples
def realData(loc,batch):
    """
    loc is the random location or mean around which samples are centred
    """
    """
    Generate numbers to right of the random point
    """
    xr = np.arange(loc,loc+(0.1*batch/2),0.1)
    xr = xr[0:int(batch/2)]
    """
    Generate numbers to left of the random point
    """
    xl = np.arange(loc-(0.1*batch/2),loc,0.1)
```

The complete code for this step can be found at https://packt.live/3fe02j3.

The function we are defining here comprises code that was used to generate the **sine()** wave dataset in *Exercise 7.01, Generating a Data Distribution from a Known Function*. The inputs to this function are the random number and the batch size. Once the random number is provided, the series is generated with the same process we followed in *Exercise 7.01, Generating a Data Distribution from a Known Function*. We also generate the labels for the real data distribution, which will be 1. The final return value will be the two features and the label.

3. Let's define a function called **fakeInputs** to generate inputs for the generator function (this is *function 1*, which we explained in the *Functions to Generate Fake Samples* section):

```
# Function to generate inputs for generator function


def fakeInputs(batch,infeats):
    """
    Sample data points equal to (batch x input feature size)
    from a random distribution
    """
    genInput = rar.dn(infeats * batch)
    # Reshape the input
    X = genInput.reshape(batch ,infeats)
    return X
```

In this function, we're generating random numbers in the format we want **([batch size , input features])**. This function generates the fake data that was sampled from the random distribution as the return value.

4. Next, we'll be defining a function that will return a generator model:

```
# Function for the generator model
def genModel(infeats,outfeats):
    #Defining the Generator model
    Genmodel = Sequential()
    Genmodel.add(Dense(32,activation = 'linear',\
                       kernel_initializer='he_uniform',\
                       input_dim=infeats))
    Genmodel.add(Dense(32,activation = 'relu',\
                       kernel_initializer='he_uniform'))
    Genmodel.add(Dense(64,activation = 'elu',\
                       kernel_initializer='he_uniform'))
    Genmodel.add(Dense(32,activation = 'elu',\
                       kernel_initializer='he_uniform'))
    Genmodel.add(Dense(32,activation = 'selu',\
                       kernel_initializer='he_uniform'))
    Genmodel.add(Dense(outfeats,activation = 'selu'))
    return Genmodel
```

This is the same model that we implemented in *Exercise 7.02, Building a Generative Network*. The return value for this function will be the generator model.

5. The following function will be used to create fake samples using the generator model:

```
# Function to create fake samples using the generator model
def fakedataGenerator(Genmodel,batch,infeats):
    # first generate the inputs to the model
    genInputs = fakeInputs(batch,infeats)
    """
    use these inputs inside the generator model
    to generate fake distribution
    """
    X_fake = Genmodel.predict(genInputs)
    # Generate the labels of fake data set
    y_fake = np.zeros((batch,1))
    return X_fake,y_fake
```

In the preceding code, we are implementing *function 3*, which we covered in the *Functions to Generate Fake Samples* section. As you can see, we call the generator model we defined in *Step 4* as input, along with the batch size and the input features. The return values for this function are the fake data that's generated, along with its label (**0**).

6. Now, let's define the parameters to be used in the functions we have just created:

```
"""
Define the arguments like batch size,input feature size
and output feature size
"""
batch = 128
infeats = 10
outfeats = 2
```

7. Let's build the discriminator model using the following code:

```
# Define the discriminator model
Discmodel = Sequential()
Discmodel.add(Dense(16, activation='relu',\
                    kernel_initializer = 'he_uniform',\
```

```
                              input_dim=outfeats))
Discmodel.add(Dense(16,activation='relu' ,\
                       kernel_initializer = 'he_uniform'))
Discmodel.add(Dense(16,activation='relu' ,\
                       kernel_initializer = 'he_uniform'))
Discmodel.add(Dense(1,activation='sigmoid'))
# Compiling the model
Discmodel.compile(loss='binary_crossentropy',\
                       optimizer='adam', metrics=['accuracy'])
```

The mode of construction for the discriminator model is similar to what we did in the generator network. Please note that the activation function for the last layer will be a sigmoid as we need a probability regarding whether the output is a rea network or a fake network.

8. Print the summary of the discriminator network:

```
# Print the summary of the discriminator model
Discmodel.summary()
```

You should get the following output:

```
Model: "sequential"
```

Layer (type)	Output Shape	Param #
dense (Dense)	(None, 16)	48
dense_1 (Dense)	(None, 16)	272
dense_2 (Dense)	(None, 16)	272
dense_3 (Dense)	(None, 1)	17

```
Total params: 609
Trainable params: 609
Non-trainable params: 0
```

Figure 7.9: Model summary

From the summary, we can see the size of the network based on the architecture we defined. We can see that the first three dense layers have 16 neurons each, which we defined in *Step 7* when we built the discriminator network. The final layer will only have one output as this is a sigmoid layer. This outputs the probability of whether the data distribution is real (**1**) or fake (**0**).

9. Invoke the generator model function to be used in the training process:

```
# Calling the Generator model function
Genmodel = genModel(infeats,outfeats)
Genmodel.summary()
```

You should get the following output:

```
Model: "sequential_1"
_____
Layer (type)                 Output Shape              Param #
=================================================================
dense_4 (Dense)              (None, 32)                352
_____
dense_5 (Dense)              (None, 32)                1056
_____
dense_6 (Dense)              (None, 64)                2112
_____
dense_7 (Dense)              (None, 32)                2080
_____
dense_8 (Dense)              (None, 32)                1056
_____
dense_9 (Dense)              (None, 2)                 66
=================================================================
Total params: 6,722
Trainable params: 6,722
Non-trainable params: 0
_____
```

Figure 7.10: Model summary

You will notice that the architecture is the same as what we developed in *Exercise 7.02, Building a Generative Network*.

10. Now, we need to define the number of epochs to train the network for, as follows:

```
# Defining the number of epochs
nEpochs = 20000
```

11. Now, let's start training the discriminator network:

Exercise7.03.ipynb

```
# Train the discriminator network
for i in range(nEpochs):
    # Generate the random number for generating real samples
    loc = np.random.normal(3,1,1)
    """
    Generate samples equal to the bath size
    from the real distribution
    """
```

```
x_real, y_real = realData(loc,batch)
#Generate fake samples using the fake data generator function
x_fake, y_fake = fakedataGenerator(Genmodel,batch,infeats)
```

The complete code for this step can be found at https://packt.live/3fe02j3.

Here, we iterate the training of the model on both the real and fake data for 20,000 epochs. The number of epochs is arrived at after some level of experimentation. We should try this out with different values for the number of epochs until we get some good accuracy figures. For every 4,000 epochs, we print the accuracy of the model on both the real dataset and the fake dataset. The printing frequency is arbitrary and is based on the number of plots you want to see to check the progress of the training process. After training, you will see that the discriminator achieves very good accuracy levels.

You should get an output similar to the following:

```
Real accuracy:0.265625,Fake accuracy:0.59375
Real accuracy:1.0,Fake accuracy:0.828125
Real accuracy:1.0,Fake accuracy:0.90625
Real accuracy:1.0,Fake accuracy:0.9453125
Real accuracy:1.0,Fake accuracy:0.9453125
```

> **NOTE**
>
> Since we are working with random values here, the output you get may vary from the one you see here. It will also vary with every run.

From the accuracy levels, we can see that the discriminator was very good (accuracy = 1) at identifying the real dataset initially and shows relatively poor accuracy levels for the fake dataset. After around 4,000 epochs, we can see that the discriminator has become good at identifying both the fake and real datasets as both the accuracies are near 1.0.

> **NOTE**
>
> To access the source code for this specific section, please refer to https://packt.live/3fe02j3
>
> You can also run this example online at https://packt.live/2ZYiYMG.
> You must execute the entire Notebook in order to get the desired result.

In this exercise, we defined different helper functions and also built the discriminator function. Finally, we trained the discriminator model on real data and fake data. At the end of the training process, we saw that the discriminator learned to discriminate between the real dataset and fake dataset really well. Having trained the discriminator network, it's now time to move on to the climax, which is building the GAN.

PROCESS 4 – IMPLEMENTING THE GAN

We have finally arrived at the moment we have been waiting for all this while. In the previous three processes, we have been progressively building all the building blocks for the GAN, such as the fake data generator, real data generator, generator network, and discriminator network. The GAN is, in fact, the integration of all these building blocks. The real game in the GAN is the process in which we integrate these components with each other. Let's address this right away.

INTEGRATING ALL THE BUILDING BLOCKS

When building the discriminator network, we generated real samples and fake samples and fed them to the discriminator during training. The training process made the discriminator "smart", which enabled it to correctly identify what is fake and what is real. In probability terms, this would mean that when the discriminator gets a fake sample, it will predict a probability close to "0" and when the sample is real, it will predict a probability close to "1". However, getting the discriminator to be smart is not our end objective. Our end objective is to get the generator model smart so that it starts generating examples that look like real samples and, in the process, fools the discriminator. This can be achieved by training the generator and updating its parameters (that is, the weights and bias) to enable it to generate samples that look like real samples. However, there is still a problem, because in the generator network, we did not include an optimizer step and therefore the generator network by itself cannot be trained. The way to get around this problem is by building another network (let's call it **Ganmodel**) that connects the generator and discriminator in sequence and then include an optimizer function in the new network so that it goes and updates the parameters of its constituents when backpropagation happens. In terms of pseudocode, this network will look something like this:

```
Ganmodel = Sequential()
# First adding the generator model
Ganmodel.add(Genmodel)
"""
```

```
Next adding the discriminator model
without training the parameters
"""

Ganmodel.add(Discmodel)
# Compile the model for loss to optimise the Generator model
Ganmodel.compile(loss='binary_crossentropy',optimizer = 'adam')
```

In this model, the generator model will generate fake samples that are fed into the discriminator model, which in turn will then generate a probability as to whether the example is fake or real. Based on the label of the example, it will have a certain loss that will be propagated through the discriminator to the generator, updating the parameters of both the models. In other words, based on the loss, the backpropagation algorithm will update each parameter based on the gradient of the parameter with respect to the loss. So, this will solve our problem of not having defined an optimizer function for the generator.

However, there is one more catch to this network. Our discriminator network has already been trained and was made really smart when we trained the discriminator network separately. We don't want to train the discriminator model again in this new network and make it smarter. This can be solved by defining that we don't want to train the discriminator parameters in the network. With this new change, the **Ganmodel** would look as follows:

```
# First define that discriminator model cannot be trained
Discmodel.trainable = False
Ganmodel = Sequential()
# First adding the generator model
Ganmodel.add(Genmodel)
"""

Next adding the discriminator model
without training the parameters
"""

Ganmodel.add(Discmodel)
# Compile the model for loss to optimise the Generator model
Ganmodel.compile(loss='binary_crossentropy',optimizer = 'adam')
```

By making **Discmodel.trainable = False**, we're telling the network that we don't want to update the parameters of the discriminator network during backpropagation. So, the discriminator network will act as a conduit to pass on the error during the backpropagation stage to the generator network.

If you think all our problems have been solved, you are in for a rude awakening. We know that when the discriminator model is presented with a fake distribution, it will predict the probability to a value very close to **0**. We also know that the labels of the fake dataset are also **0**. So, in terms of loss, there would be very minimal loss being propagated back to the generator. With such a minuscule loss, the subsequent update to the parameters of the generator model will also be very minuscule. This will not enable the generator to generate samples that are like the real samples. The generator will only be able to learn if a large loss is generated and propagated to it so that its parameters are updated in the direction of real parameters. So, how do we get the loss to be high? What if, instead of defining the labels of the fake samples as **0**, we define them as **1**? If we do this, the discriminator model, as usual, will predict a probability close to 0 for fake examples. However, we now have a situation where the loss function would be large because the labels are 1. When this large loss function gets propagated back to the generator network, the parameters will be updated significantly, which will enable it to be smarter. Subsequently, what will happen is the generator will start generating samples that look more like the real samples, and they would meet our objective.

This concept can be explained with the following figure. Here, we can see that at the initial level of training, the probability for the fake data is close to zero (**0.01**) and the label that we've given for the fake data is **1**. This will ensure that we get a large loss that gets backpropagated to the generator network:

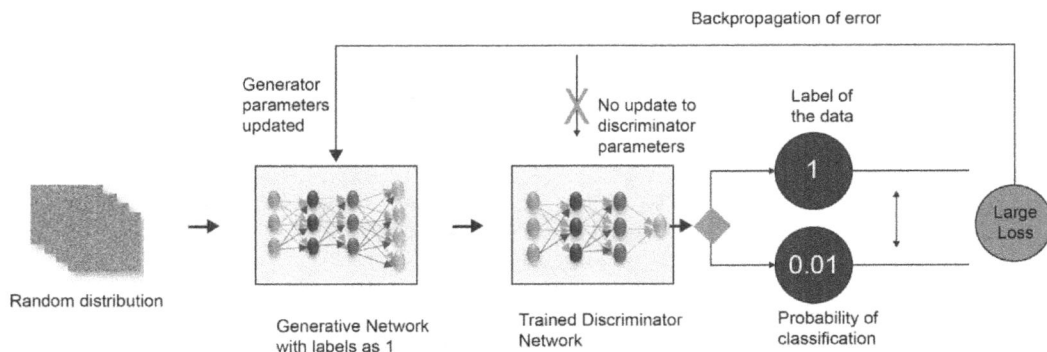

Figure 7.11: GAN process

Now that we have seen the dynamics of the GAN model, let's tie all the pieces together to define the process we will follow in order to build the GAN.

PROCESS FOR BUILDING THE GAN

The complete process for the GAN is all about tying together the pieces we have built into a logical order. We will use all the functions we built when we defined the discriminator function. In addition, we will also make new functions; for instance, a function for the discriminator network and another function for the GAN model. All these functions will be called at specific points to make the GAN model. The end-to-end process will be as follows:

1. Define the function to generate a real data distribution. This function is the same function we developed in *Exercise 7.03, Implementing the Discriminator Network* for the discriminator network.

2. Define the three functions that were created for generating fake samples. These are a function for generating fake inputs, a function for the generator network, and a function for generating fake samples and labels. All these functions are the same as the ones we developed in *Exercise 7.03, Implementing the Discriminator Network* for the discriminator network.

3. Create a new function for the discriminator network, just like we created in *Exercise 7.03, Implementing the Discriminator Network*. This function will have the output features (2) as its input as both the real dataset and fake dataset have two features. This function will return the discriminator model.

4. Create a new function for the GAN model as per the pseudocode we developed in the previous section (*Process 4 – Building a GAN*). This function will have the generator model and the discriminator model as its inputs.

5. Start the training process.

THE TRAINING PROCESS

The training process here is similar to the process we implemented in *Exercise 7.03, Implementing the Discriminator Network* for the discriminator network. The steps for the training process are as follows:

1. Generate a random number and then generate a batch of real samples and its labels using the function to generate real samples.

2. Generate a batch of fake samples and its labels using the third function we described regarding the functions for generating fake samples. The third function will use both the other functions to generate the fake samples.

3. Train the discriminator model using the **train_on_batch()** function using the batch of real samples and fake samples.

4. Generate another batch of fake inputs to train the GAN model. These fake samples are generated using *function 1* in the fake sample generation process.

5. Generate the labels for the fake samples that are intended to fool the discriminator. These labels will be 1s instead of 0s.

6. Train the GAN model using the **train_on_batch()** function using the fake samples and its labels, as described in *Steps 4* and *5*.

7. *Steps 1* to *6* are repeated for the number of epochs we want the training to run for. This is done through a **for** loop over the number of epochs.

8. At every intermediate step, we calculate the accuracy of the model on the fake samples and real samples using the **evaluate()** function. The accuracy of the model is also printed.

9. We also generate output plots at certain epochs.

Now that we have seen the complete process behind training a GAN, let's dive into *Exercise 7.04, Implementing the GAN*, which implements this process.

EXERCISE 7.04: IMPLEMENTING THE GAN

In this exercise, we will build and train the GAN by implementing the process we discussed in the previous section. Follow these steps to complete this exercise:

1. Open a new Jupyter Notebook and name it **Exercise 7.04**. Import the following library packages:

```
# Import the required library functions
import tensorflow as tf
import numpy as np
from numpy.random import randn
from tensorflow.keras.models import Sequential
from tensorflow.keras.layers import Dense
from matplotlib import pyplot
```

2. Let's create a function to generate the real samples:

Exercise7.04.ipynb

```
# Function to generate real samples
def realData(loc,batch):
    """
    loc is the random location or mean
```

```
around which samples are centred
"""
# Generate numbers to right of the random point
xr = np.arange(loc,loc+(0.1*batch/2),0.1)
xr = xr[0:int(batch/2)]
# Generate numbers to left of the random point
xl = np.arange(loc-(0.1*batch/2),loc,0.1)
```

The complete code for this step can be found on https://packt.live/3iJJHVS

The function we're creating here follows the same process we implemented in *Exercise 7.01*, *Generating a Data Distribution from a Known Function*. The inputs to this function are the random number and the batch size. We get the real data distribution with both our features, along with the label for the real data distribution as return values, from this function. The return values from this function are the real dataset and its label.

3. Here, let's define the function to generate inputs for the generator network:

```
# Function to generate inputs for generator function
def fakeInputs(batch,infeats):
"""
    Sample data points equal to (batch x input feature size)
 from a random distribution
    """
    genInput = randn(infeats * batch)
    # Reshape the input
    X = genInput.reshape(batch ,infeats)
    return X
```

This function generates the fake data that was sampled from the random distribution as output.

4. Now, let's go ahead and define the function for building the generator network:

```
# Function for the generator model
def genModel(infeats,outfeats):
    # Defining the Generator model
    Genmodel = Sequential()
    Genmodel.add(Dense(32,activation = 'linear',\
                      kernel_initializer='he_uniform',\
                      input_dim=infeats))
    Genmodel.add(Dense(32,activation = 'relu',\
                      kernel_initializer='he_uniform'))
    Genmodel.add(Dense(64,activation = 'elu',\
                      kernel_initializer='he_uniform'))
```

```
        Genmodel.add(Dense(32,activation = 'elu',\
                       kernel_initializer='he_uniform'))
        Genmodel.add(Dense(32,activation = 'selu',\
                       kernel_initializer='he_uniform'))
        Genmodel.add(Dense(outfeats,activation = 'selu'))
        return Genmodel
```

This is the same function we built in *Exercise 7.02, Building a Generative Network*. This function returns the generator model.

5. In this step, we will define the function that will create fake samples using the generator network:

```
# Function to create fake samples using the generator model
def fakedataGenerator(Genmodel,batch,infeats):
    # first generate the inputs to the model
    genInputs = fakeInputs(batch,infeats)
    """
    use these inputs inside the generator model
    to generate fake distribution
    """
    X_fake = Genmodel.predict(genInputs)
    # Generate the labels of fake data set
    y_fake = np.zeros((batch,1))
    return X_fake,y_fake
```

The function we are defining here takes the random data distribution as input (to the generator network we defined in the previous step) and generates the fake distribution. The label for the fake distribution, which is 0, is also generated within the function. In other words, the outputs from this function are the fake dataset and its label.

6. Now, let's define the parameters that we will be using within the different functions:

```
"""
Define the arguments like batch size,input feature size
and output feature size
"""
batch = 128
infeats = 10
outfeats = 2
```

7. Next, let's build the discriminator model as a function:

```
# Discriminator model as a function
def discModel(outfeats):
    Discmodel = Sequential()
    Discmodel.add(Dense(16, activation='relu',\
                        kernel_initializer = 'he_uniform',\
                        input_dim=outfeats))
    Discmodel.add(Dense(16,activation='relu' ,\
                        kernel_initializer = 'he_uniform'))
    Discmodel.add(Dense(16,activation='relu' ,\
                        kernel_initializer = 'he_uniform'))
    Discmodel.add(Dense(1,activation='sigmoid'))
    # Compiling the model
    Discmodel.compile(loss='binary_crossentropy',\
                      optimizer='adam',metrics=['accuracy'])
    return Discmodel
```

The network architecture will be like the one we developed in *Exercise 7.03, Implementing the Discriminator Network*. This function will return the discriminator.

8. Print the summary of the discriminator network:

```
# Print the summary of the discriminator model
Discmodel = discModel(outfeats)
Discmodel.summary()
```

You should get the following output:

```
Model: "sequential"

Layer (type)                 Output Shape              Param #
=================================================================
dense (Dense)                (None, 16)                48

dense_1 (Dense)              (None, 16)                272

dense_2 (Dense)              (None, 16)                272

dense_3 (Dense)              (None, 1)                 17
=================================================================
Total params: 609
Trainable params: 609
Non-trainable params: 0
```

Figure 7.12: Discriminator model summary

This output is the same as the one we received for the network we implemented in *Exercise 7.03, Implementing the Discriminator Network*, where we defined the discriminator function.

9. Invoke the generator model function for use in the training process:

```
# Calling the Generator model function
Genmodel = genModel(infeats,outfeats)
Genmodel.summary()
```

You should get the following output:

```
Model: "sequential_10"

Layer (type)                Output Shape              Param #
=================================================================
dense_34 (Dense)            (None, 32)                352
_____
dense_35 (Dense)            (None, 32)                1056
_____
dense_36 (Dense)            (None, 64)                2112
_____
dense_37 (Dense)            (None, 32)                2080
_____
dense_38 (Dense)            (None, 32)                1056
_____
dense_39 (Dense)            (None, 2)                 66
=================================================================
Total params: 6,722
Trainable params: 6,722
Non-trainable params: 0
_____
```

Figure 7.13: Generator model summary

You will notice that the architecture is the same as what we developed in *Exercise 7.02, Building a Generative Network*.

10. Before we begin training, let's visualize the fake data distribution. For this, we generate the fake dataset using the **fakedataGenerator()** function and then visualize it using **pyplot**:

```
# Let us visualize the initial fake data
x_fake, _ = fakedataGenerator(Genmodel,batch,infeats)
# Plotting the fake data using pyplot
pyplot.scatter(x_fake[:, 0], x_fake[:, 1], color='blue')
# Adding x and y labels
```

```
pyplot.xlabel('Feature 1 of the distribution')
pyplot.ylabel('Feature 2 of the distribution')
pyplot.show()
```

You should get an output similar to the following. Please note that data generation is stochastic in nature (random) and that you might not get the same plot:

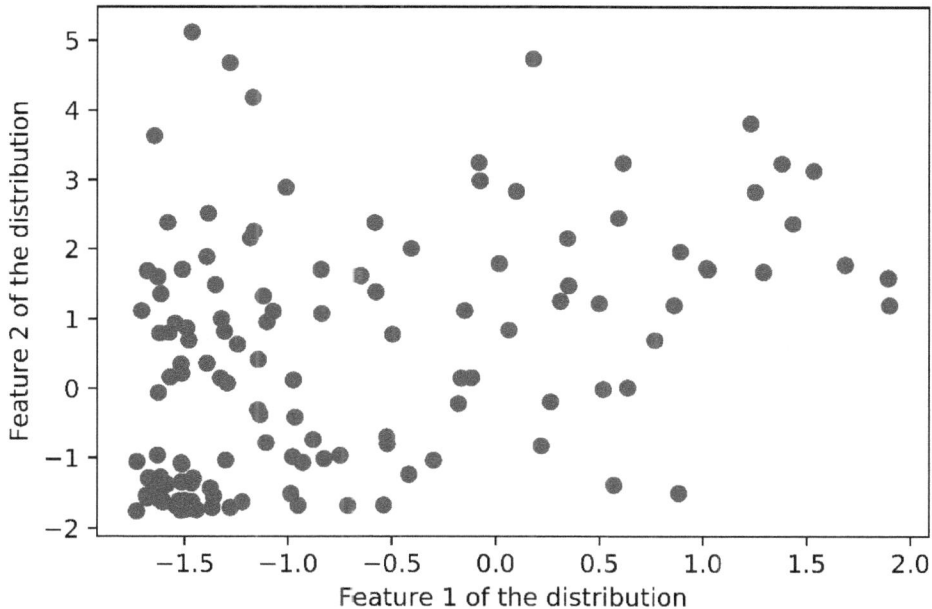

Figure 7.14: Plot from the fake input distribution

From the preceding plot, you can see that the data distribution is quite random. We need to convert this random data into a form similar to the sine wave, which was our real data distribution.

11. Now, let's define the GAN model as a function. This function is similar to the pseudocode we developed in *Process 4*, where we defined the GAN. The GAN is a wrapper model around the generator model and the discriminator model. Please note that we define the discriminator model as **not trainable** within this function:

```
"""
Define the combined generator and discriminator model,
for updating the generator
"""
```

```
def ganModel(Genmodel,Discmodel):
    # First define that discriminator model cannot be trained
    Discmodel.trainable = False
    Ganmodel = Sequential()
    # First adding the generator model
    Ganmodel.add(Genmodel)
    """
    Next adding the discriminator model
    without training the parameters
    """
    Ganmodel.add(Discmodel)
    # Compile the model for loss to optimise the Generator model
    Ganmodel.compile(loss='binary_crossentropy',optimizer = 'adam')
    return Ganmodel
```

This function will return the GAN model.

12. Now, let's invoke the GAN function. Please note that the inputs to the GAN model are the previously defined generator model and the discriminator model:

```
# Initialise the gan model
gan_model = ganModel(Genmodel,Discmodel)
```

13. Print the summary of the GAN model:

```
# Print summary of the GAN model
gan_model.summary()
```

You should get the following output:

```
Model: "sequential_2"

_____
Layer (type)                 Output Shape              Param #
=================================================================
sequential_1 (Sequential)    (None, 2)                 6722
_____
sequential (Sequential)      (None, 1)                 609
=================================================================
Total params: 7,331
Trainable params: 6,722
Non-trainable params: 609
_____
```

Figure 7.15: Summary of the GAN model

Note that the parameters of each layer of the GAN model are equivalent to the parameters of the generator and discriminator models. The GAN model is just a wrapper around these two models we defined earlier.

14. Let's define the number of epochs to train the network:

```
# Defining the number of epochs
nEpochs = 20000
```

15. Now, we start the process of training the network:

Exercise7.04.ipynb

```
# Train the GAN network
for i in range(nEpochs):
    # Generate the random number for generating real samples
    loc = np.random.normal(3,1,1)
    """
    Generate samples equal to the bath size
    from the real distribution
    """
    x_real, y_real = realData(loc,batch)
    #Generate fake samples using the fake data generator function
    x_fake, y_fake = fakedataGenerator(Genmodel,batch,infeats)
    # train the  discriminator on the real samples
    Discmodel.train_on_batch(x_real, y_real)
    # train the discriminator on the fake samples
    Discmodel.train_on_batch(x_fake, y_fake)
```

The complete code for this step can be found at https://packt.live/3iIJHVS

It needs to be noted here that the training of the discriminator model with the fake and real samples and the training of the GAN model happens concurrently. The only difference is that training the GAN model proceeds without updating the parameters of the discriminator model. The other thing to note is that, inside the GAN, the labels for the fake samples would be 1. This is to generate large loss terms that will be backpropagated through the discriminator network to update the generator parameters.

> **NOTE:**
>
> Please note that the third line of code from the bottom (`filename = 'GAN_Training_Plot%03d.png' % (i)`) saves a plot once every 2,000 epochs. The plots will be saved in the same folder that your Jupyter Notebook is located in. You can also specify the path you want to save the plots at. This can be done as follows:
>
> `filename = 'D:/Project/GAN_Training_Plot%03d.png' % (i)`
>
> You can access the plots that were generated through this exercise at https://packt.live/2W1FjaI.

You should get an output similar to the one shown here. Since the predictions are stochastic in nature (that is to say, they're random), you might not get the same plots shown in this example. Your values may vary; however, they will be similar to what's shown here:

```
Real accuracy:0.2421875,Fake accuracy:0.0234375
Real accuracy:0.625,Fake accuracy:0.609375
Real accuracy:0.6484375,Fake accuracy:0.9609375
Real accuracy:0.84375,Fake accuracy:0.734375
Real accuracy:0.3671875,Fake accuracy:0.734375
Real accuracy:0.53125,Fake accuracy:0.703125
Real accuracy:0.578125,Fake accuracy:0.640625
Real accuracy:0.640625,Fake accuracy:0.8203125
Real accuracy:0.515625,Fake accuracy:0.7109375
Real accuracy:0.5625,Fake accuracy:0.859375
```

From the preceding output, you can see that the real dataset accuracy levels are progressively going down and that the fake dataset's accuracy is going up. In ideal situations, the accuracy of the discriminator network has to be around the 0.5 level, which indicates that the discriminator is really confused as to whether a sample is fake or real. Now, let's look at some of the plots that were generated at different epoch levels as to how the data points are converging to look like the real function. The following plot is the distribution of the random data point before it was fed into the GAN (*Step 10*):

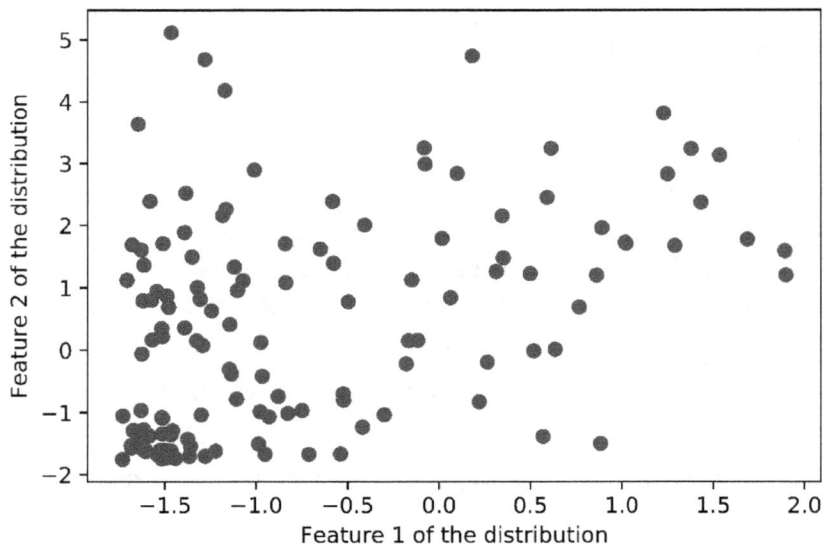

Figure 7.16: Plot from the fake input distribution

Notice the distribution of the data where the data points are mostly centered on a mean of 0. This is because the random points are generated from a normal distribution that has a mean of 0 and a standard deviation of 1. Now, using the raw data, let's study the progression of the fake dataset as the generator is trained.

> **NOTE**
>
> To access the source code for this specific section, please refer to https://packt.live/3iIJHVS.
>
> You can also run this example online at https://packt.live/3gF5DPW. You must execute the entire Notebook in order to get the desired result.

The three plots shown below map the progression of the fake data distribution vis-a-vis the real data distribution. The x axis represents feature 1, while the y axis represents feature 2. In the plots, the red points pertain to the data from the real distribution and the blue plots pertain to the data from the fake distribution. From the following plot, we can see that at epoch **2000**, the fake plots are within the domain; however, they are not aligned to the shape of the real data distribution.

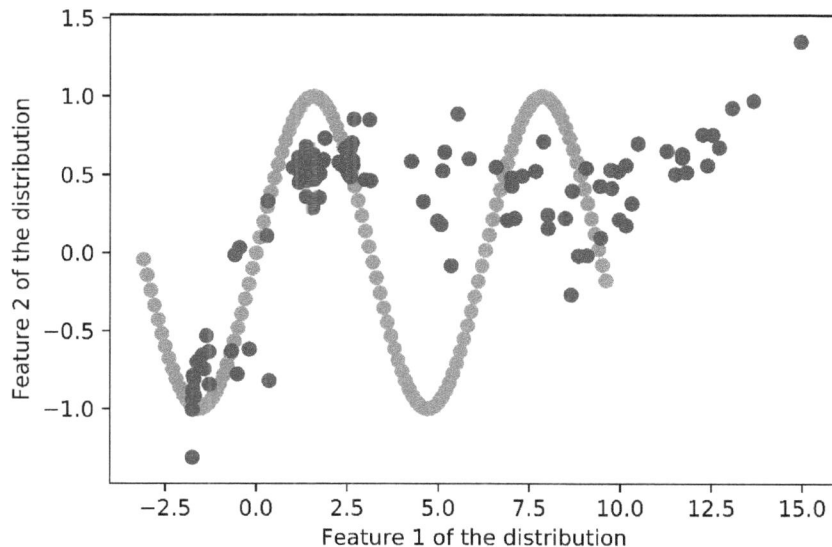

Figure 7.17: Plot of fake data distribution vis-à-vis the real data distribution at epoch 2000

By epoch **10000**, which is when the generator has been trained almost halfway, there is a consolidation nearer to the real data distribution:

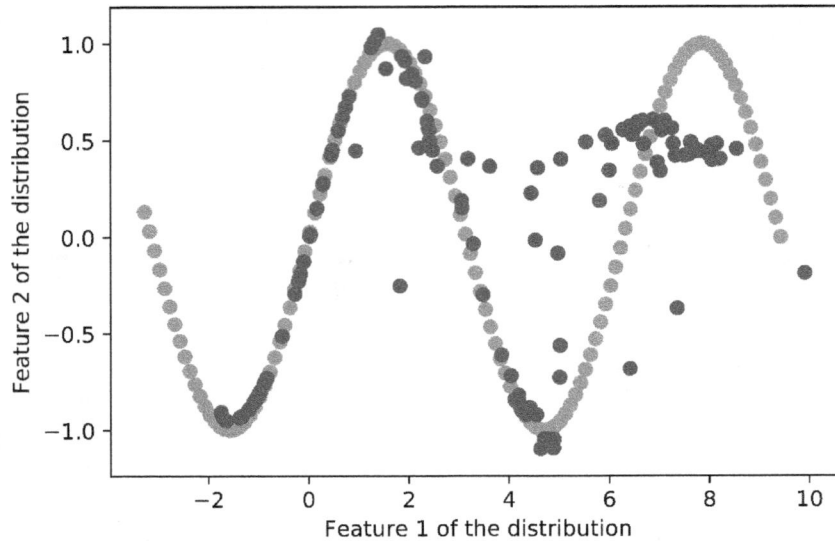

Figure 7.18: Plot of fake data distribution vis-à-vis the real data distribution at epoch 10000

By epoch **18000**, we can see that most of the points are aligned to the real data distribution, which is an indicator that the GAN has been trained reasonably well.

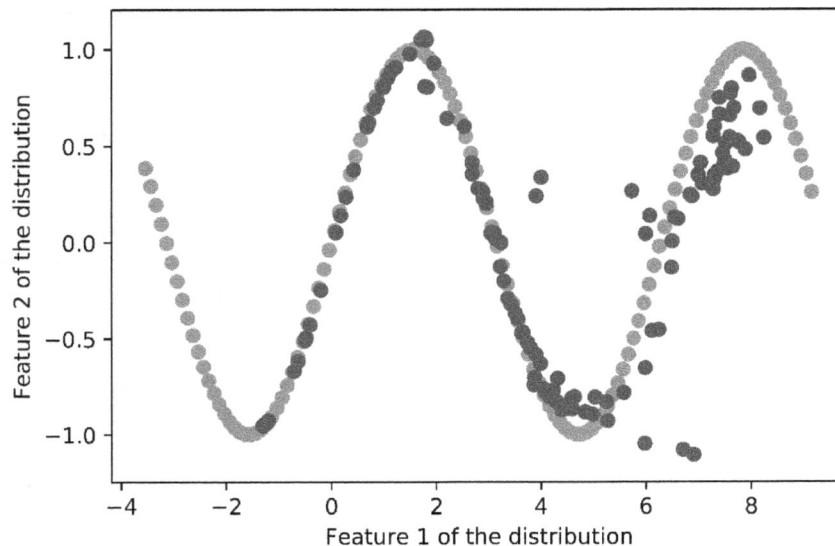

Figure 7.19: Plot of fake data distribution vis-à-vis the real data distribution at epoch 18000

However, you can see that the data points after **x = 4** have a lot more noise than the ones on the left. One reason for this could be the random data distribution we generated before we trained the *GAN(Step 10)* contains data that is distributed predominantly between **−2** and **4**. Such data is aligning well to the target distribution (sine wave) within the same range and is a little wobbly around the target distribution to the right of **x = 4**. However, you should also note that getting 100% alignment to the target distribution is an extremely difficult proposition that would involve experimenting with different model architectures and more experiments. We encourage you to experiment and be innovative with different components within the architecture to get the distribution more aligned.

> **NOTE**
>
> The results we have gotten in the above exercise will vary every time we run the code.

This brings us to the end of the complete process of progressively building a GAN. Through a series of exercises, we have learned what a GAN is, its constituents, and how all of them are tied together to train a GAN. We will take what we've learned forward and develop more advanced GANs using different datasets.

DEEP CONVOLUTIONAL GANS

In the previous sections, where we implemented a GAN, we made use of an architecture based on the **Multi-Layer Perceptron (MLP)**. As you may recall from the previous chapters, MLPs have fully connected layers. This implies that all the neurons in each layer have connections to all the neurons of the subsequent layer. For this reason, MLPs are also called fully connected layers. The GAN that we developed in the previous section can also be called a **Fully Connected GAN (FCGAN)**. In this section, we will learn about another architecture called **Deep Convolutional GANs (DCGANS)**. As the name implies, this is based on the **Convolutional Neural Network (CNN)** architecture that you learned about in *Chapter 4, Deep Learning for Text – Embeddings*. Let's revisit some of the building blocks of DCGANs.

BUILDING BLOCKS OF DCGANS

Most of the building blocks of DCGANs are similar to what you learned about when you were introduced to CNNs in *Chapter 3, Image Classification with Convolutional Neural Networks*. Let's revisit some of the important ones.

Convolutional Layers

As you learned in *Chapter 3, Image Classification with Convolutional Neural Networks*, convolutional operations involve filters or kernels moving over the input image to generate a set of feature maps. The convolutional layer can be implemented in Keras using the following line of code:

```
from tensorflow.keras import Sequential
model = Sequential()
model.add(Conv2D(64, kernel_size=(5, 5),\
                strides=(2,2), padding='same'))
```

> **NOTE**
>
> The above code block is solely meant to explain how the code is implemented. It may not result in a desirable output when run in its current form. For now, try to understand the syntax completely; we will be putting this code into practice soon.

In the first part of the preceding code, the **Sequential()** class is imported from the **tensorflow.keras** module. It is then instantiated to a variable model in the second line of code. The convolutional layer is added to the **Sequential()** class by defining the number of filters, kernel size, the required strides, and the padding indicators. In the preceding line of code, 64 indicates the number of feature maps. A **kernel_size** value of **(5,5)** indicates the size of the filters that will be convolved over the input image to generate the feature maps. The **strides** value of **(2,2)** indicates that the filters will move two cells at a time, both horizontally and vertically, in the process of generating the feature maps. **padding = 'same'** indicates that we want the output of the convolutional operation to be of the same size as the input.

> **NOTE:**
>
> The choice of architecture to use, such as the number of filters, size of kernels, stride, and more, is an art and can be mastered with lots of experimentation on the domain.

Activation Functions

In the previous section, we implemented some activation functions such as ReLU, ELU, SELU, and linear. In this section, we will be introduced to another activation function called LeakyReLU. LeakyReLU is another variation of ReLU. Unlike ReLU, which doesn't allow any negative values, LeakyReLU allows a small non-zero gradient that is controlled by a factor, α. This factor, α, controls the slope of the gradient for the negative values.

Upsampling Operation

In a CNN, an image gets down-sampled to lower dimensions by operations such as max pooling and convolutional operations. However, in a GAN, the dynamics of a generator network operate in a direction opposite to the convolutional operation; that is, from lower or coarser dimensions, we have to transform an image to a denser form (that is, with more dimensions). One way to do that is through an operation called **UpSampling**. In this operation, the input dimensions are doubled. Let's understand this operation in more detail using a small example.

The following code can be used to import the required library files. The function that's specific for **UpSampling** is **UpSampling2D** from **keras.layers**:

```
from tensorflow.keras.models import Sequential
from tensorflow.keras.layers import UpSampling2D
```

The following code creates a simple model that takes an array of shape **(3,3,1)** as input in the **UpSampling** layer:

```
# A model for UpSampling2d
model = Sequential()
model.add(UpSampling2D(input_shape=(3,3,1)))
model.summary()
```

The output will be as follows:

```
Model: "sequential_165"

_____
Layer (type)                    Output Shape              Param #
=================================================================
up_sampling2d_18 (UpSampling (None, 6, 6, 1)              0
=================================================================
Total params: 0
Trainable params: 0
Non-trainable params: 0
_____
```

Figure 7.20: Model summary for UpSampling2D

From the summary, we can see that the output has been doubled to **(None, 6,6,1)**, wherein the middle two dimensions have been doubled. To understand what change this makes to an array of shape **(3,3,1)**, we will need to define an array of size **(3,3)**, as follows:

```
# Defining an array of shape (3,3)
import numpy as np

X = np.array([[1,2,3],[4,5,6],[7,8,9]])
X.shape
```

The output will be as follows:

```
(3, 3)
```

The array we've defined has only two dimensions. However, the input to the model we defined needs four dimensions, where the dimensions are in the order **(examples, width, height, channels)**. We can create the additional dimensions using the **reshape()** function, as follows:

```
# Reshaping the array
X = X.reshape((1,3,3,1))
X.shape
```

The output will be as follows:

```
(1, 3, 3, 1)
```

We can use the following code to make some predictions with the **UpSampling** model we created and observe the dimensions of the resultant array:

```
# Predicting with the model
y = model.predict(X)
# Printing the output shape
y[0,:,:,0]
```

The output will be as follows:

```
array([[1., 1., 2., 2., 3., 3.],
       [1., 1., 2., 2., 3., 3.],
       [4., 4., 5., 5., 6., 6.],
       [4., 4., 5., 5., 6., 6.],
       [7., 7., 8., 8., 9., 9.],
       [7., 7., 8., 8., 9., 9.]], dtype=float32)
```

Figure 7.21: Output shape of the unsampled model

From the preceding output, we can see how the resultant array has been transformed. As we can see, each of the inputs has been doubled to get the resultant array. We will be using the **UpSampling** method in *Exercise 7.05, Implementing the DCGAN*.

Transpose Convolution

Transpose convolution is different from the **UpSampling** method we just saw. **UpSampling** was more or less a naïve doubling of the input values. However, transpose convolutions have weights that are learned during the training phase. Transpose convolutions work similarly to convolutional operations but in reverse. Instead of reducing the dimensions, transpose convolutions expand the dimensions of the input through a combination of the kernel size and its strides. As learned in *Chapter 3, Image Processing with Convolutional Neural Networks*, strides are the step sizes where we convolve or move the filters over the image to get an output. We also control the output of transpose convolutions with the **padding = 'same'** parameter, just like we do in convolutional operations.

Let's take a look at a code example of how transpose convolutions work.

First, we will need to import the necessary library files. The function that's specific to transpose convolution operations is **Conv2DTranspose** from **keras.layers**:

```
from tensorflow.keras.models import Sequential
from tensorflow.keras.layers import Conv2DTranspose
```

Now, we can create a simple model that takes an image of shape **(3,3,1)** in the transpose convolution layer:

```
# A model for transpose convolution
model = Sequential()
model.add(Conv2DTranspose(1, (4,4), (2,2),\
          input_shape=(3,3,1),padding='same'))
model.summary()
```

In the transpose convolution layer, the first parameter **(1)** is the number of filters. The second one **(4,4)** is the size of kernel and the last one **(2,2)** is the strides. With **padding = 'same'**, the output will not be dependent on the size of the kernel but will be multiples of the stride and the input dimension. The summary that will be generated by the preceding code will be as follows:

```
Model: "sequential_187"

_____
Layer (type)                    Output Shape              Param #
================================================================
conv2d_transpose_21 (Conv2DT (None, 6, 6, 1)                17
================================================================
Total params: 17
Trainable params: 17
Non-trainable params: 0

_____
```

Figure 7.22: Summary of the model

From the summary, we can see that the output has been doubled to **(None, 6,6,1)**, which would work like the multiplying the strides by the input dimensions (None, 2 × 3, 2 × 3, 1).

Now, let's see what changes occur to a real array of shape **(1,3,3,1)**. Remember that we also created this array earlier:

```
# Defining an array of shape (3,3)
X = np.array([[1,2,3],[4,5,6],[7,8,9]])
X = X.reshape((1,3,3,1))
X.shape
```

The output is as follows:

```
(1, 3, 3, 1)
```

To generate the transposed array, we need to make some predictions using the transpose convolution model we created. By printing the shape, we can also observe the dimensions of the resultant array:

```
 # Predicting with the model
y = model.predict(X)
# Printing the shape
print(y.shape)
# Printing the output shape
y[0,:,:,0]
```

The output will be as follows:

```
(1, 6, 6, 1)
array([[[-0.90538687,  0.        ,  -1.8107737 ,  0.        ,  -2.7161605 ,
          0.        ],
        [ 0.        ,  0.        ,  0.        ,  0.        ,  0.        ,
          0.        ],
        [-3.6215475 ,  0.        ,  -4.526934 ,  0.        ,  -5.432321 ,
          0.        ],
        [ 0.        ,  0.        ,  0.        ,  0.        ,  0.        ,
          0.        ],
        [-6.337708 ,  0.        ,  -7.243095 ,  0.        ,  -8.148481 ,
          0.        ],
        [ 0.        ,  0.        ,  0.        ,  0.        ,  0.        ,
          0.        ]]], dtype=float32)
```

Figure 7.23: Transformed array

> **NOTE**
>
> The output you get may vary from the one we have shown above.

From the preceding output, we can see how the resultant array has been transformed. The values in the generated array are the end result of the dynamics between the weights of the kernel on the input image.

Now that we have seen some of the basic building blocks of a DCGAN, we'll go ahead and build it in the next exercise.

GENERATING HANDWRITTEN IMAGES USING DCGANS

Now, we will try to generate a data distribution similar to the data pertaining to handwritten digits using a DCGAN. We will be using the MNIST handwritten digits dataset as the real dataset. This dataset has a training set of 60,000 examples, all of which are handwritten images of digits from 0 to 9. The implementation process for this GAN will be similar to *Exercise 7.04, Implementing the GAN*, where we implemented the GAN for the known function. Let's look at the steps we will follow for this problem statement.

First, we'll need to define the function that will be used to generate a real data distribution:

```
# Get the MNIST data
    (X_train, _), (_, _) = mnist.load_data()
```

The preceding function will generate the real data distribution from the MNIST dataset. The train and test sets can be generated using the **mnist.load_data()** function. Using this function, we get all the related datasets in the form **(X_train,y_train)**, **(X_test,y_test)**. Since we only require the **X_train** data, we do not store the other datasets in variables.

The MNIST data is two-dimensional; that is, (width, height). Since we require three-dimensional data (width, height, channel) for convolutional operations, we need to create the third dimension as 1 using the **np.newaxis** function. Please note that the first dimensions will be the number of examples:

```
# Reshaping the input data to include channel
    X = X_train[:,:,:,np.newaxis]
```

```
# Generating a batch of data
    imageBatch = X[np.random.randint(0, X.shape[0], size=batch)]
```

The other process is to generate batches of the training data. To generate batches of data, we sample some integers between 0 and the number of examples in the training set. The sample's size will be equal to the batch size we want. This is implemented as follows:

```
# Generating a batch of data
    imageBatch = X[np.random.randint(0, X.shape[0], size=batch)]
```

We will only be returning the **X** variable. The labels that are batches of 1s will be separately generated during the training process.

Then, we need to define the three functions that will be used for generating fake samples. These are a function for generating fake inputs, a function for the generator network, and a function for generating fake samples and labels. Most of these functions are the same as what we developed in *Exercise 7.04*, *Implementing the GAN*. The generator model will be constructed as a convolutional model with intermittent use of **Up-Sampling/Converse2Dtranspose** operations.

Next, we need to create a new function for the discriminator network. This discriminator model will, again, be a convolutional model with the final layer as a sigmoid layer where we output a probability, that is, the probability of an image being real or fake. The input dimensions to the discriminator model will be the dimensions of the images generated from MNIST and the fake images, which will be (batch size, 28,28,1).

The GAN model will be the same as the one we created in *Exercise 7.04, Implementing the GAN*. This function will have the generator model and the discriminator model as its inputs.

THE TRAINING PROCESS

The training process will be similar to the process we implemented in *Exercise 7.04, Implementing the GAN*. The steps for the training process are as follows:

1. Generate a batch of MNIST data using the function to generate a real dataset.

2. Generate a batch of fake samples using *function 3* described in the functions for generating fake samples.

3. Concatenate the real samples and fake samples into one DataFrame. This will be the input variable for the discriminator model.

4. The labels will be a series of 1s and 0s corresponding to the real data and fake data that was concatenated earlier.

5. Train the discriminator model using the **train_on_batch()** function using the **X** variable and the labels.

6. Generate another batch of fake inputs for training the GAN model. These fake samples are generated using *function 1* in the fake sample generation process.

7. Generate the labels for the fake samples that are intended to fool the discriminator. These labels will be 1s instead of 0s.

8. Train the GAN model using the **train_on_batch()** function using the fake samples and its labels, as described in *Steps 6* and *7*.

9. *Steps 1* to *8* are repeated for the number of epochs we want the training to run for. This is done through a **for** loop over the number of epochs.

10. At every intermediate step, we calculate the accuracy of the discriminator model.

11. We also generate output plots at certain epochs.

Now that we have seen the complete process behind training a DCGAN, let's dive into the next exercise, which implements this process.

EXERCISE 7.05: IMPLEMENTING THE DCGAN

In this exercise, we will build and train the DCGAN on the MNIST dataset. We will use the MNIST dataset as the real data distribution. We will then generate fake data from a random distribution. After that, we will train the GAN to generate data that is similar to the MNIST dataset's. Follow these steps to complete this exercise:

1. Open a new Jupyter Notebook and name it **Exercise 7.05**. Import the following library packages and the MNIST dataset:

```
# Import the required library functions
import numpy as np
import matplotlib.pyplot as plt
from matplotlib import pyplot
import tensorflow as tf
from tensorflow.keras.layers import Input
from tensorflow.keras.initializers import RandomNormal
from tensorflow.keras.models import Model, Sequential
from tensorflow.keras.layers \
import Reshape, Dense, Dropout, Flatten,Activation
from tensorflow.keras.layers import LeakyReLU,BatchNormalization
from tensorflow.keras.layers \
import Conv2D, UpSampling2D,Conv2DTranspose
from tensorflow.keras.datasets import mnist
from tensorflow.keras.optimizers import Adam
```

2. Define the function that will be used to generate real datasets. The real dataset is generated from the MNIST data:

> **NOTE:**
>
> Alternatively, you can download the MNIST dataset from
> https://packt.live/2X4xeCL

```
# Function to generate real data samples
def realData(batch):
    # Get the MNIST data
    (X_train, _), (_, _) = mnist.load_data()
```

```
# Reshaping the input data to include channel
X = X_train[:,:,:,np.newaxis]
# normalising the data
X = (X.astype('float32') - 127.5)/127.5
# Generating a batch of data
imageBatch = X[np.random.randint(0, X.shape[0], size=batch)]
return imageBatch
```

The return value from this function is the batch of MNIST data. Note that we normalize the input data by subtracting **127.5**, which is half the maximum pixel values (255), and divide by the same amount. This will help with converging the solution faster.

3. Now, let's generate a set of images from the MNIST dataset:

```
# # Generating a batch of images
mnistData = realData(25)
```

4. Next, let's visualize the plots using **matplotlib**:

```
# Plotting the image
for j in range(5*5):
    pyplot.subplot(5,5,j+1)
    # turn off axis
    pyplot.axis('off')
    pyplot.imshow(mnistData[j,:,:,0],cmap='gray_r')
```

You should get an output similar to the following:

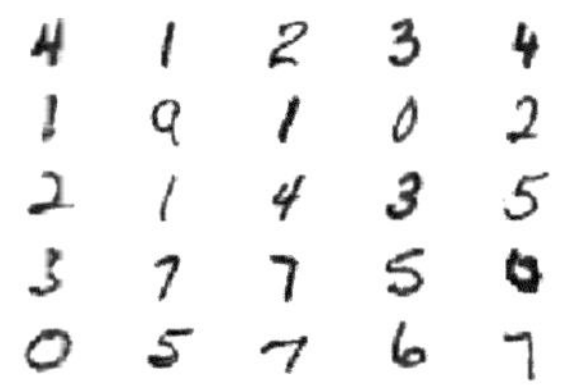

Figure 7.24: Visualized data – digits from the dataset

From the output, we can see the visualization of some of the digits. We can see that the image is centrally positioned within a white background.

NOTE

The digits that are visualized when you run the code will differ from the ones we've shown here.

5. Define the function to generate inputs for the generator network. The fake data
    will be random data points generated from a uniform distribution:

```
# Function to generate inputs for generator function
def fakeInputs(batch,infeats):
    #Generate random noise data with shape (batch,input features)
    x_fake = np.random.uniform(-1,1,size=[batch,infeats])
    return x_fake
```

6. Let's define the function for building the generator network. Building the
    generator network is similar to building any CNN network. In this generator
    network, we will use the **UpSampling** method:

Exercise7.05.ipynb

```
# Function for the generator model
def genModel(infeats):
    # Defining the Generator model
    Genmodel = Sequential()
    Genmodel.add(Dense(512,input_dim=infeats))
    Genmodel.add(Activation('relu'))
    Genmodel.add(BatchNormalization())
    # second layer of FC => RElu => BN layers
    Genmodel.add(Dense(7*7*64))
    Genmodel.add(Activation('relu'))
```

The complete code for this step can be found on https://packt.live/2ZPg8cJ.

In the model, we can see the progressive use of the transpose convolution
operation. The initial input has the dimensions of 100. This is progressively
increased to the desired image size of batch size x 28 x 28 through a series of
transpose convolution operations.

7. Next, we define the function to create fake samples. In this function, we only
    return the **X** variable:

```
# Function to create fake samples using the generator model
def fakedataGenerator(Genmodel,batch,infeats):
    # first generate the inputs to the model
    genInputs = fakeInputs(batch,infeats)
    """
    use these inputs inside the generator model
    to generate fake distribution
    """
    X_fake = Genmodel.predict(genInputs)

    return X_fake
```

The return value from this function is the fake dataset.

8. Define the parameters that we will use, along with the summary of the generator network:

```
# Define the arguments like batch size and input feature
batch = 128
infeats = 100
Genmodel = genModel(infeats)
Genmodel.summary()
```

You should get the following output:

```
Layer (type)                    Output Shape           Param #
=================================================================
dense (Dense)                   (None, 512)            51712
_____
activation (Activation)         (None, 512)            0
_____
batch_normalization_v2 (Batc    (None, 512)            2048
_____
dense_1 (Dense)                 (None, 3136)           1608768
_____
activation_1 (Activation)       (None, 3136)           0
_____
batch_normalization_v2_1 (Ba    (None, 3136)           12544
_____
reshape (Reshape)               (None, 7, 7, 64)       0
_____
conv2d_transpose (Conv2DTran    (None, 14, 14, 32)     51232
_____
activation_2 (Activation)       (None, 14, 14, 32)     0
_____
batch_normalization_v2_2 (Ba    (None, 14, 14, 32)     128
_____
conv2d_transpose_1 (Conv2DTr    (None, 28, 28, 1)      801
_____
activation_3 (Activation)       (None, 28, 28, 1)      0
=================================================================
Total params: 1,727,233
```

Figure 7.25 Summary of the model

From the summary, please note how the dimension of the input changes with each transpose convolution operation. Finally, we get an output that is equal in dimension to the real data set: **(None,28 ,28,1)**.

9. Let's use the generator function to generate a fake sample before training:

```
# Generating a fake sample and printing the shape
fake = fakedataGenerator(Genmodel,batch,infeats)
fake.shape
```

You should get the following output:

```
(128, 28, 28, 1)
```

10. Now, let's plot the generated fake sample:

```
# Plotting the fake sample

plt.imshow(fake[1, :, :, 0], cmap='gray_r')

plt.xlabel('Fake Sample Image')
```

You should get an output similar to the one shown here:

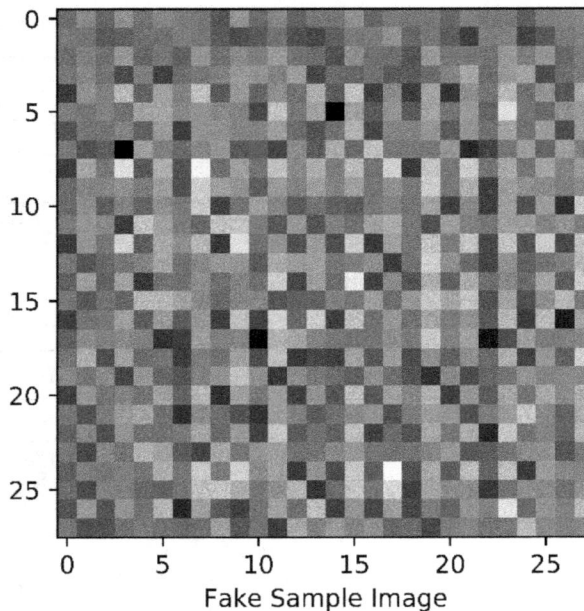

Figure 7.26: Plot of the fake sample image

This is the plot of the fake sample before training. After training, we want samples like these to look like the MNIST samples we visualized earlier in this exercise.

11. Now, let's build the discriminator model as a function. The network will be a CNN network like the one you learned about in *Chapter 3, Image Classification with Convolutional Neural Networks*:

Exercise7.05.ipynb

```
# Descriminator model as a function
def discModel():
    Discmodel = Sequential()
    Discmodel.add(Conv2D(32,kernel_size=(5,5),strides=(2,2), \
                       padding='same',input_shape=(28,28,1)))
    Discmodel.add(LeakyReLU(0.2))
    # second layer of convolutions
    Discmodel.add(Conv2D(64, kernel_size=(5,5), \
                       strides=(2, 2), padding='same'))
```

The complete code for this step can be found on https://packt.live/2ZPg8cJ.

In the discriminator network, we have included all the necessary layers, such as the convolutional operations and LeakyReLU activations. Please note that the last layer is a sigmoid layer as we want the output as a probability of the sample to be real (1) or fake (0).

12. Print the summary of the discriminator network:

```
# Print the summary of the discriminator model
Discmodel = discModel()
Discmodel.summary()
```

You should get the following output:

Layer (type)	Output Shape	Param #
conv2d (Conv2D)	(None, 14, 14, 32)	832
leaky_re_lu (LeakyReLU)	(None, 14, 14, 32)	0
conv2d_1 (Conv2D)	(None, 7, 7, 64)	51264
leaky_re_lu_1 (LeakyReLU)	(None, 7, 7, 64)	0
flatten (Flatten)	(None, 3136)	0
dense_2 (Dense)	(None, 512)	1606144
leaky_re_lu_2 (LeakyReLU)	(None, 512)	0
dense_3 (Dense)	(None, 1)	513

```
Total params: 1,658,753
Trainable params: 1,658,753
Non-trainable params: 0
```

Figure 7.27: Summary of the model architecture

The preceding screenshot shows the summary of the model architecture. This is based on the different layers we implemented using the **Sequential** class. For example, we can see that the first layer has 32 filter maps, the second layer has 64 filter maps, and the last layer has one output that corresponds to the sigmoid activation.

13. Next, define the GAN model as a function:

```
"""
Define the combined generator and discriminator model,
for updating the generator
"""
def ganModel(Genmodel,Discmodel):
    # First define that discriminator model cannot be trained
    Discmodel.trainable = False
    Ganmodel = Sequential()
    # First adding the generator model
    Ganmodel.add(Genmodel)
    """
    Next adding the discriminator model
    without training the parameters
    """
    Ganmodel.add(Discmodel)
    # Compilise the model for loss to optimise the Generator model
    Ganmodel.compile(loss='binary_crossentropy',\
                    optimizer = 'adam')
    return Ganmodel
```

The structure of the GAN model is similar to the one we developed in *Exercise 7.04, Implementing the GAN*.

14. Now, it's time to invoke the GAN function. Please note that the inputs to the GAN model are the previously defined generator and discriminator models:

```
# Initialise the gan model
gan_model = ganModel(Genmodel,Discmodel)
# Print summary of the GAN model
gan_model.summary()
```

From the preceding code, we can see that the inputs to the GAN model are the previously defined generator and discriminator models. You should get the following output:

```
Layer (type)                  Output Shape            Param #
=================================================================
sequential (Sequential)       (None, 28, 28, 1)       1727233
_____
sequential_1 (Sequential)     (None, 1)               1658753
=================================================================
Total params: 3,385,986
Trainable params: 1,719,873
Non-trainable params: 1,666,113
```

Figure 7.28: Model summary

Please note that the parameters of each layer of the GAN model are equivalent to the parameters of the generator and discriminator models. The GAN model is just a wrapper around the models we defined earlier.

15. Define the number of epochs to train the network:

```
# Defining the number of epochs
nEpochs = 5000
```

16. Now, let's train the GAN:

> **NOTE:**
>
> Before you run the code that follows, make sure you have a folder titled **handwritten** in the same path as your Jupyter Notebook.

`Exercise7.05.ipynb`

```
# Train the GAN network
for i in range(nEpochs):
    """
    Generate samples equal to the bath size
    from the real distribution
    """
    x_real = realData(batch)
    #Generate fake samples using the fake data generator function
    x_fake = fakedataGenerator(Genmodel,batch,infeats)
    # Concatenating the real and fake data
    X = np.concatenate([x_real,x_fake])
    #Creating the dependent variable and initializing them as '0'
    Y = np.zeros(batch * 2)
```

The full code for this step can be found at https://packt.live/2ZPg8cJ.

From the preceding code, we can see that the training of the discriminator model with the fake and real samples and the training of the GAN model happens concurrently. The only difference is the training of the GAN model proceeds without updating the parameters of the discriminator model. The other thing to note is that, inside the GAN, the labels for the fake samples would be 1 to generate large loss terms that will be backpropagated through the discriminator network to update the generator parameters. We also display the predicted probability of the GAN for every 10 epochs. When calculating the probability, we combine a sample of real data and fake data and then take the mean of the predicted probability. We also save a copy of the generated image.

> **NOTE:**
>
> We'll analyze the plots that will be generated in the section that follows.

You should get an output similar to the following:

```
Discriminator probability:0.6213402152061462
Discriminator probability:0.7360671758651733
Discriminator probability:0.6130768656730652
Discriminator probability:0.5046337842941284
Discriminator probability:0.5005484223365784
Discriminator probability:0.50015789270401
Discriminator probability:0.5000558495521545
Discriminator probability:0.5000174641609192
Discriminator probability:0.5000079274177551
Discriminator probability:0.4999823570251465
Discriminator probability:0.5000027418136597
Discriminator probability:0.5000032186508179
Discriminator probability:0.5000043511390686
Discriminator probability:0.5000077486038208
```

> **NOTE**
>
> The output for the preceding code may not be an exact match with what you get when you run the code.

From the predicted probability of the test data, we can see that the values are hovering around the **.55** mark. This is an indication that the discriminator is confused about whether the image is fake or real. If the discriminator were sure that an image was real, it would predict a probability close to 1, while it would predict a probability close to 0 if it were sure that the image was fake. In our case, we can see that the probability is around the .55 mark, which indicates that the generator is learning to generate images similar to the real images. This has confused the discriminator. *A value close to 50% accuracy for the discriminator is the desired value.*

17. Now, let's generate fake images after the training process and visualize them:

```
# Images predicted after training
x_fake = fakedataGenerator(Genmodel,25,infeats)
# Visualizing the plots
for j in range(5*5):
    pyplot.subplot(5,5,j+1)
    # turn off axis
    pyplot.axis('off')
    pyplot.imshow(x_fake[j,:,:,0],cmap='gray_r')
```

The output will be as follows:

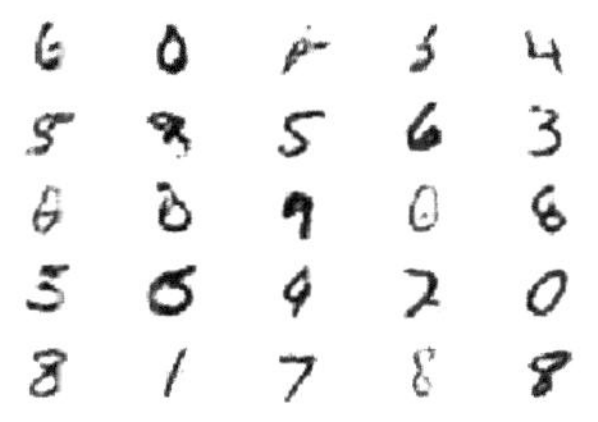

Figure 7.29: Predicted image post training

We can see that the generated images from the trained generator model closely resonate with the real handwritten digits.

> **NOTE**
>
> To access the source code for this specific section, please refer to https://packt.live/2ZPg8cJ.
>
> This section does not currently have an online interactive example, and will need to be run locally.

In this exercise, we developed a GAN to generate distributions similar to MNIST handwritten digits. In the section that follows, we will analyze the images that were generated at each epoch during this exercise.

ANALYSIS OF SAMPLE PLOTS

Now, let's look at the output sample plots from the previous exercise to see what the generated images look like. By completing the previous exercise, these should have been saved in the same path where your Jupyter Notebook is located, under a subfolder called **handwritten**:

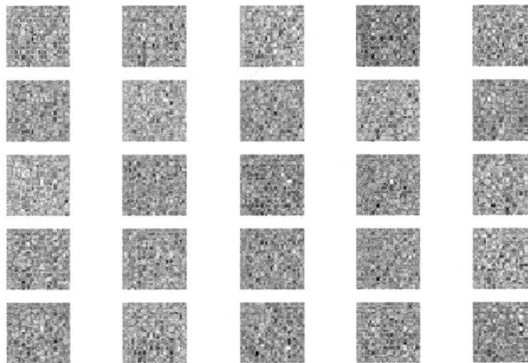

Figure 7.30: Sample plot after 10 iterations

The preceding images are those that were generated after 10 iterations. We can see that these images look more like random noise. However, we can also see that there are traces of white patches forming within the image, which indicates the GAN is learning some of the features of the real image:

Figure 7.31: Sample plot after 500 iterations

The preceding images are the plots after 500 iterations. From these images, we can see some semblance of the real image. We can see that the white background of the real images is being formed. We can also see the distribution getting aggregated at the center of the image:

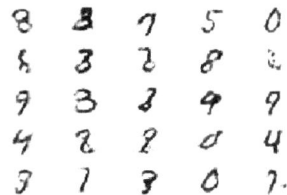

Figure 7.32: Sample plot after 2,000 iterations

The preceding image is after 2,000 iterations. We can see that many digits have started to form; for example, 8, 5 ,3 ,9 ,4, 7, 0, and so on. We can also see that the dark shades of the images have started to become more pronounced. Now, let's look at the images that were generated during the last iteration:

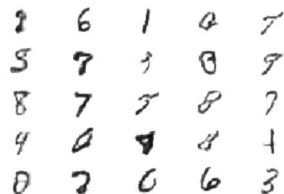

Figure 7.33: Sample plot after 5,000 iterations

A question to ask at this stage is, are these images perfect? Absolutely not. Would running the training for more epochs improve the results further? Not necessarily. Getting those perfect images would entail hours of training and experimentation with different model architectures. You can take this as a challenge to improve the output through your choices of architecture and the parameters within the model.

GANs are a really active area of research and the possibilities they are opening up point to the direction of computers slowly becoming creative. However, there are some common problems when implementing GANs. Let's look at some of them.

COMMON PROBLEMS WITH GANS

GANs are difficult networks to train and stabilize. There are different failure modes for GANs. Let's get a perspective of some of the common failure modes.

MODE COLLAPSE

A very common failure mode of GANs, especially on multi-modal data, is a situation called **mode collapse**. This refers to a situation where the generator learns only some specific variety of the distribution within the data. For example, in an MNIST data distribution, if the GAN generates only one particular digit (say, 5) after training, then this is a case of mode collapse.

One way to combat mode collapse is to group data according to the different classes and train the discriminator accordingly. This will give the discriminator the ability to identify different modes that are present in the data.

CONVERGENCE FAILURE

Another prominent failure mode in GANs is convergence failure. In this failure mode, the network fails to converge with the loss as it never settles during the training phase. Some methods that researchers have used to get over this problem include adding noise to discriminatory networks and penalizing discriminator weights through regularization techniques.

Notwithstanding the numerous challenges inherent in training and building GANs, it still remains one of the most active areas of research within the deep learning community. The promises and the applications that are made possible by GANs are what make this area one of the most sought-after domains in deep learning. Now that we have laid some of the foundations for GANs, let's use what we've learned to build a GAN for a different dataset.

ACTIVITY 7.01: IMPLEMENTING A DCGAN FOR THE MNIST FASHION DATASET

In this activity, you will implement a DCGAN to generate images similar to the ones found in the MNIST fashion dataset. The MNIST fashion dataset is similar to the handwritten digital images dataset that you implemented in *Exercise 7.05*, *Implementing the DCGAN*. This dataset consists of grayscale images of 10 different fashion accessories and comprises 60,000 training samples. The following is a sample of the images included in this dataset:

Figure 7.34: Sample of the MNIST fashion dataset

The objective of this activity is to build a GAN and generate images similar to the fashion dataset. The high-level steps for this activity will be similar to *Exercise 7.05*, *Implementing the DCGAN*, where you implemented a DCGAN for handwritten digits. You will be completing this activity in two parts, first by creating the relevant functions and then by training the model.

Generating Key Functions: Here, you will be creating the required functions, such as the generator function and the discriminator function:

1. Define the function that will generate a real data distribution. This function has to generate the real data distribution from the MNIST fashion dataset. The fashion dataset can be imported using the following code:

    ```
    from tensorflow.keras.datasets import fashion_mnist
    ```

    The training set can be generated using the **fashion_mnist.load_data()** function.

    > **NOTE:**
    >
    > Alternatively, you can download the dataset from https://packt.live/2X4xeCL.

2. Define the three functions that will be used to generate fake samples; that is, the function for generating fake inputs, the function for the generator network, and the function for generating fake samples and labels. Use *Converse2Dtranspose* operations within the generator function.

3. Create a new function for the discriminator network.

4. Create the GAN model. You can take cues from *Exercise 7.05, Implementing the DCGAN*, on how to do this.

The Training Process: You will follow a process similar to the one in *Exercise 7.05, Implementing the DCGAN*:

1. Generate a batch of MNIST data using the function for generating a real dataset.

2. Generate a batch of fake samples using the third function described in the functions for generating fake samples.

3. Concatenate the real samples and fake samples into one DataFrame and generate their labels.

4. Train the discriminator model using the **train_on_batch()** function using the **X** variable and the labels.

5. Generate another batch of fake inputs for training the GAN model, along with their labels.

6. Train the GAN model using the **train_on_batch()** function using the fake samples and their labels.

7. Repeat the training for around 5,000 epochs.

8. At every intermediate step, calculate the accuracy of the discriminator model.

The discriminator probabilities you'll get should be around **0.5**. The expected output will be a generated image that looks similar to the one shown here:

Figure 7.35: Expected output for this activity

> **NOTE:**
>
> The detailed steps for this activity, along with the solutions and additional commentary, are presented on page 426.

SUMMARY

You have come a long way from being introduced to one of the most promising areas in deep learning. Let's revisit some of the concepts that we learned about in this chapter.

We started this chapter by understanding what GANs are and their major applications. We then went on to understand the various building blocks of GANs, such as the real datasets, fake datasets, the discriminator operation, the generator operation, and the GAN operation.

We executed a problem statement to progressively build a **fully connected GAN (FCGAN)** to solve a real function. In the process of building the GAN, we also implemented exercises for creating real datasets, creating fake datasets, creating a generator network, creating a discriminator network, and finally combining all these individual components to create the GAN. We visualized the different plots and understood how the generated data distribution mimicked the real data distribution.

In the next section, we understood the concept of DCGANs. We also visited some of the unique concepts in DCGANs such as upsampling and transpose convolutions. We implemented a GAN for the MNIST digital handwritten images and visualized the fake images we generated using a DCGAN. Finally, we also implemented a DCGAN for the MNIST fashion dataset in an activity.

Having laid the foundations, the next question is, where do we go from here? GANs are a large area by itself and there's quite a lot of buzz around it these days. To start with, it would be good to tweak the models you have already learned by tweaking the architecture and activation functions and trying out other parameters such as batch normalization. After playing around with different variations of the current models, it will be time to explore other networks such as the **Least Squares GAN (LSGAN)** and **Wasserstein GAN (WGAN)**. Then, there is this large playing field of conditional GANs such as the **Conditional GAN (cGan)**, InfoGAN, **Auxiliary Classifier GAN (AC-GAN)**, and **Semi-Supervised GAN (SGAN)**. Once you've done this, you'll have set the stage for advanced topics such as CycleGAN, BigGAN, and StyleGAN.

This chapter also brings down the curtain on the amazing journey you've made throughout this book. First, you were introduced to what deep learning is and the different use cases that are possible with deep learning. Subsequently, you learned the basics of neural networks, which are the foundations of deep learning. From there, you went on to learn about advanced techniques such as CNNs, which are the workhorses for use cases such as image recognition. Along with that, you learned about recurrent neural networks, which can be used for sequence data. Finally, you were introduced to GANs, a class of networks that's making lots of waves within the domain. Having equipped yourself with this set of tools now is the time to apply your learning to your domain. The possibilities and opportunities are endless. All we need to do is consolidate our current learning and move ahead step by step. I wish you all the best on your journey in scaling new peaks in the deep learning domain.

APPENDIX

CHAPTER 1: BUILDING BLOCKS OF DEEP LEARNING

ACTIVITY 1.01: SOLVING A QUADRATIC EQUATION USING AN OPTIMIZER

SOLUTION

Let's solve the following quadratic equation:

$$x^2 - 10x + 25 = 0$$

Figure 1.29: Quadratic equation to be solved

We already know that the solution to this quadratic equation is **x=5**.

We can use an optimizer to solve this. For the optimizer, **x** is the variable and the cost function is the left-hand side expression, which is as follows:

$$x^2 - 10x + 25$$

Figure 1.30: Left-hand side expression

The optimizer will find the value of **x** for which the expression is the minimum – in this case, it is **0**. Please note that this will work only for quadratic equations that are perfect squares, such as in this case. The left-hand side expression is a perfect square that can be explained with the following equation:

$$(x - 5)^2$$

Figure 1.31: Perfect square

Now, let's look at the code for solving this:

1. Open a new Jupyter Notebook and rename it *Activity 1.01*.

2. Import **tensorflow**:

```
import tensorflow as tf
```

3. Create the variable **x** and initialize it to 0.0:

```
x=tf.Variable(0.0)
```

4. Construct the **loss** function as a **lambda** function:

    ```
    loss=lambda:abs(x**2-10*x+25)
    ```

5. Create an instance of an optimizer with a learning rate of **.01**:

    ```
    optimizer=tf.optimizers.Adam(.01)
    ```

6. Run the optimizer through 10,000 iterations. You can start with a smaller number such as 1,000 and keep increasing the number of iterations until you get the solution:

    ```
    for i in range(10000):
        optimizer.minimize(loss,x)
    ```

7. Print the value of **x**:

    ```
    tf.print(x)
    ```

    The output is as follows:

    ```
    4.99919891
    ```

This is the solution to our quadratic equation. It may be noted that, irrespective of the number of iterations, you will never get a perfect 5.

> **NOTE**
>
> To access the source code for this specific section, please refer to https://packt.live/3gBTFGA.
>
> You can also run this example online at https://packt.live/2Dqa2Id. You must execute the entire Notebook in order to get the desired result.

CHAPTER 2: NEURAL NETWORKS

ACTIVITY 2.01: BUILD A MULTILAYER NEURAL NETWORK TO CLASSIFY SONAR SIGNALS

SOLUTION

Let's see how the solution looks. Remember—this is one solution, but there could be many variations:

1. Import all the required libraries:

```
import tensorflow as tf
import pandas as pd
from sklearn.preprocessing import LabelEncoder
# Import Keras libraries
from tensorflow.keras.models import Sequential
from tensorflow.keras.layers import Dense
```

2. Load and examine the data:

```
df = pd.read_csv('sonar.csv')
df.head()
```

The output is:

bute_8	attribute_9	attribute_10	...	attribute_52	attribute_53	attribute_54	attribute_55	attribute_56	attribute_57	attribute_58	attribute_59	attribute_60	Class
0.1601	0.3109	0.2111	...	0.0027	0.0065	0.0159	0.0072	0.0167	0.0180	0.0084	0.0090	0.0032	Rock
0.3481	0.3337	0.2872	...	0.0084	0.0089	0.0048	0.0094	0.0191	0.0140	0.0049	0.0052	0.0044	Rock
0.3771	0.5598	0.6194	...	0.0232	0.0166	0.0095	0.0180	0.0244	0.0316	0.0164	0.0095	0.0078	Rock
0.1276	0.0598	0.1264	...	0.0121	0.0036	0.0150	0.0085	0.0073	0.0050	0.0044	0.0040	0.0117	Rock
0.2467	0.3564	0.4459	...	0.0031	0.0054	0.0105	0.0110	0.0015	0.0072	0.0048	0.0107	0.0094	Rock

Figure 2.37: Contents of sonar.csv

Observe that there are 60 features, and the target has two values—Rock and Mine.

This means that this is a binary classification problem. Let's prepare the data before we build the neural network.

3. Separate the features and the labels:

```
X_input = df.iloc[:, :-1]
Y_label = df['Class'].values
```

In this code, **X_input** is selecting all the rows of all the columns except the **Class** column, and **Y_label** is just selecting the **Class** column.

4. Labels are in text format. We need to encode them as numbers before we can use them with our model:

```
labelencoder_Y = LabelEncoder()
Y_label = labelencoder_Y.fit_transform(Y_label)
Y_label = Y_label.reshape([208, 1])
```

The **reshape** function at the end will convert the labels into matrix format, which is expected by the model.

5. Build the multilayer model with Keras:

```
model = Sequential()
model.add(Dense(300,input_dim=60, activation = 'relu'))
model.add(Dense(200, activation = 'relu'))
model.add(Dense(100, activation = 'relu'))
model.add(Dense(1, activation = 'sigmoid'))
```

You can experiment with the number of layers and neurons, but the last layer can only have one neuron with a sigmoid activation function, since this is a binary classifier.

6. Set the training parameters:

```
model.compile(optimizer='adam',loss='binary_crossentropy', \
                metrics=['accuracy'])
```

7. Train the model:

```
model.fit(X_input, Y_label, epochs=30)
```

The truncated output will be somewhat similar to the following:

```
Train on 208 samples
Epoch 1/30
208/208 [==============================] - 0s 205us/sample -
loss:
  0.1849 - accuracy: 0.9038
Epoch 2/30
208/208 [==============================] - 0s 220us/sample -
loss:
  0.1299 - accuracy: 0.9615
Epoch 3/30
```

```
208/208 [==============================] - 0s 131us/sample -
loss:
   0.0947 - accuracy: 0.9856
Epoch 4/30
208/208 [==============================] - 0s 151us/sample -
loss:
   0.1046 - accuracy: 0.9712
Epoch 5/30
208/208 [==============================] - 0s 171us/sample -
loss:
   0.0952 - accuracy: 0.9663
Epoch 6/30
208/208 [==============================] - 0s 134us/sample -
loss:
   0.0777 - accuracy: 0.9856
Epoch 7/30
208/208 [==============================] - 0s 129us/sample -
loss:
   0.1043 - accuracy: 0.9663
Epoch 8/30
208/208 [==============================] - 0s 142us/sample -
loss:
   0.0842 - accuracy: 0.9712
Epoch 9/30
208/208 [==============================] - 0s 155us/sample -
loss:
   0.1209 - accuracy: 0.9423
Epoch 10/30
208/208 [==============================] - ETA: 0s - loss:
   0.0540 - accuracy: 0.98 - 0s 334us/sample - los
```

8. Let's evaluate the trained model and examine its accuracy:

```
model.evaluate(X_input, Y_label)
```

The output is as follows:

```
208/208 [==============================] - 0s 128us/sample -
loss:
  0.0038 - accuracy: 1.0000
 [0.003758653004367191, 1.0]
```

As you can see, we have been able to successfully train a multilayer binary neural network and get 100% accuracy within 30 epochs.

> **NOTE**
>
> To access the source code for this specific section, please refer to https://packt.live/38EMoDi.
>
> You can also run this example online at https://packt.live/2W2sygb. You must execute the entire Notebook in order to get the desired result.

CHAPTER 3: IMAGE CLASSIFICATION WITH CONVOLUTIONAL NEURAL NETWORKS (CNNS)

ACTIVITY 3.01: BUILDING A MULTICLASS CLASSIFIER BASED ON THE FASHION MNIST DATASET

SOLUTION

1. Open a new Jupyter Notebook.

2. Import **tensorflow.keras.datasets.fashion_mnist**:

   ```
   from tensorflow.keras.datasets import fashion_mnist
   ```

3. Load the Fashion MNIST dataset using **fashion_mnist.load_data()** and save the results to **(features_train, label_train), (features_test, label_test)**:

   ```
   (features_train, label_train), (features_test, label_test) = \
   fashion_mnist.load_data()
   ```

4. Print the shape of the training set:

   ```
   features_train.shape
   ```

   The output will be as follows:

   ```
   (60000, 28, 28)
   ```

   The training set is composed of **60000** images of size **28** by **28**. We will need to reshape it and add the channel dimension.

5. Print the shape of the testing set:

   ```
   features_test.shape
   ```

   The output will be as follows:

   ```
   (10000, 28, 28)
   ```

   The testing set is composed of **10000** images of size **28** by **28**. We will need to reshape it and add the channel dimension

6. Reshape the training and testing sets with the dimensions **(number_rows, 28, 28, 1)**:

```
features_train = features_train.reshape(60000, 28, 28, 1)
features_test = features_test.reshape(10000, 28, 28, 1)
```

7. Create three variables called **batch_size**, **img_height**, and **img_width** that take the values **16**, **28**, and **28**, respectively:

```
batch_size = 16
img_height = 28
img_width = 28
```

8. Import **ImageDataGenerator** from **tensorflow.keras.preprocessing**:

```
from tensorflow.keras.preprocessing.image \
import ImageDataGenerator
```

9. Create an **ImageDataGenerator** called **train_img_gen** with data augmentation: **rescale=1./255, rotation_range=40, width_shift_range=0.1, height_shift_range=0.1, shear_range=0.2, zoom_range=0.2, horizontal_flip=True, fill_mode='nearest'**:

```
train_img_gen = ImageDataGenerator(rescale=1./255, \
                                   rotation_range=40, \
                                   width_shift_range=0.1, \
                                   height_shift_range=0.1, \
                                   shear_range=0.2, \
                                   zoom_range=0.2, \
                                   horizontal_flip=True, \
                                   fill_mode='nearest')
```

10. Create an **ImageDataGenerator** called **val_img_gen** with rescaling (by dividing by 255):

```
val_img_gen = ImageDataGenerator(rescale=1./255)
```

11. Create a data generator called **train_data_gen** using **.flow()** and specify the batch size, features, and labels from the training set:

```
train_data_gen = train_img_gen.flow(features_train, \
                                     label_train, \
                                     batch_size=batch_size)
```

12. Create a data generator called **val_data_gen** using **.flow()** and specify the batch size, features, and labels from the testing set:

```
val_data_gen = train_img_gen.flow(features_test, \
                                   label_test, \
                                   batch_size=batch_size)
```

13. Import **numpy** as **np**, **tensorflow** as **tf**, and **layers** from **tensorflow.keras**:

```
import numpy as np
import tensorflow as tf
from tensorflow.keras import layers
```

14. Set **8** as the seed for **numpy** and **tensorflow** using **np.random_seed()** and **tf.random.set_seed()**:

```
np.random.seed(8)
tf.random.set_seed(8)
```

15. Instantiate a **tf.keras.Sequential()** class into a variable called **model** with the following layers: A convolution layer with **64** kernels of shape **3**, **ReLU** as the activation function, and the necessary input dimensions; a max pooling layer; a convolution layer with **128** kernels of shape **3** and **ReLU** as the activation function; a max pooling layer; a flatten layer; a fully connected layer with **128** units and **ReLU** as the activation function; a fully connected layer with **10** units and **softmax** as the activation function.

The code should be as follows:

```
model = tf.keras.Sequential\
        ([layers.Conv2D(64, 3, activation='relu', \
                        input_shape=(img_height, \
                                     img_width ,1)), \
                        layers.MaxPooling2D(), \
                        layers.Conv2D(128, 3, \
                                      activation='relu'), \
                        layers.MaxPooling2D(),\
                        layers.Flatten(), \
                        layers.Dense(128, \
                                     activation='relu'), \
                        layers.Dense(10, \
                                     activation='softmax')])
```

16. Instantiate a **tf.keras.optimizers.Adam()** class with **0.001** as the learning rate and save it to a variable called optimizer:

```
optimizer = tf.keras.optimizers.Adam(0.001)
```

17. Compile the neural network using .**compile()** with **loss='sparse_ categorical_crossentropy', optimizer=optimizer, metrics=['accuracy']**:

```
model.compile(loss='sparse_categorical_crossentropy', \
              optimizer=optimizer, metrics=['accuracy'])
```

18. Fit the neural networks with **fit_generator()** and provide the train and validation data generators, **epochs=5**, the steps per epoch, and the validation steps:

```
model.fit_generator(train_data_gen, \
                    steps_per_epoch=len(features_train) \
                                    // batch_size, \
                    epochs=5, \
                    validation_data=val_data_gen, \
                    validation_steps=len(features_test) \
                                     // batch_size)
```

The expected output will be as follows:

```
Epoch 1/5
3750/3750 [==============================] - 55s 15ms/step - loss: 0.8060 - accuracy: 0.6972 - val_loss:
0.6738 - val_accuracy: 0.7481
Epoch 2/5
3750/3750 [==============================] - 52s 14ms/step - loss: 0.5892 - accuracy: 0.7788 - val_loss:
0.5588 - val_accuracy: 0.7911
Epoch 3/5
3750/3750 [==============================] - 54s 14ms/step - loss: 0.5211 - accuracy: 0.8041 - val_loss:
0.4928 - val_accuracy: 0.8170
Epoch 4/5
3750/3750 [==============================] - 53s 14ms/step - loss: 0.4864 - accuracy: 0.8162 - val_loss:
0.4896 - val_accuracy: 0.8186
Epoch 5/5
3750/3750 [==============================] - 7022s 2s/step - loss: 0.4590 - accuracy: 0.8271 - val_loss:
0.4696 - val_accuracy: 0.8334
```

Figure 3.30: Model training log

We trained our CNN on five epochs, and we achieved accuracy scores of **0.8271** on the training set and **0.8334** on the validation set, respectively. Our model is not overfitting much and achieved quite a high score. The accuracy is still increasing after five epochs, so we may get even better results if we keep training it. This is something you may try by yourself.

> **NOTE**
>
> To access the source code for this specific section, please refer to https://packt.live/2ObmA8t.
>
> You can also run this example online at https://packt.live/3fiyyJi. You must execute the entire Notebook in order to get the desired result.

ACTIVITY 3.02: FRUIT CLASSIFICATION WITH TRANSFER LEARNING

SOLUTION

1. Open a new Jupyter Notebook.

2. Import **tensorflow** as **tf**:

```
import tensorflow as tf
```

3. Create a variable called **file_url** containing the link to the dataset:

```
file_url = 'https://github.com/PacktWorkshops'\
           '/The-Deep-Learning-Workshop'\
           '/raw/master/Chapter03/Datasets/Activity3.02'\
           '/fruits360.zip'
```

> **NOTE**
>
> In the aforementioned step, we are using the dataset stored at
> https://packt.live/3eePQ8G. If you have stored the dataset at any
> other URL, please change the highlighted path accordingly.

4. Download the dataset using **tf.keras.get_file** with **'fruits360.zip'**, **origin=file_url**, **extract=True** as parameters and save the result to a variable called **zip_dir**:

```
zip_dir = tf.keras.utils.get_file('fruits360.zip', \
                                  origin=file_url, \
                                  extract=True)
```

5. Import the **pathlib** library:

```
import pathlib
```

6. Create a variable called **path** containing the full path to the **fruits360_filtered** directory using **pathlib.Path(zip_dir).parent**:

```
path = pathlib.Path(zip_dir).parent / 'fruits360_filtered'
```

7. Create two variables called **train_dir** and **validation_dir** that take the full paths to the train (**Training**) and validation (**Test**) folders, respectively:

```
train_dir = path / 'Training'
validation_dir = path / 'Test'
```

8. Create two variables called **total_train** and **total_val** that will get the number of images for the training and validation sets, that is, **11398** and **4752**:

```
total_train = 11398
total_val = 4752
```

9. Import **ImageDataGenerator** from **tensorflow.keras.
    preprocessing**:

```
from tensorflow.keras.preprocessing.image \
import ImageDataGenerator
```

10. Create an **ImageDataGenerator** called **train_img_gen** with data
    augmentation: **rescale=1./255, rotation_range=40, width_
    shift_range=0.1, height_shift_range=0.1, shear_
    range=0.2, zoom_range=0.2, horizontal_flip=True, fill_
    mode='nearest'**:

```
train_img_gen = ImageDataGenerator(rescale=1./255, \
                                   rotation_range=40, \
                                   width_shift_range=0.1, \
                                   height_shift_range=0.1, \
                                   shear_range=0.2, \
                                   zoom_range=0.2, \
                                   horizontal_flip=True, \
                                   fill_mode='nearest')
```

11. Create an **ImageDataGenerator** called **val_img_gen** with rescaling (by
    dividing by 255):

```
val_img_gen = ImageDataGenerator(rescale=1./255)
```

12. Create four variables called **batch_size**, **img_height**, **img_width**, and
    channel that take the values **16**, **100**, **100**, and **3**, respectively:

```
batch_size=16
img_height = 100
img_width = 100
channel = 3
```

13. Create a data generator called **train_data_gen** using **.flow_from_
    directory()** and specify the batch size, training folder, and target size:

```
train_data_gen = train_image_generator.flow_from_directory\
                 (batch_size=batch_size, \
                  directory=train_dir, \
                  target_size=(img_height, img_width))
```

14. Create a data generator called **val_data_gen** using **.flow_from_directory()** and specify the batch size, validation folder, and target size:

```
val_data_gen = validation_image_generator.flow_from_directory\
                (batch_size=batch_size, \
                directory=validation_dir, \
                target_size=(img_height, img_width))
```

15. Import **numpy** as **np**, **tensorflow** as **tf**, and **layers** from **tensorflow.keras**:

```
import numpy as np
import tensorflow as tf
from tensorflow.keras import layers
```

16. Set **8** as the seed for **numpy** and **tensorflow** using **np.random_seed()** and **tf.random.set_seed()**:

```
np.random.seed(8)
tf.random.set_seed(8)
```

17. Import **VGG16** from **tensorflow.keras.applications**:

```
from tensorflow.keras.applications import VGG16
```

18. Instantiate a **VGG16** model into a variable called **base_model** with the following parameters:

```
base_model = VGG16(input_shape=(img_height, \
                                img_width, channel), \
                                weights='imagenet', \
                                include_top=False)
```

19. Set this model to non-trainable using the **.trainable** attribute:

```
base_model.trainable = False
```

20. Print the summary of this **VGG16** model:

```
base_model.summary()
```

The expected output will be as follows:

```
block3_conv3 (Conv2D)         (None, 25, 25, 256)        590080

block3_pool (MaxPooling2D)    (None, 12, 12, 256)        0

block4_conv1 (Conv2D)         (None, 12, 12, 512)        1180160

block4_conv2 (Conv2D)         (None, 12, 12, 512)        2359808

block4_conv3 (Conv2D)         (None, 12, 12, 512)        2359808

block4_pool (MaxPooling2D)    (None, 6, 6, 512)          0

block5_conv1 (Conv2D)         (None, 6, 6, 512)          2359808

block5_conv2 (Conv2D)         (None, 6, 6, 512)          2359808

block5_conv3 (Conv2D)         (None, 6, 6, 512)          2359808

block5_pool (MaxPooling2D)    (None, 3, 3, 512)          0
=================================================================
Total params: 14,714,688
Trainable params: 0
Non-trainable params: 14,714,688
```

Figure 3.31: Model summary

This output shows us the architecture of **VGG16**. We can see that there are **14,714,688** parameters in total, but there is no trainable parameter. This is expected as we have frozen all the layers of this model.

21. Create a new model using **tf.keras.Sequential()** by adding the base model to the following layers: **Flatten()**, **Dense(1000, activation='relu')**, and **Dense(120, activation='softmax')**. Save this model to a variable called **model**:

```
model = tf.keras.Sequential([base_model, \
                             layers.Flatten(), \
                             layers.Dense(1000, \
                                          activation='relu'), \
                             layers.Dense(120, \
                                          activation='softmax')])
```

22. Instantiate a **tf.keras.optimizers.Adam()** class with **0.001** as the learning rate and save it to a variable called **optimizer**:

```
optimizer = tf.keras.optimizers.Adam(0.001)
```

23. Compile the neural network using **.compile()** with **loss='categorical_crossentropy', optimizer=optimizer, metrics=['accuracy']**:

```
model.compile(loss='categorical_crossentropy', \
              optimizer=optimizer, metrics=['accuracy'])
```

24. Fit the neural networks with **fit_generator()** and provide the train and validation data generators, **epochs=5**, the steps per epoch, and the validation steps. This model may take a few minutes to train:

```
model.fit_generator(train_data_gen, \
                    steps_per_epoch=len(features_train) \
                                    // batch_size, \
                    epochs=5, \
                    validation_data=val_data_gen, \
                    validation_steps=len(features_test) \
                                     // batch_size)
```

The expected output will be as follows:

```
Train for 712 steps, validate for 297 steps
Epoch 1/5
712/712 [==============================] - 340s 478ms/step - loss: 2.0376 - accuracy: 0.5071 - val_loss: 1.2164 - val
_accuracy: 0.6431
Epoch 2/5
712/712 [==============================] - 343s 482ms/step - loss: 0.6580 - accuracy: 0.8043 - val_loss: 0.6205 - val
_accuracy: 0.8136
Epoch 3/5
712/712 [==============================] - 343s 481ms/step - loss: 0.4190 - accuracy: 0.8742 - val_loss: 0.5638 - val
_accuracy: 0.8293
Epoch 4/5
712/712 [==============================] - 339s 477ms/step - loss: 0.3450 - accuracy: 0.8902 - val_loss: 0.4557 - val
_accuracy: 0.8731
Epoch 5/5
712/712 [==============================] - 331s 465ms/step - loss: 0.2854 - accuracy: 0.9106 - val_loss: 0.3776 - val
_accuracy: 0.8920

<tensorflow.python.keras.callbacks.History at 0x15af12a90>
```

Figure 3.32: Expected output

Here, we used transfer learning to customize a pretrained **VGG16** model on ImageNet so that it fits our fruit classification dataset. We replaced the head of the model with our own fully connected layers and trained these layers on five epochs. We achieved an accuracy score of **0.9106** for the training set and **0.8920** for the testing set. These are quite remarkable results given the time and hardware used to train this model. You can try to fine-tune this model and see whether you can achieve an even better score.

> **NOTE**
>
> To access the source code for this specific section, please refer to https://packt.live/2DsVRCI.
>
> This section does not currently have an online interactive example, and will need to be run locally.

CHAPTER 4: DEEP LEARNING FOR TEXT – EMBEDDINGS

ACTIVITY 4.01: TEXT PREPROCESSING OF THE 'ALICE IN WONDERLAND' TEXT

SOLUTION

You need to perform the following steps:

> **NOTE**
>
> Before commencing this activity, make sure you have defined the **alice_raw** variable as demonstrated in the section titled *Downloading Text Corpora Using NLTK*.

1. Change the data to lowercase and separate into sentences:

```
txt_sents = tokenize.sent_tokenize(alice_raw.lower())
```

2. Tokenize the sentences:

```
txt_words = [tokenize.word_tokenize(sent) for sent in txt_sents]
```

3. Import **punctuation** from the **string** module and **stopwords** from NLTK:

```
from string import punctuation
stop_punct = list(punctuation)
from nltk.corpus import stopwords
stop_nltk = stopwords.words("english")
```

4. Create a variable holding the contextual stop words **--** and **said**:

```
stop_context = ["--", "said"]
```

5. Create a master list for the stop words to remove words that contain terms from punctuation, NLTK stop words, and contextual stop words:

```
stop_final = stop_punct + stop_nltk + stop_context
```

6. Define a function to drop these tokens from any input sentence (tokenized):

```
def drop_stop(input_tokens):
    return [token for token in input_tokens \
            if token not in stop_final]
```

7. Remove the terms in **stop_final** from the tokenized text:

```
alice_words_nostop = [drop_stop(sent) for sent in txt_words]
print(alice_words_nostop[:2])
```

Here's what the first two sentences look like:

```
[['alice', "'s", 'adventures', 'wonderland', 'lewis', 'carroll',
'1865', 'chapter', 'i.', 'rabbit-hole', 'alice', 'beginning',
'get', 'tired', 'sitting', 'sister', 'bank', 'nothing', 'twice',
'peeped', 'book', 'sister', 'reading', 'pictures', 'conversations',
"'and", 'use', 'book', 'thought', 'alice', "'without", 'pictures',
'conversation'], ['considering', 'mind', 'well', 'could', 'hot',
'day', 'made', 'feel', 'sleepy', 'stupid', 'whether', 'pleasure',
'making', 'daisy-chain', 'would', 'worth', 'trouble', 'getting',
'picking', 'daisies', 'suddenly', 'white', 'rabbit', 'pink', 'eyes',
'ran', 'close']]
```

8. Using the **PorterStemmer** algorithm from NLTK, perform stemming on the result. Print out the first five sentences of the result:

```
from nltk.stem import PorterStemmer
stemmer_p = PorterStemmer()
alice_words_stem = [[stemmer_p.stem(token) for token in sent] \
                    for sent in alice_words_nostop]
print(alice_words_stem[:5])
```

The output will be as follows:

```
[['alic', "'s", 'adventur', 'wonderland', 'lewi', 'carrol', '1865',
'chapter', 'i.', 'rabbit-hol', 'alic', 'begin', 'get', 'tire',
'sit', 'sister', 'bank', 'noth', 'twice', 'peep', 'book', 'sister',
'read', 'pictur', 'convers', "'and", 'use', 'book', 'thought',
'alic', "'without", 'pictur', 'convers'], ['consid', 'mind', 'well',
'could', 'hot', 'day', 'made', 'feel', 'sleepi', 'stupid', 'whether',
'pleasur', 'make', 'daisy-chain', 'would', 'worth', 'troubl', 'get',
'pick', 'daisi', 'suddenli', 'white', 'rabbit', 'pink', 'eye', 'ran',
'close'], ['noth', 'remark', 'alic', 'think', 'much', 'way', 'hear',
'rabbit', 'say', "'oh", 'dear'], ['oh', 'dear'], ['shall', 'late']]
```

> **NOTE**
>
> To access the source code for this specific section, please refer to https://packt.live/2VVNEgf.
>
> You can also run this example online at https://packt.live/38Gr54r.
> You must execute the entire Notebook in order to get the desired result.

ACTIVITY 4.02: TEXT REPRESENTATION FOR ALICE IN WONDERLAND

SOLUTION

You need to perform the following steps:

1. From *Activity 4.01, Text Preprocessing Alice in Wonderland*, print the first three sentences from the result after stop word removal. This is the data you will work with:

```
print(alice_words_nostop[:3])
```

The output is as follows:

```
[['alice', "'s", 'adventures', 'wonderland', 'lewis', 'carroll',
'1865', 'chapter', 'i.', 'rabbit-hole', 'alice', 'beginning',
'get', 'tired', 'sitting', 'sister', 'bank', 'nothing', 'twice',
'peeped', 'book', 'sister', 'reading', 'pictures', 'conversations',
"'and", 'use', 'book', 'thought', 'alice', "'without", 'pictures',
'conversation'], ['considering', 'mind', 'well', 'could', 'hot',
'day', 'made', 'feel', 'sleepy', 'stupid', 'whether', 'pleasure',
'making', 'daisy-chain', 'would', 'worth', 'trouble', 'getting',
'picking', 'daisies', 'suddenly', 'white', 'rabbit', 'pink', 'eyes',
'ran', 'close'], ['nothing', 'remarkable', 'alice', 'think', 'much',
'way', 'hear', 'rabbit', 'say', "'oh", 'dear']]
```

2. Import **word2vec** from Gensim and train your word embeddings with default parameters:

```
from gensim.models import word2vec
model = word2vec.Word2Vec(alice_words_nostop)
```

3. Find the **5** terms most similar to **rabbit**:

```
model.wv.most_similar("rabbit", topn=5)
```

The output is as follows:

```
[('alice', 0.9963310360908508),
 ('little', 0.9955872463226318),
 ('went', 0.9955698251724243),
 ("'s", 0.9955658312658691),
 ('would', 0.9954401254653931)]
```

4. Using a **window** size of **2**, retrain the word vectors:

```
model = word2vec.Word2Vec(alice_words_nostop, window=2)
```

5. Find the terms most similar to **rabbit**:

```
model.wv.most_similar("rabbit", topn=5)
```

The output will be as follows:

```
[('alice', 0.9491485357284546),
 ("'s", 0.9364748001098633),
 ('little', 0.9345826506614685),
 ('large', 0.9341927170753479),
 ('duchess', 0.9341296553611755)]
```

6. Retrain word vectors using the Skip-gram method with a window size of **5**:

```
model = word2vec.Word2Vec(alice_words_nostop, window=5, sg=1)
```

7. Find the terms most similar to **rabbit**:

```
model.wv.most_similar("rabbit", topn=5)
```

The output will be as follows:

```
[('gardeners', 0.9995723366737366),
 ('end', 0.9995588064193726),
 ('came', 0.9995309114456177),
 ('sort', 0.9995298385620117),
 ('upon', 0.9995272159576416)]
```

8. Find the representation for the phrase **white rabbit** by averaging the vectors for **white** and **rabbit**:

```
v1 = model.wv['white']
v2 = model.wv['rabbit']
res1 = (v1+v2)/2
```

9. Find the representation for **mad hatter** by averaging the vectors for **mad** and **hatter**:

```
v1 = model.wv['mad']
v2 = model.wv['hatter']
res2 = (v1+v2)/2
```

10. Find the cosine similarity between these two phrases:

```
model.wv.cosine_similarities(res1, [res2])
```

This gives us the following value:

```
array([0.9996213], dtype=float32)
```

11. Load the pre-trained GloVe embeddings of size 100D using the formatted keyed vectors:

```
from gensim.models.keyedvectors import KeyedVectors
glove_model = KeyedVectors.load_word2vec_format\
("glove.6B.100d.w2vformat.txt", binary=False)
```

12. Find representations for **white rabbit** and **mad hatter**:

```
v1 = glove_model['white']
v2 = glove_model['rabbit']
res1 = (v1+v2)/2

v1 = glove_model['mad']
v2 = glove_model['hatter']
res2 = (v1+v2)/2
```

13. Find the **cosine** similarity between the two phrases. Has the cosine similarity changed?

```
glove_model.cosine_similarities(res1, [res2])
```

The following is the output of the preceding code:

```
array([0.4514577], dtype=float32)
```

Here, we can see that the cosine similarity between the two phrases "**mad hatter**" and "**white rabbit**" is far lower from the GloVe model. This is because the GloVe model hasn't seen the terms together in its training data as much as they appear in the book. In the book, the terms **mad** and **hatter** appear together a lot because they form the name of an important character. In other contexts, of course, we con't see **mad** and **hatter** together as often.

> **NOTE**
>
> To access the source code for this specific section, please refer to https://packt.live/2VVNEgf
>
> This section does not currently have an online interactive example, and will need to be run locally.

CHAPTER 5: DEEP LEARNING FOR SEQUENCES

ACTIVITY 5.01: USING A PLAIN RNN MODEL TO PREDICT IBM STOCK PRICES

SOLUTION

1. Import the necessary libraries, load the **.csv** file, reverse the index, and plot the time series (the **Close** column) for visual inspection:

```
import pandas as pd, numpy as np
import matplotlib.pyplot as plt
inp0 = pd.read_csv("IBM.csv")
inp0 = inp0.sort_index(ascending=False)
inp0.plot("Date", "Close")
plt.show()
```

The output will be as follows, with the closing price plotted on the *Y-axis*:

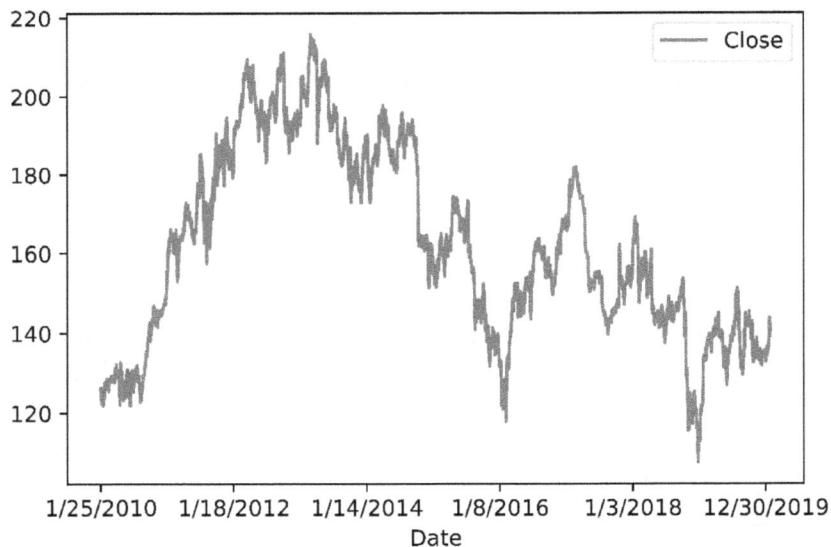

Figure 5.40: The trend for IBM stock prices

2. Extract the values for **Close** from the DataFrame as a **numpy** array and plct them using **matplotlib**:

```
ts_data = inp0.Clcse.values.reshape(-1,1)
plt.figure(figsize=[14,5])
plt.plot(ts_data)
plt.show()
```

The resulting trend is as follows, with the index plotted on the *X-axis*:

Figure 5.41: The stock price data visualized

3. Assign the final 25% data as test data and the first 75% as train data:

```
train_recs = int(len(ts_data) * 0.75)
train_data = ts_data[:train_recs]
test_data = ts_data[train_recs:]
len(train_data), len(test_data)
```

The output will be as follows:

```
(1888, 630)
```

4. Using **MinMaxScaler** from **sklearn**, scale the train and test data:

```
from sklearn.preprocessing import MinMaxScaler
scaler = MinMaxScaler()
train_scaled = scaler.fit_transform(train_data)
test_scaled = scaler.transform(test_data)
```

5. Using the **get_lookback** function we defined earlier in this chapter (refer to the *Preparing the Data for Stock Price Prediction* section), get the lookback data for the train and test sets using a lookback period of 10:

```
look_back = 10
trainX, trainY = get_lookback(train_scaled, look_back=look_back)
testX, testY = get_lookback(test_scaled, look_back= look_back)
trainX.shape, testX.shape
```

The output will be as follows:

```
((1888, 10), (630, 10))
```

6. From Keras, import all the necessary layers for employing plain RNNs (**SimpleRNN, Activation, Dropout, Dense,** and **Reshape**) and 1D convolutions (Conv1D). Also, import the **mean_squared_error** metric from **sklearn**:

```
from tensorflow.keras.models import Sequential
from tensorflow.keras.layers import SimpleRNN, Activation, Dropout,
Dense, Reshape, Conv1D
from sklearn.metrics import mean_squared_error
```

7. Build a model with a 1D convolution layer (5 filters of size 3) and an RNN layer with 32 neurons. Add 25% dropout after the RNN layer. Print the model's summary:

```
model_comb = Sequential()
model_comb.add(Reshape((look_back,1), \
                       input_shape = (look_back,)))
model_comb.add(Conv1D(5, 3, activation='relu'))
model_comb.add(SimpleRNN(32))
model_comb.add(Dropout(0.25))
model_comb.add(Dense(1))
model_comb.add(Activation('linear'))
model.summary()
```

The output will be as follows:

```
Layer (type)                 Output Shape             Param #
=================================================================
reshape_4 (Reshape)          (None, 10, 1)               0
_____
conv1d_3 (Conv1D)            (None, 8, 5)                20
_____
simple_rnn_3 (SimpleRNN)     (None, 32)                1216
_____
dropout (Dropout)            (None, 32)                  0
_____
dense_4 (Dense)              (None, 1)                  33
_____
activation_4 (Activation)    (None, 1)                   0
=================================================================
Total params: 1,269
Trainable params: 1,269
Non-trainable params: 0
_____
```

Figure 5.42: Summary of the model

8. Compile the model with the **mean_squared_error** loss and the **adam** optimizer. Fit this on the train data in five epochs, with a validation split of 10% and a batch size of 1:

```
model_comb.compile(loss='mean_squared_error', \
                   optimizer='adam')
model_comb.fit(trainX, trainY, epochs=5, \
               batch_size=1, verbose=2, \
               validation_split=0.1)
```

The output will be as follows:

```
Train on 1699 samples, validate on 189 samples
Epoch 1/5
1699/1699 - 7s - loss: 0.0101 - val_loss: 4.5731e-04
Epoch 2/5
1699/1699 - 6s - loss: 0.0047 - val_loss: 0.0013
Epoch 3/5
1699/1699 - 6s - loss: 0.0032 - val_loss: 4.4642e-04
Epoch 4/5
1699/1699 - 6s - loss: 0.0027 - val_loss: 4.2862e-04
Epoch 5/5
1699/1699 - 6s - loss: 0.0019 - val_loss: 3.4329e-04
```

Figure 5.43: Training and validation loss

9. Using the **get_model_perf** method, print the RMSE of the model:

```
get_model_perf(model_comb)
```

The output will be as follows:

```
Train RMSE: 0.03 RMSE
Test RMSE: 0.03 RMSE
```

10. Plot the predictions – the entire view, as well as the zoomed-in view:

```
%matplotlib notebook
plt.figure(figsize=[10,5])
plot_pred(model_comb)
```

We should see the following plot of predictions (dotted lines) versus the actuals (solid lines):

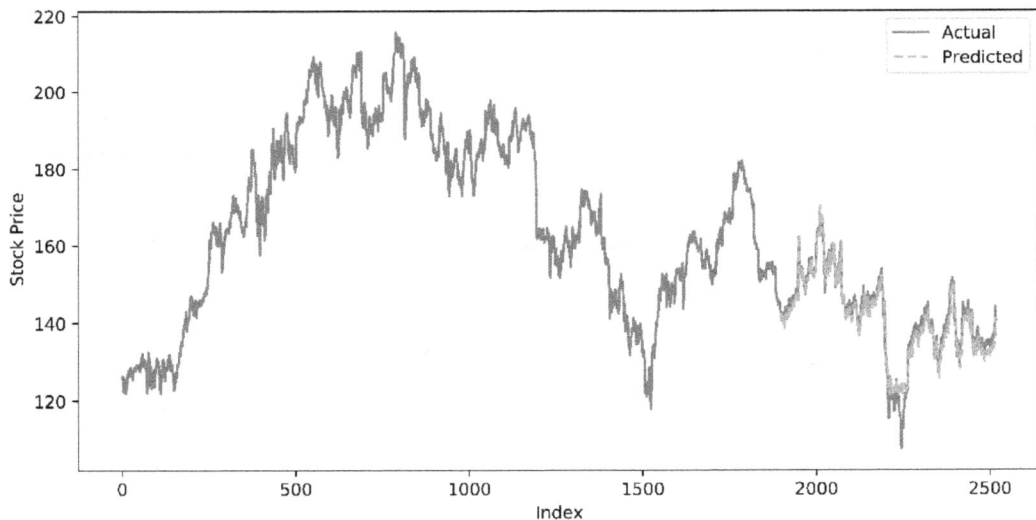

Figure 5.44: Predictions versus actuals

The zoomed-in view is as follows:

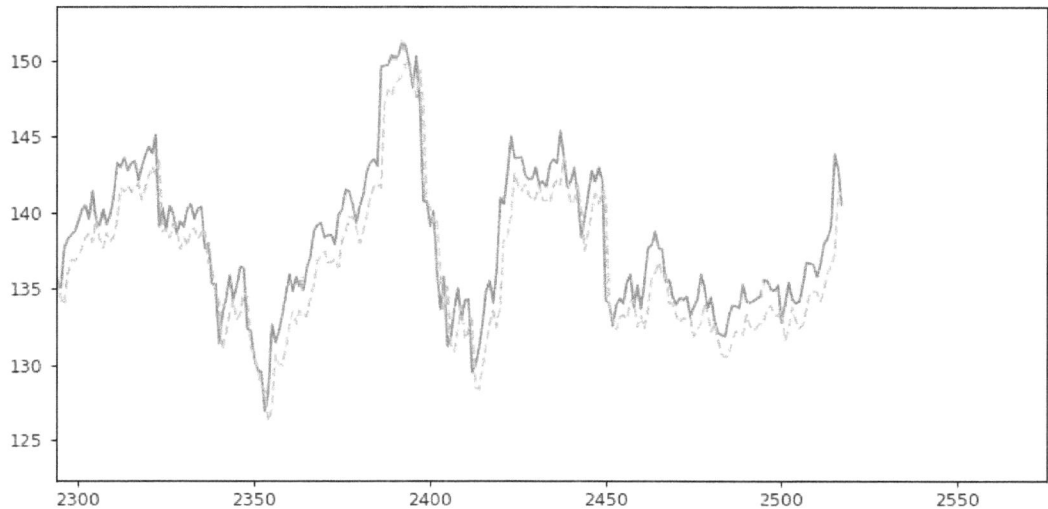

Figure 5.45: Predictions (dotted lines) versus actuals (solid lines) – detailed view

We can see that the model does a great job of catching the finer patterns and does extremely well at predicting the daily stock price.

> **NOTE**
>
> To access the source code for this specific section, please refer to https://packt.live/2ZctArW.
>
> You can also run this example online at https://packt.live/38EDOEA. You must execute the entire Notebook in order to get the desired result.

CHAPTER 6: LSTMS, GRUS, AND ADVANCED RNNS

ACTIVITY 6.01: SENTIMENT ANALYSIS OF AMAZON PRODUCT REVIEWS

SOLUTION

1. Read in the data files for the **train** and **test** sets. Examine the shapes of the datasets and print out the top **5** records from the **train** data:

```
import pandas as pd, numpy as np
import matplotlib.pyplot as plt
%matplotlib inline
train_df = pd.read_csv("Amazon_reviews_train.csv")
test_df = pd.read_csv("Amazon_reviews_test.csv")
print(train_df.shape, train_df.shape)
train_df.head(5)
```

The dataset's shape and header are as follows:

```
(25000, 2) (25000, 2)
```

	review_text	label
0	Stuning even for the non-gamer: This sound tra...	1
1	The best soundtrack ever to anything.: I'm rea...	1
2	Amazing!: This soundtrack is my favorite music...	1
3	Excellent Soundtrack: I truly like this soundt...	1
4	Remember, Pull Your Jaw Off The Floor After He...	1

Figure 6.26: First five records from the train dataset

2. For convenience, when it comes to processing, separate the raw text and the labels for the **train** and **test** sets. You should have **4** variables, as follows: **train_raw** comprising raw text for the train data, **train_labels** with labels for the train data, **test_raw** containing raw text for the test data, and **test_labels** comprising Labels for the test data. Print the first two reviews from the **train** text.

```
train_raw = train_df.review_text.values
train_labels = train_df.label.values
```

```
test_raw = test_df.review_text.values
test_labels = test_df.label.values
train_raw[:2]
```

The preceding code results in the following output:

```
array(['Stuning even for the non-gamer: This sound track was beautiful! It paints th
e senery in your mind so well I would recomend it even to people who hate vid. game
music! I have played the game Chrono Cross but out of all of the games I have ever p
layed it has the best music! It backs away from crude keyboarding and takes a freshe
r step with grate guitars and sculful orchestras. It would impress anyone who cares
to listen! ^_^',
       "The best soundtrack ever to anything.: I'm reading a lot of reviews saying t
hat this is the best 'game soundtrack' and I figured that I'd write a review to disa
gree a bit. This in my opinino is Yasunori Mitsuda's ultimate masterpiece. The music
is timeless and I'm been listening to it for years now and its beauty simply refuses
to fade.The price tag on this is pretty staggering I must say, but if you are going
to buy any cd for this much money, this is the only one that I feel would be worth e
very penny."],
      dtype=object)
```

Figure 6.27: Raw text from the train dataset

3. Normalize the case and tokenize the test and train texts using NLTK's **word_tokenize** (after importing it, of course – hint: use a list comprehension for cleaner code). Download **punkt** from **nltk** if you haven't used the tokenizer before. Print the first review from the train data to check if the tokenization worked.

```
import nltk
nltk.download('punkt')
from nltk.tokenize import word_tokenize
train_tokens = [word_tokenize(review.lower()) \
                for review in train_raw]
test_tokens = [word_tokenize(review.lower()) \
               for review in test_raw]
print(train_tokens[0])
```

The tokenized data gets printed as follows:

```
['stuning', 'even', 'for', 'the', 'non-gamer', ':', 'this', 'sound', 'track', 'was',
'beautiful', '!', 'it', 'paints', 'the', 'senery', 'in', 'your', 'mind', 'so', 'wel
l', 'i', 'would', 'recomend', 'it', 'even', 'to', 'people', 'who', 'hate', 'vid',
'.', 'game', 'music', '!', 'i', 'have', 'played', 'the', 'game', 'chrono', 'cross',
'but', 'out', 'of', 'all', 'of', 'the', 'games', 'i', 'have', 'ever', 'played', 'i
t', 'has', 'the', 'best', 'music', '!', 'it', 'backs', 'away', 'from', 'crude', 'key
boarding', 'and', 'takes', 'a', 'fresher', 'step', 'with', 'grate', 'guitars', 'an
d', 'soulful', 'orchestras', '.', 'it', 'would', 'impress', 'anyone', 'who', 'care
s', 'to', 'listen', '!', '^_^']
```

Figure 6.28: Tokenized review from the `train` dataset

4. Import any stop words (built in to NLTK) and punctuation from the string module. Define a function (**drop_stop**) to remove these tokens from any input tokenized sentence. Download **stopwords** from NLTK if you haven't used it before:

```
from string import punctuation
stop_punct = list(punctuation)
nltk.download("stopwords")
from nltk.corpus import stopwords
stop_nltk = stopwords.words("english")
stop_final = stop_punct + stop_nltk
def drop_stop(input_tokens):
    return [token for token in input_tokens \
            if token not in stop_final]
```

5. Using the defined function (**drop_stop**), remove the redundant stop words from the **train** and the **test** texts. Print the first review of the processed **train** texts to check whether the function worked:

```
train_tokens_no_stop = [drop_stop(sent) \
                        for sent in train_tokens]
test_tokens_no_stop = [drop_stop(sent) \
                       for sent in test_tokens]
print(train_tokens_no_stop[0])
```

We'll get the following output:

```
['stuning', 'even', 'non-gamer', 'sound', 'track', 'beautiful',
 'paints', 'senery', 'mind', 'well', 'would', 'recomend', 'even',
 'people', 'hate', 'vid', 'game', 'music', 'played', 'game',
 'chrono', 'cross', 'games', 'ever', 'played', 'best', 'music',
 'backs', 'away', 'crude', 'keyboarding', 'takes', 'fresher',
 'step', 'grate', 'guitars', 'soulful', 'orchestras', 'would',
 'impress', 'anyone', 'cares', 'listen', '^_^']
```

6. Using **PorterStemmer** from NLTK, stem the tokens for both the **train** and **test** data:

```
from nltk.stem import PorterStemmer
stemmer_p = PorterStemmer()
train_tokens_stem = [[stemmer_p.stem(token) for token in sent] \
                        for sent in train_tokens_no_stop]
test_tokens_stem = [[stemmer_p.stem(token) for token in sent] \
                        for sent in test_tokens_no_stop]
print(train_tokens_stem[0])
```

The result should be printed as follows:

```
['stune', 'even', 'non-gam', 'sound', 'track', 'beauti', 'paint',
 'seneri', 'mind', 'well', 'would', 'recomend', 'even', 'peopl',
 'hate', 'vid', 'game', 'music', 'play', 'game', 'chrono', 'cross',
 'game', 'ever', 'play', 'best', 'music', 'back', 'away', 'crude',
 'keyboard', 'take', 'fresher', 'step', 'grate', 'guitar', 'soul',
 'orchestra', 'would', 'impress', 'anyon', 'care', 'listen', '^_^']
```

7. Create the strings for each of the **train** and **text** reviews. This will help us work with the utilities in Keras to create and pad the sequences. Create the **train_texts** and **test_texts** variables. Print the first review from the processed **train** data to confirm this:

```
train_texts = [" ".join(txt) for txt in train_tokens_stem]
test_texts = [" ".join(txt) for txt in test_tokens_stem]
print(train_texts[0])
```

The result of the preceding code is as follows:

```
stune even non-gam sound track beauti paint seneri mind well would
recommend even peopl hate vid game music play game chrono cross game
ever play best music back away crude keyboard take fresher step grate
guitar soul orchestra would impress anyon care listen ^_^
```

8. From Keras' preprocessing utilities for text (**keras.preprocessing.text**), import the **Tokenizer** module. Define a vocabulary size of **10000** and instantiate the tokenizer with this vocabulary:

```
from tensorflow.keras.preprocessing.text import Tokenizer
vocab_size = 10000
tok = Tokenizer(num_words=vocab_size)
```

9. Fit the tokenizer on the **train** texts. This works just like **CountVectorizer** did in *Chapter 4, Deep Learning for Text – Embeddings*, and trains the vocabulary. After fitting, use the **texts_to_sequences** method of the tokenizer on the **train** and **test** sets to create the sequences for them. Print the sequence for the first review in the train data:

```
tok.fit_on_texts(train_texts)
train_sequences = tok.texts_to_sequences(train_texts)
test_sequences = tok.texts_to_sequences(test_texts)
print(train_sequences[0])
```

The encoded sequence is as follows:

```
[22, 514, 7161, 85, 190, 184, 1098, 283, 20, 11, 1267, 22,
 56, 370, 9682, 114, 41, 71, 114, 8166, 1455, 114, 51, 71,
 29, 41, 58, 182, 2931, 2153, 75, 8167, 816, 2666, 829, 719,
 3871, 11, 483, 120, 268, 110]
```

10. We need to find the optimal length of the sequences to process the model. Get the length of the reviews from the **train** set into a list and plot a histogram of the lengths:

```
seq_lens = [len(seq) for seq in train_sequences]
plt.hist(seq_lens)
plt.show()
```

The distribution of the lengths is as follows:

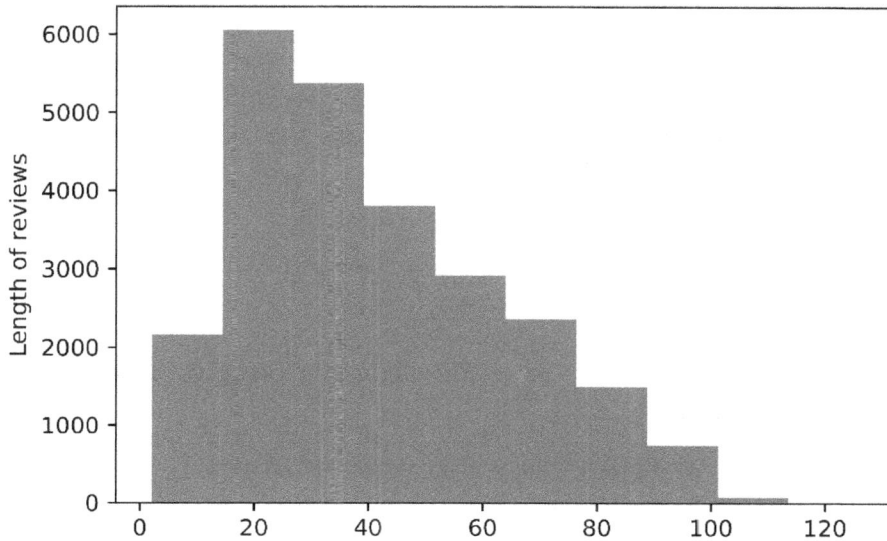

Figure 6.29: Histogram of text lengths

11. The data is now in the same format as the IMDb data we used in this chapter. Using a sequence length of **100** (define the **maxlen = 100** variable), use the **pad_sequences** method from the **sequence** module in Keras' preprocessing utilities (**keras.preprocessing.sequence**) to limit the sequences to **100** for both the **train** and **test** data. Check the shape of the result for the train data:

```
maxlen = 100
from tensorflow.keras.preprocessing.sequence import pad_sequences
X_train = pad_sequences(train_sequences, maxlen=maxlen)
X_test = pad_sequences(test_sequences, maxlen=maxlen)
X_train.shape
```

The shape is as follows:

```
(25000, 100)
```

12. To build the model, import all the necessary layers from Keras (**embedding**, **spatial dropout**, **LSTM**, **dropout**, and **dense**) and import the **Sequential** model. Initialize the **Sequential** model:

```
from tensorflow.keras.models import Sequential

from tensorflow.keras.layers import Dense, Embedding,
SpatialDropout1D, Dropout, GRU, LSTM
model_lstm = Sequential()
```

13. Add an embedding layer with **32** as the vector size (**output_dim**). Add a spatial dropout of **40%**:

```
model_lstm.add(Embedding(vocab_size, output_dim=32))
model_lstm.add(SpatialDropout1D(0.4))
```

14. Build a stacked LSTM model with **2** layers that have **64** cells each. Add a dropout layer with **40%** dropout:

```
model_lstm.add(LSTM(64, return_sequences=True))
model_lstm.add(LSTM(64, return_sequences=False))
model_lstm.add(Dropout(0.4))
```

15. Add a dense layer with **32** neurons with **relu** activation, then a **50%** dropout layer, followed by another dense layer of **32** neurons with **relu** activation, and follow this up with another dropout layer with **50%** dropout:

```
model_lstm.add(Dense(32, activation='relu'))
model_lstm.add(Dropout(0.5))
model_lstm.add(Dense(32, activation='relu'))
model_lstm.add(Dropout(0.5))
```

16. Add a final dense layer with a single neuron with **sigmoid activation** and compile the model. Print the model summary:

```
model_lstm.add(Dense(1, activation='sigmoid'))
model_lstm.compile(loss='binary_crossentropy', \
                   optimizer='rmsprop', \
                   metrics=['accuracy'])
model_lstm.summary()
```

The summary of the model will be as follows:

Layer (type)	Output Shape	Param #
embedding_3 (Embedding)	(None, None, 32)	320000
spatial_dropout1d_2 (Spatial	(None, None, 32)	0
lstm_4 (LSTM)	(None, None, 64)	24832
lstm_5 (LSTM)	(None, 64)	33024
dropout_5 (Dropout)	(None, 64)	0
dense_5 (Dense)	(None, 32)	2080
dropout_6 (Dropout)	(None, 32)	0
dense_6 (Dense)	(None, 32)	1056
dropout_7 (Dropout)	(None, 32)	0
dense_7 (Dense)	(None, 1)	33

```
Total params: 381,025
Trainable params: 381,025
Non-trainable params: 0
```

Figure 6.30: Stacked LSTM model summary

17. Fit the model on the training data with a **20%** validation split and a batch size of **128**. Train for **5 epochs**:

```
history_lstm = model_lstm.fit(X_train, train_labels, \
                              batch_size=128, \
                              validation_split=0.2, \
                              epochs = 5)
```

We will get the following training output:

```
Train on 20000 samples, validate on 5000 samples
Epoch 1/5
20000/20000 [==============================] - 86s 4ms/sample - loss: 0.5599 - accuracy: 0.7080 - val_los
s: 0.4052 - val_accuracy: 0.8186
Epoch 2/5
20000/20000 [==============================] - 71s 4ms/sample - loss: 0.3692 - accuracy: 0.8594 - val_los
s: 0.3954 - val_accuracy: 0.8478
Epoch 3/5
20000/20000 [==============================] - 71s 4ms/sample - loss: 0.3056 - accuracy: 0.8870 - val_los
s: 0.3774 - val_accuracy: 0.8466
Epoch 4/5
20000/20000 [==============================] - 71s 4ms/sample - loss: 0.2759 - accuracy: 0.8983 - val_los
s: 0.3677 - val_accuracy: 0.8586
Epoch 5/5
20000/20000 [==============================] - 69s 3ms/sample - loss: 0.2508 - accuracy: 0.9074 - val_los
s: 0.3447 - val_accuracy: 0.8518
```

Figure 6.31: Stacked LSTM model training output

18. Make a prediction on the test set using the **predict_classes** method of the model. Then, print out the confusion matrix:

```
from sklearn.metrics import accuracy_score, confusion_matrix
test_pred = model_lstm.predict_classes(X_test)
print(confusion_matrix(test_labels, test_pred))
```

We will get the following result:

```
[[10226,  1931],
 [ 1603, 11240]]
```

19. Using the **accuracy_score** method from **scikit-learn**, calculate the accuracy of the test set.

```
print(accuracy_score(test_labels, test_pred))
```

The accuracy we get is:

```
0.85864
```

As we can see, the accuracy score is around **86%**, and looking at the confusion matrix (output of *step 18*), the model does a decent job of predicting both classes well. We got this accuracy without doing any hyperparameter tuning. You can tweak the hyperparameters to get significantly higher accuracy.

> **NOTE**
>
> To access the source code for this specific section, please refer to https://packt.live/3fpo0YI
>
> You can also run this example online at https://packt.live/2Wi75QH. You must execute the entire Notebook in order to get the desired result.

CHAPTER 7: GENERATIVE ADVERSARIAL NETWORKS

ACTIVITY 7.01: IMPLEMENTING A DCGAN FOR THE MNIST FASHION DATASET

SOLUTION

1. Open a new Jupyter Notebook and name it **Activity 7.01**. Import the following library packages:

```
# Import the required library functions
import numpy as np
import matplotlib.pyplot as plt
from matplotlib import pyplot
import tensorflow as tf
from tensorflow.keras.layers import Input
from tensorflow.keras.initializers import RandomNormal
from tensorflow.keras.models import Model, Sequential
from tensorflow.keras.layers \
import Reshape, Dense, Dropout, Flatten,Activation
from tensorflow.keras.layers import LeakyReLU,BatchNormalization
from tensorflow.keras.layers import Conv2D,
UpSampling2D,Conv2DTranspose
from tensorflow.keras.datasets import fashion_mnist
from tensorflow.keras.optimizers import Adam
```

2. Create a function that will generate real data samples from the fashion MNIST data:

```
# Function to generate real data samples
def realData(batch):
    # Get the MNIST data
    (X_train, _), (_, _) = fashion_mnist.load_data()
    # Reshaping the input data to include channel
    X = X_train[:,:,:,np.newaxis]
    # normalising the data to be between 0 and 1
    X = (X.astype('float32') - 127.5)/127.5
    # Generating a batch of data
    imageBatch = X[np.random.randint(0, X.shape[0], \
                                     size=batch)]

    return imageBatch
```

The output from this function is the batch of MNIST data. Please note that we normalize the input data by subtracting **127.5**, which is half the max pixel value, and dividing by the same value. This will help in converging the solution faster.

3. Now, let's generate a set of images from the MNIST dataset:

```
# Generating a set of  sample images
fashionData = realData(25)
```

You should get the following output:

```
Downloading data from https://storage.googleapis.com/tensorflow/tf-keras-datasets/train-labels-idx1-ubyte.gz
32768/29515 [==============================] - 0s 8us/step
Downloading data from https://storage.googleapis.com/tensorflow/tf-keras-datasets/train-images-idx3-ubyte.gz
26427392/26421880 [==============================] - 17s 1us/step
Downloading data from https://storage.googleapis.com/tensorflow/tf-keras-datasets/t10k-labels-idx1-ubyte.gz
8192/5148 [==============================] - 0s 0us/step
Downloading data from https://storage.googleapis.com/tensorflow/tf-keras-datasets/t10k-images-idx3-ubyte.gz
4423680/4422102 [==============================] - 8s 2us/step
```

Figure 7.36: Generating images from MNIST

4. Now, let's visualize the images with **matplotlib**:

```
# for j in range(5*5):
    pyplot.subplot(5,5,j-1)
    # turn off axis
    pyplot.axis('off')
    pyplot.imshow(fashionData[j,:,:,0],cmap='gray_r')
```

You should get an output similar to the one shown here:

Figure 7.37: Plotted images

From the output, we can see the visualization of several fashion articles. We can see that the images are centrally located within a white background. This are the images that we'll try to recreate.

5. Now, let's define the function to generate inputs for the generator network. The inputs are random data points that are generated from a random uniform distribution:

```
# Function to generate inputs for generator function
def fakeInputs(batch,infeats):
    # Generate random noise data with shape (batch,input features)
    x_fake = np.random.uniform(-1,1,size=[batch,infeats])
    return x_fake
```

This function generates the fake data that was sampled from the random distribution as the output.

6. Let's define the function for building the generator network:

`Activity7.01.ipynb`

```
# Function for the generator model
def genModel(infeats):
    # Defining the Generator model
    Genmodel = Sequential()
    Genmodel.add(Dense(512,input_dim=infeats))
    Genmodel.add(Activation('relu'))
    Genmodel.add(BatchNormalization())
    # second layer of FC => RElu => BN layers
    Genmodel.add(Dense(7*7*64))
    Genmodel.add(Activation('relu'))
    Genmodel.add(BatchNormalization())
```

The complete code for this step can be found at https://packt.live/3fpobDm

Building the generator network is similar to building any CNN network. In this generator network, we will use the transpose convolution method for upsampling images. In this model, we can see the progressive use of the transpose convolution. The initial input starts with a dimension of 100, which is our input feature. The dimension of the MNIST dataset is batch size x 28 x 28. Therefore, we have upsampled the data twice to get the output as batch size x 28 x 28.

7. Next, we define the function that will be used to create fake samples:

```
# Function to create fake samples using the generator model
def fakedataGenerator(Genmodel,batch,infeats):
    # first generate the inputs to the model
    genInputs = fakeInputs(batch,infeats)
    """
    use these inputs inside the generator model \
    to generate fake distribution
    """
    X_fake = Genmodel.predict(genInputs)

    return X_fake
```

In this function, we only return the **X** variable. The output from this function is the fake dataset.

8. Define the parameters that we will use in many of the functions, along with the summary of the generator network:

```
# Define the arguments like batch size and input feature
batch = 128
infeats = 100
Genmodel = genModel(infeats,)
Genmodel.summary()
```

You should get the following output:

```
Model: "sequential"

_____
Layer (type)                 Output Shape              Param #
=================================================================
dense (Dense)                (None, 512)               51712
_____
activation (Activation)      (None, 512)               0
_____
batch_normalization_v2 (Batc (None, 512)               2048
_____
dense_1 (Dense)              (None, 3136)              1608768
_____
activation_1 (Activation)    (None, 3136)              0
_____
batch_normalization_v2_1 (Ba (None, 3136)              12544
_____
reshape (Reshape)            (None, 7, 7, 64)          0
_____
conv2d_transpose (Conv2DTran (None, 14, 14, 32)        51232
_____
activation_2 (Activation)    (None, 14, 14, 32)        0
_____
batch_normalization_v2_2 (Ba (None, 14, 14, 32)        128
_____
conv2d_transpose_1 (Conv2DTr (None, 28, 28, 1)         801
_____
activation_3 (Activation)    (None, 28, 28, 1)         0
=================================================================
Total params: 1,727,233
Trainable params: 1,719,873
Non-trainable params: 7,360
_____
```

Figure 7.38: Summary of the generative model

From the summary, please note how the dimension of the input noise changes with each transpose convolution operation. Finally, we get an output that is equal in dimension to the real dataset, (None,28 ,28,1).

9. Let's use the generator function to generate a fake sample before training:

```
# Generating a fake sample and printing the shape
fake = fakedataGenerator(Genmodel,batch,infeats)
fake.shape
```

You should get the following output:

```
(128, 28, 28, 1)
```

10. Now, let's plot the generated fake sample:

```
# Plotting the fake sample

plt.imshow(fake[1, :, :, 0], cmap='gray_r')
```

You should get an output similar to the following:

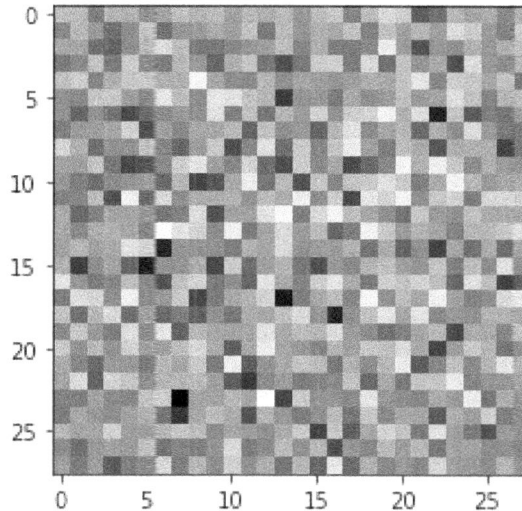

Figure 7.39: Output of the fake sample

This is the plot of the fake sample before training. After training, we want samples like these to look like the MNIST fashion samples we visualized earlier in this activity.

11. Build the discriminator model as a function. The network architecture will be similar to a CNN architecture:

Activity7.01.ipynb

```
# Descriminator model as a function
def discModel():
    Discmodel = Sequential()
    Discmodel.add(Conv2D(32,kernel_size=(5,5),strides=(2,2),\
                padding='same',input_shape=(28,28,1)))
    Discmodel.add(LeakyReLU(0.2))
    # second layer of convolutions
    Discmodel.add(Conv2D(64, kernel_size=(5,5), strides=(2, 2), \
                padding='same'))
    Discmodel.add(LeakyReLU(0.2))
```

The full code for this step can be found at https://packt.live/3-pobDm

In the discriminator network, we have included all the necessary layers, such as the convolutional operations and **LeakyReLU**. Please note that the last layer is a sigmoid layer as we want the output as a probability of whether the sample is real (1) or fake (0).

12. Print the summary of the discriminator network:

```
# Print the summary of the discriminator model
Discmodel = discModel()
Discmodel.summary()
```

You should get the following output:

```
Model: "sequential_1"
_____
Layer (type)                 Output Shape              Param #
=================================================================
conv2d (Conv2D)              (None, 14, 14, 32)        832
_____
leaky_re_lu (LeakyReLU)      (None, 14, 14, 32)        0
_____
conv2d_1 (Conv2D)            (None, 7, 7, 64)          51264
_____
leaky_re_lu_1 (LeakyReLU)    (None, 7, 7, 64)          0
_____
flatten (Flatten)            (None, 3136)              0
_____
dense_2 (Dense)              (None, 512)               1606144
_____
leaky_re_lu_2 (LeakyReLU)    (None, 512)               0
_____
dense_3 (Dense)              (None, 1)                 513
=================================================================
Total params: 1,658,753
Trainable params: 1,658,753
Non-trainable params: 0
```

Figure 7.40: Discriminator model summary

13. Define the GAN model as a function:

```
# Define the combined generator and discriminator model, for updating
the generator
def ganModel(Genmodel,Discmodel):
    # First define that discriminator model cannot be trained
    Discmodel.trainable = False
    Ganmodel = Sequential()
    # First adding the generator model
```

```
    Ganmodel.add(Genmodel)
    """

    Next adding the discriminator model
    without training the parameters
    """

    Ganmodel.add(Discmodel)
    """

    Compile the model for loss to optimise the Generator model
    """

    Ganmodel.compile(loss='binary_crossentropy',\
                     optimizer = 'adam')
    return Ganmodel
```

The structure of the GAN model is similar to the one we developed in *Exercise 7.05, Implementing the DCGAN*.

14. Now, it's time to invoke the GAN function:

```
# Initialise the GAN model
gan_model = ganModel(Genmodel,Discmodel)
# Print summary of the GAN model
gan_model.summary()
```

Please note that the inputs to the GAN model are the previously defined generator model and the discriminator model. You should get the following output:

```
Model: "sequential_2"

_____
Layer (type)                 Output Shape              Param #
=================================================================
sequential (Sequential)      (None, 28, 28, 1)         1727233

_____
sequential_1 (Sequential)    (None, 1)                 1658753
=================================================================
Total params: 3,385,986
Trainable params: 1,719,873
Non-trainable params: 1,666,113
```

Figure 7.41: GAN model summary

Please note that the parameters of each layer of the GAN model are equivalent to the parameters of the generator and discriminator models. The GAN model is just a wrapper around the two models we defined earlier.

15. Define the number of epochs to train the network on using the following code:

```
# Defining the number of epochs
nEpochs = 5000
```

16. Now, we can start the process of training the network:

> **NOTE:**
>
> Before you run the following code, make sure you have created a folder called **output** in the same path as your Jupyter Notebook.

Activity7.01.ipynb

```
# Train the GAN network
for i in range(nEpochs):
    """
    Generate samples equal to the batch size
    from the real distribution
    """
    x_real = realData(batch)
    #Generate fake samples using the fake data generator function
    x_fake = fakedataGenerator(Genmodel,batch,infeats)
    # Concatenating the real and fake data
    X = np.concatenate([x_real,x_fake])
    #Creating the dependent variable and initializing them as '0'
    Y = np.zeros(batch * 2)
```

The complete code for this step can be found on https://packt.live/3fpobDm

It needs to be noted here that the training of the discriminator model with the fake and real samples and the training of the GAN model happens concurrently. The only difference is the training of the GAN model proceeds without updating the parameters of the discriminator model. The other thing to note is that, inside the GAN, the labels for the fake samples would be 1 to generate large loss terms that will be backpropagated through the discriminator network to update the generator parameters. We also display the predicted probability of the GAN for every 50 epochs. When calculating the probability, we combine a sample of real data and a sample of fake data and then take the mean of the predicted probability. We also save a copy of the generated image.

You should get an output similar to the following:

```
Discriminator probability:0.5276428461074829
Discriminator probability:0.5038391351699829
Discriminator probability:0.47621315717697144
```

```
Discriminator probability:0.48467564582824707
Discriminator probability:0.5270703434944153
Discriminator probability:0.5247280597686768
Discriminator probability:0.5282968282699585
```

Let's also look at some of the plots that were generated from the training process at various epochs:

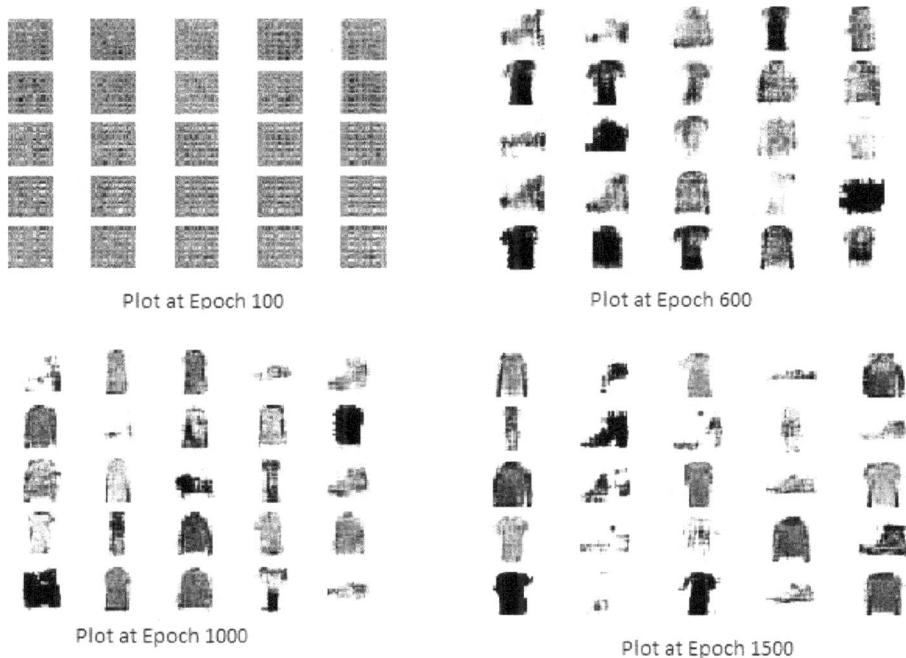

Plot at Epoch 100   Plot at Epoch 600

Plot at Epoch 1000   Plot at Epoch 1500

Figure 7.42: Images generated during the training process

From the preceding plots, we can see the progression of the training process. We can see that by epoch 100, the plots were mostly noise. By epoch 600, the forms of the fashion articles started to become more pronounced. At epoch 1,500, we can see that the fake images are looking very similar to the fashion dataset.

NOTE:

You can take a closer look at these images by going to https://packt.live/2W1Fjal.

17. Now, let's look at the images that were generated after training:

```
# Images generated after training
x_fake = fakedataGenerator(Genmodel,25,infeats)
# Displaying the plots
for j in range(5*5):
pyplot.subplot(5,5,j+1)
    # turn off axis
    pyplot.axis('off')
    pyplot.imshow(x_fake[j,:,:,0],cmap='gray_r')
```

You should get an output similar to the following:

Figure 7.43: Images generated after the training process

From the training accuracy levels, you can see that the accuracy of the discriminator model hovers around the .50 range, which is the desired range. The purpose of the generator is to create fake images that look like real ones. When the generator generates images that look very similar to real images, the discriminator gets confused as to whether the image has been generated from the real distribution or fake distribution. This phenomenon manifests in an accuracy level of around 50% for the discriminator, which is the desired level.

> **NOTE**
>
> To access the source code for this specific section, please refer to https://packt.live/3fpobDr.
>
> This section does not currently have an online interactive example, and will need to be run locally.

INDEX

www.ingramcontent.com/pod-product-compliance
Lightning Source LLC
Chambersburg PA
CBHW080128220326
41598CB00032B/4998